BREEDING BIRDS *of* ONTARIO
NIDIOLOGY AND DISTRIBUTION
VOLUME 1: NONPASSERINES

BREEDING BIRDS *of* ONTARIO
NIDIOLOGY AND DISTRIBUTION
VOLUME 1: NONPASSERINES

George K. Peck

and

Ross D. James

A Life Sciences Miscellaneous Publication
of the Royal Ontario Museum, Toronto

ROYAL ONTARIO MUSEUM
PUBLICATIONS IN LIFE SCIENCES

The Royal Ontario Museum publishes three series in the Life Sciences.

LIFE SCIENCES CONTRIBUTIONS, a numbered series of original scientific publications including monographic works.

LIFE SCIENCES OCCASIONAL PAPERS, a numbered series of original scientific publications, primarily short and usually of taxonomic significance.

LIFE SCIENCES MISCELLANEOUS PUBLICATIONS, an unnumbered series of publications of varied subject matter and format.

All manuscripts considered for publication are subject to the scrutiny and editorial policies of the Life Sciences Editorial Board, and to review by persons outside the Museum staff who are authorities in the particular field involved.

Canadian Cataloguing in Publication Data

Peck, George K.
Breeding birds of Ontario: nidiology and distribution

(Life sciences miscellaneous publication, ISSN 0082-5093)
Includes bibliographies and indexes.
Contents: v. 1. Nonpasserines.
ISBN 0-88854-288-7 (v. 1)

1. Birds - Ontario - Eggs and nests. 2. Birds - Ontario - Breeding - Maps. 3. Birds - Ontario - Breeding. I. James, Ross D., 1943- II. Royal Ontario Museum. III. Title. IV. Series: Life Sciences miscellaneous publications.

QL685.5.06P32 598.29713 C82-095259-1
QL1.T6539

Cover: Semipalmated Plover, *Charadrius semipalmatus* Bonaparte (p. 163). Drawing by Ross D. James.

100 Queen's Park, Toronto, Canada M5S 2C6
ISBN 0-8854-288-7
ISSN 0082-5093
PRINTED AND BOUND IN CANADA BY THE ALGER PRESS

To James L. Baillie
whose friendship and guidance were greatly valued and whose knowledge, enthusiasm, and love of birds were a constant source of inspiration.

GEORGE K. PECK is a veterinarian and a research associate with the Department of Ornithology of the Royal Ontario Museum. Dr. Peck has been in charge of compiling nest record cards and preparing annual reports for the Ontario Nest Records Scheme (ONRS) since 1966. He has travelled throughout North America to study and photograph breeding birds, and fieldwork in Ontario has taken him to almost every region of the province. He is a skilled photographer and has an extensive collection of colour slides of the nests, eggs, young, and habitats of many breeding birds of North America. His appointment as a research associate in 1977 was in recognition of his extensive contribution to studies in ornithology at the ROM.

ROSS D. JAMES is associate curator in the Department of Ornithology of the Royal Ontario Museum. He began work with the museum in 1966 while pursuing graduate studies in avian biology. While conducting research and fieldwork for his studies, he has travelled to many parts of North and Central America, including most regions of Ontario. He was appointed assistant curator in the Department of Ornithology in 1973. He was senior author of *Annotated Checklist of Birds of Ontario*, published by the Museum in 1976. He is a photographer, and also prepared the drawings for this volume.

Contents

Figures

Nidiology and Distribution

Introduction

Ontario, Canada's second largest province, covers a huge area of 1 068 620 km² (412 582 m² and extends 1624 km (1015 mi) from east to west, and 1736 km (1085 mi) from north to south, through 15 degrees of latitude (41°45′N to 56°44′N) and almost 21 degrees of longitude (74°21′W to 95°05′W). Its area is one half again that of the state of Texas, almost twice that of France, and five times that of the British Isles. In Ontario an area of ca 43 500 000 ha (107 500 000 acres) is covered by forests, and two-thirds of the province is the Canadian Shield of Precambrian rock.

The climate is varied and is especially influenced in the north by the cooling effect of James and Hudson bays, and in the south and west it is moderated and warmed by the Great Lakes. The extremes of climate in the province are perhaps best exemplified if one compares the climate on Point Pelee and Pelee Island, with their cactus, Carolinian Forest, and southern breeding-bird species, to the climate on the northern seacoasts, with their tundra, permafrost, and arctic breeding-bird species.

The various forest and physiographic regions in the province are illustrated in Figure 1. They exert a profound influence on the distribution of some breeding birds, and a knowledge of the extent of these regions is necessary in order to understand more fully why some species are found in specific areas of Ontario and not in others. The features of these regions influence to a large extent the environmental conditions and may determine whether a particular species can find suitable nesting habitat.

The **Deciduous Forest** region (Carolinian Zone) in the extreme south of the province, the northern continuation of the widespread deciduous forests of the eastern United States, is dominated by broad-leaved tree species (Fig. 145), many of which reach their northern limits in this area. Coniferous trees are few and scattered in the region. In recent times most of this region has been cleared for farming (Fig. 151), and wooded areas are confined largely to farm wood lots and to areas too rocky to farm.

The **Great Lakes-St. Lawrence Forest** region, which extends right across the province and includes a portion west of Lake Superior, is characterized by a mixture of deciduous and coniferous trees (Fig. 155). Local predominance of either type of tree depends upon climate, drainage, exposure, and soil conditions; the proportion of coniferous trees increases gradually towards the north. South and east of the Canadian Shield in this region, as well as in some areas of the Shield, most of the forests have been cleared for agricultural purposes. In the more rocky areas of the Shield, farms are often very poor (Fig. 157) and abandoned fields are numerous; however, some sections, such as the aspen parklands of western Rainy River, are more productive.

The **Boreal Forest** region of northern Ontario, the largest forest region in the province, is usually divided into three main sections (Rowe, 1959): Boreal Forest, Hudson Bay Lowland Forest, and Forest-Tundra. Farming is all but absent in this region; it is confined to a few small areas.

The Boreal Forest section is dominated by spruce trees (Fig. 159). Birch and poplar are the only broad-leaved trees that grow in significant numbers, and they are only of local distribution. Logging is the major disturbance in this area.

The Hudson Bay Lowland Forest section, which covers almost the same area as the Hudson Bay Lowland physiographic region, is a transition forest mainly of subarctic appearance, with extensive areas of swamp and muskeg (Fig. 163). Much of the forest is reduced to a short, open cover of spruce and tamarack.

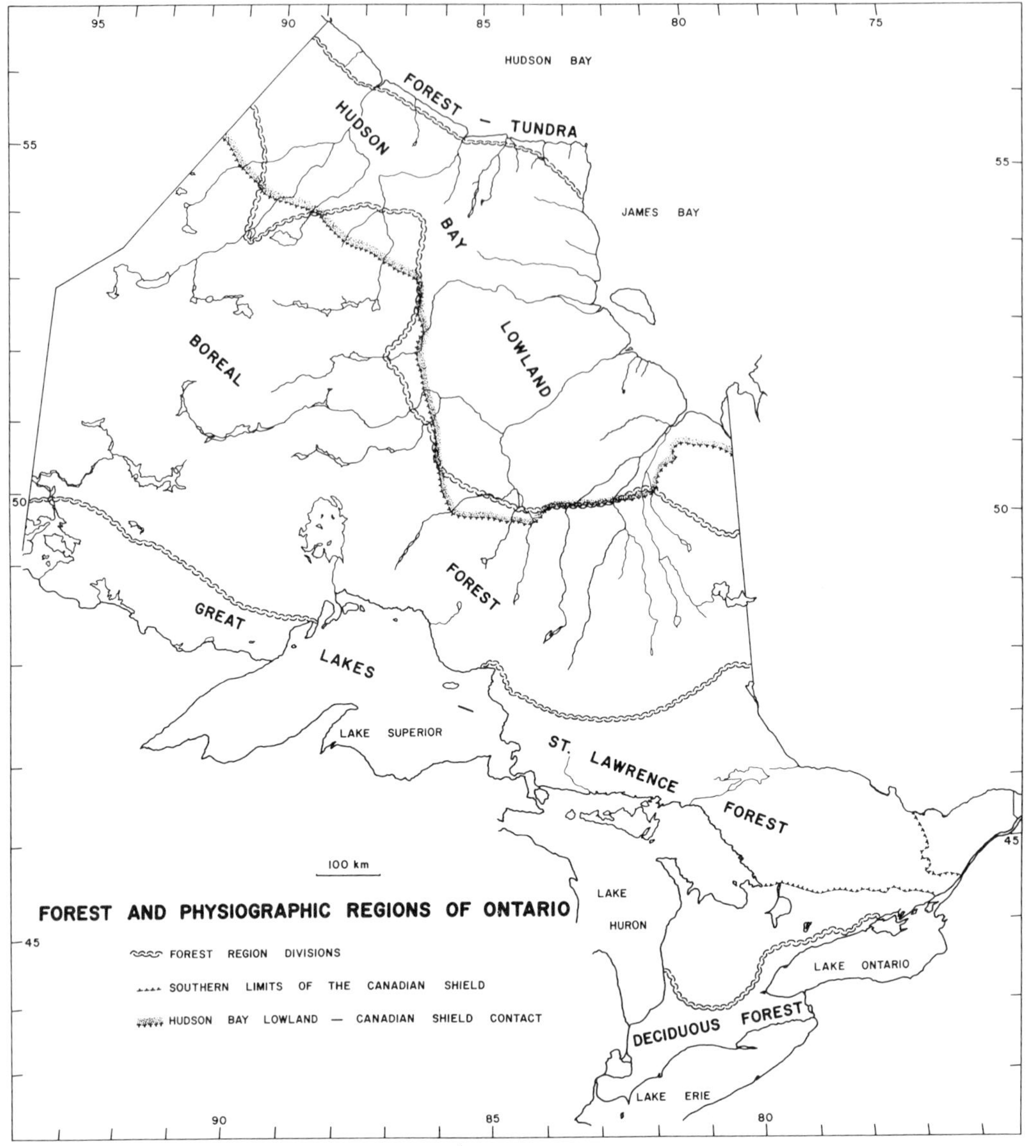

Fig. 1 Forest and physiographic regions of Ontario.

The Forest-Tundra section is a narrow strip along the north coast, most of which features tundra vegetation only, which is characteristic of regions much farther north in Canada (Figs. 167, 169). The severe climate largely restricts trees to stunted stands inland along the lake and river shores. The cold waters along the coasts of Hudson Bay and the northern portion of James Bay maintain arctic conditions, which attract several birds species ordinarily found farther north.

The **Hudson Bay Lowland** physiographic region occupies about one-quarter of the province. It is a flat, poorly drained plain underlain by palaeozoic rock (Hustich, 1957), with a maximum elevation of ca 150 m (492 ft), at its junction with the Canadian Shield in the Albany

River basin. Outcrops of bedrock are few and are mainly confined to the Shield-Lowland contact zone and to eroded river channels.

The **Canadian Shield,** the largest physiographic region of Ontario, surrounds the Hudson Bay Lowland in the north, includes the rest of northern Ontario, and extends into much of southern Ontario, as far as southern Georgian Bay in the west and Kingston and the St. Lawrence River in the east. Outcrops of Precambrian rock are common because much of the overlying material was stripped away by glacial action. The presence of the Shield exerts a marked influence on the breeding distribution of many species. For example, lakes in this region not only are low in nutrients and productivity, but also often have abrupt rocky shorelines with little soil or vegetation to provide the attractive marshy breeding habitats so prevalent in the south.

South of the Canadian Shield, and including the eastern lowlands between the Ottawa and St. Lawrence rivers, southern Ontario is developed predominantly as an agricultural area. Rich glacial tills over palaeozoic limestone bedrock have created excellent farming conditions and, in areas of low relief, support numerous productive marshes. Human activity, in the form of forest removal and urbanization, has been the major influence on birdlife in this region and has encouraged some species and restricted others. The ranges of a number of the many species that still breed in this agricultural region are rather suddenly curtailed to the north by the presence of the Canadian Shield.

The literature and records dealing with the breedings birds of Ontario have increased enormously since the publication of a two-part study, *The Distribution of Breeding Birds in Ontario*, in 1936 and 1937 by James L. Baillie and Paul Harrington. This study listed 210 breeding species in the province. In 1956 the Ontario Nest Records Scheme (ONRS) was initiated, and it now comprises almost 80 000 nest cards. Approximately two-thirds of these cards represent relatively recent records contributed since 1956 by about 775 contributors. The remaining cards are of historical nesting records gleaned from sources such as the following: the nest and egg collection of the Royal Ontario Museum (ROM); unpublished records in field notes, manuscripts, and lists in ROM files; *Contributions of the Royal Ontario Museum of Zoology*; articles in *Audubon Field Notes, American Birds, The Auk, Canadian Field-Naturalist, Ontario Field Biologist, Ontario Natural Science Bulletin, Transactions of the Royal Canadian Institute, Biological Science Bulletin of the University of Toronto, The Wilson Bulletin*, and a number of naturalist club bulletins; and ornithological books such as *Life Histories of North American Birds* (Bent 1919–1968), *Birds of the Detroit-Windsor Area* (Kelley et al., 1963), *History of the Birds of Kingston, Ontario* (Quilliam, 1973), *Birds of the Oshawa-Lake Scugog Region, Ontario* (Tozer and Richards, 1974), and *The Birds of Simcoe County, Ontario* (Devitt, 1967). Thus the ONRS is certainly the most comprehensive source of information on the breeding biology of the birds of Ontario.

In 1976 R. D. James, P. L. McLaren, and J. C. Barlow published *Annotated Checklist of the Birds of Ontario*, which listed the current status and gave a brief outline of the distribution of all birds that had been documented in Ontario. However, the breeding ranges and egg dates of many species included in that work differ from those given in the present volume because of the continuing accumulation of new data. In *Breeding Birds of Ontario* we endeavour to correlate and summarize all breeding and nesting records from the ONRS and from other available sources, published and unpublished.

With the addition of the American avocet and the black-billed magpie in 1980, the Ontario breeding-bird list totals 283 species. Nine species are considered to be hypothetical, that is, without material evidence of breeding: yellow-crowned night heron, brant, canvasback, short-billed dowitcher, western kingbird, Bohemian waxwing, northern shrike, Connecticut warbler, and pine grosbeak. The first four, the nonpasserines, are covered in this volume, in a separate section. There are a further 10 species in the breeding-bird list for which no nest has yet been found, but whose breeding has been confirmed by the collecting or photographing of flightless young: whistling swan, Ross's goose, greater scaup, bufflehead, king eider, surf scoter, pectoral sandpiper, stilt sandpiper, Hudsonian godwit, and American avocet. There are unsubstantiated reports of nesting in the ONRS files for six of the 19 species named above: yellow-crowned night heron, canvasback, bufflehead, western kingbird, Connecticut warbler, and pine grosbeak.

The present volume, which deals with the nonpasserines, covers 143 species. All available records have been considered for use in the nidiology accounts, and those as recent as 1980 have been used for some species with few nest records. The second volume will deal with the passerine species breeding in Ontario.

Some islands in James and Hudson bays are close to Ontario and are associated geographically with the province, but are under the jurisdiction of the Northwest Territories. The breeding records of some species on these islands are indicated in this volume, especially if few records exist in Ontario.

The order and nomenclature of bird species in this volume follow those of the American Ornithologists' Union check-list (1957) and associated supplements in 1973 and 1976. Plant names are based on the nomenclature used in the *Manual of Vascular Plants of Northeastern United States and Adjacent Canada* (Gleason and Cronquist, 1963) if the species were included in that work.

Records and Methods

Ontario Field Work Among the major sources of breeding-bird data for this volume are the specimens, photographs, and observations obtained on the various ROM and ONRS provincial field trips. Prominent among the trips were the ROM faunal investigations of certain provincial areas undertaken between 1923 and 1949, and the results of some of these trips are published here for the first time. Another major source of information was the observations of breeding birds and the collection of specimens obtained during explorations in 1947 and 1949 for the Geodetic Service of Canada and the Geographical Bureau and reported in *Birds of the West James Bay and Southern Hudson Bay Coasts* (Manning, 1952). Records from the provincial and national parks and reports of field work by the Ontario Ministry of Natural Resources and the Canadian Wildlife Service were also important data sources. In addition, we acquired much valuable information from the records of private field studies of many individual naturalists and from the work of bird observatories (Long Point and Prince Edward Point), conservation authorities, environmental consulting firms, and naturalists' and ornithological clubs; from nest-box projects of individuals and clubs; and from university studies.

It should be pointed out that, apart from reports of investigations carried out in areas where little previous work had been done, the bulk of the records in the ONRS reflect in their distribution a definite correlation with the distribution of contributors. Therefore, there are many more nesting and breeding records from more densely populated areas of Ontario (e.g., Durham, Halton, and York regional municipalities, and Northumberland and Simcoe counties) than from less populated parts of the province. There is a disparity in the number of records from southern and northern Ontario, which emphasizes the need for further investigations in the north.

Abbreviations The following abbreviations are used in this volume: DBH (diameter breast height), DM (District Municipality), E (egg), NMC (National Museum of Natural Sciences, National Museums of Canada), N (Nest), NWT (Northwest Territories), ONRS (Ontario Nest Records Scheme), RM (Regional Municipality), ROM (Royal Ontario Museum), ROM PR (Royal Ontario Museum Photographic Record).

Metric Conversion On almost all the ONRS cards, the various measurements recorded are given in the British Imperial system. However, since it is the metric system of measurement that is universally used in scientific publications, it has been necessary to convert these measurements to the metric system, and to record the British Imperial equivalents in parentheses. Because most measurements on the cards are estimated and because birds are not exact in nest construction and positioning, we have usually rounded off the figures when converting them to the metric system.

Nest-Card Data Processing Most of the information included in the nidiology accounts comes from the nest-record cards of the ONRS. Usually, if the nidiology information was provided on those nest cards, we have not cited further documentation. If any nidiology account here lacks information, it is because such details were not currently available.

More than 80 000 ONRS nest cards will be examined and summarized for the nidiological accounts in the two volumes of this work, as will a large number of breeding records and specimen or other material evidence of nesting or breeding from many different sources. But only information available about birds in Ontario will be included.

For the present volume, a total of 17 757 nest cards were processed. With three exceptions, every card for each species was used in the data summaries and averages. The three exceptions are killdeer (946 cards), black tern (722 cards), and mourning dove (1602 cards); because of the large number of record cards for these three species, every card was examined but a limited number of 8 to 10 cards per provincial region was actually used for data extraction. These particular cards were selected on the basis of the information they supplied (earliest and latest egg dates, stated incubation periods, multiple nest visits, unusual clutch sizes, nest locations, nest heights), and at the same time care was exercised to avoid the introduction of bias through the overselection of unusual information.

The term "records" at the beginning of the nidiological section of each species account is used synonymously for "ONRS nest-record cards". It does not necessarily indicate the number of nests considered because many nest-record cards, especially those of colonial species, list more than one nest per card. Wherever the numbers do not agree, the total numbers of nests is listed in parentheses after the total number of record cards. All the information in the nidiological section for each species, including the data in the subsections on eggs, incubation periods, and egg dates, has been derived from those nest-record cards. However, because the nest-record cards vary greatly in the amount of information they contain, the numbers of nest-record cards used to summarize clutch sizes, incubation periods, and egg dates often do not coincide with the total number of cards considered for each species.

The nidiological accounts present data in the following order: breeding habitats, nest locations, nest positions, nest heights, nest constructions, nest materials, nest sizes, clutch sizes, average clutch ranges, incubation periods, and egg dates. Since the analysis of nest-card data frequently involved the numerical listing of habitats, nest positions, tree and plant types, and nest materials, an attempt was made to list each in order of the species' preference, that is, the most frequently used first and the least frequently used last, and often to group them according to category. Incubation periods are given wherever they were stated or could be ascertained from nest cards; however, data on duration of nest building, hatching success, fledging period, and breeding success are omitted from the accounts owing to a lack of sufficient information on the cards.

Nest heights, nest sizes, clutch sizes, incubation periods, and egg dates, where sufficient numbers demanded, were averaged by the interquartile range (the middle 50 per cent) method, rather than by the conventional method of averaging, which in many cases would result in an unrealistic number, for example, an average clutch size of 3.5 eggs. The interquartile range values are derived as shown in the following example of an average-clutch-range calculation.

Data In 100 clutches ranging from 2 to 12 eggs, the number of nests with various clutch sizes were as follows:

2 eggs—3 nests
3 eggs—6 nests
4 eggs—9 nests
6 eggs—18 nests }
7 eggs—22 nests } 50 nests or 50 per cent of the total number of nests are included in these three clutch sizes
8 eggs—16 nests }
9 eggs—10 nests
11 eggs—9 nests
12 eggs—7 nests

The average clutch range includes the commonest clutch sizes, that is those that together represent at least 50 per cent of the total number of nests. In each species account the average-clutch range entry is followed by a number in parentheses, which is the total number of nests represented by the average clutch sizes, and as in the above example, this total may exceed 50 per cent of the nests considered: the total number of nests with the commonest clutch sizes is 56, or 50 per cent of all nests considered plus 6 nests. The entry would appear as follows: 6 to 8 eggs (56 nests).

The eggs subsection of each account lists all clutch sizes given on nest-record cards unless a card indicated specifically that a clutch was incomplete. Therefore, some incomplete clutch sizes may be included here because they were not identified as such; however, the use of the interquartile range method for deriving the average clutch range eliminates unusually small or large clutches, which were possibly either incomplete or the product of more than one female. When we had information that clutches beyond the size limits of the average clutch range were successfully incubated by one pair, we indicated these clutch sizes in boldface type along with those within the average clutch range.

We ascertained incubation periods from those cards with information from multiple visits that indicated completed-clutch and hatching dates. Most of the incubation periods represent the interval between the laying of the last egg and the beginning of hatching. For those species that commence incubation with the laying of the first egg (e.g., cuckoos and owls), we used the date of the laying of the first egg and date of the beginning of hatching to define the incubation period; we assumed that the first egg laid was the first to hatch.

The egg-dates subsection lists the earliest and latest dates on which viable eggs were recorded in active nests of that species and the dates for the middle 50 per cent of those nests, to indicate the height of the laying season. When a nest with viable eggs was visited more than once, we used the first and last egg dates recorded; thus the number of egg dates used in our calculations (given in parentheses) is often greater than the number of nests considered, as is illustrated in the following example: 62 nests, 2 May to 20 July (89 dates), 31 nests, 14 June to 3 July.

For colonial species, the number of colonies listed refers to the number of records available for all colony nestings, not to the number of separate colony sites. Because the numbers of different clutch sizes vary with the time of the visit to the colony (i.e., large numbers of incomplete clutches were often recorded early in the breeding season and relatively few clutches were present to record late in the season), we decided to give clutch-size percentages for some colonial species as an indication of this possible bias in average clutch sizes. Clutch-size percentages were also used for three noncolonial species (killdeer, spotted sandpiper, and great horned owl) that usually lay relatively determinate clutches, to help clarify the number of different clutch sizes recorded in the many available nest records. Egg dates for colonial species represent colonies with eggs in nests, not individual nests.

Large amounts of data, much of which have never before been published, have been summarized in this work and reveal some unique or seldom reported information about avian nesting. In some instances, this information is at variance with published data. A few examples of such information are as follows:

1) Nest habitats: interesting differences in nest habitat selection between similar species (e.g., common gallinule/American coot; Virginia rail/sora).

2) Nest location: green heron in a wood duck nesting box; belted kingfisher in a tree cavity.

3) Nest trees: common flicker preferring dead deciduous trees; yellow-bellied sapsucker preferring living deciduous trees; all woodpecker species preferring southerly exposures for cavity entrances.

4) Nest materials: red-tailed and red-shouldered hawks incorporating old passerine nests in nest structures.

5) Nest heights and sizes: extreme and average ranges based on many nests from throughout the province.

6) Clutch sizes: common loon with 4 eggs; killdeer with 7 eggs; the first published summary of average clutch sizes for the province of Ontario.

7) Colony sizes: averages presented for colonial species for which data is available.

8) Incubation periods recorded for those species for which data was available; unusual lengths of incubation (e.g., killdeer, 33 days) and of hatching (e.g., spotted sandpiper, 4 days).

9) Egg dates: extremes and averages listed for those species for which data was available.

10) Multiple clutches: indicated for those species where known.

11) Cowbird parasitism: indicated for all known hosts; unusual hosts include Virginia rail and spotted sandpiper.

Breeding Distribution By 1975 the provincial government had altered a number of county and district boundaries and created new regional and district municipalities to replace some of the former areas. The 52 provincial regions (counties, districts, and regional and district municipalities) now extant are outlined on the breeding distribution map for each species. Each map shows the distribution of nesting and breeding records currently available in those regions for the species concerned. All records from former counties and districts made before 1975 are shown on these maps in the current regions. Figures 2 and 3 outline and name each region and a number of other localities in the province. Initially, we considered plotting breeding records near the limits of each species' range and representing the range as

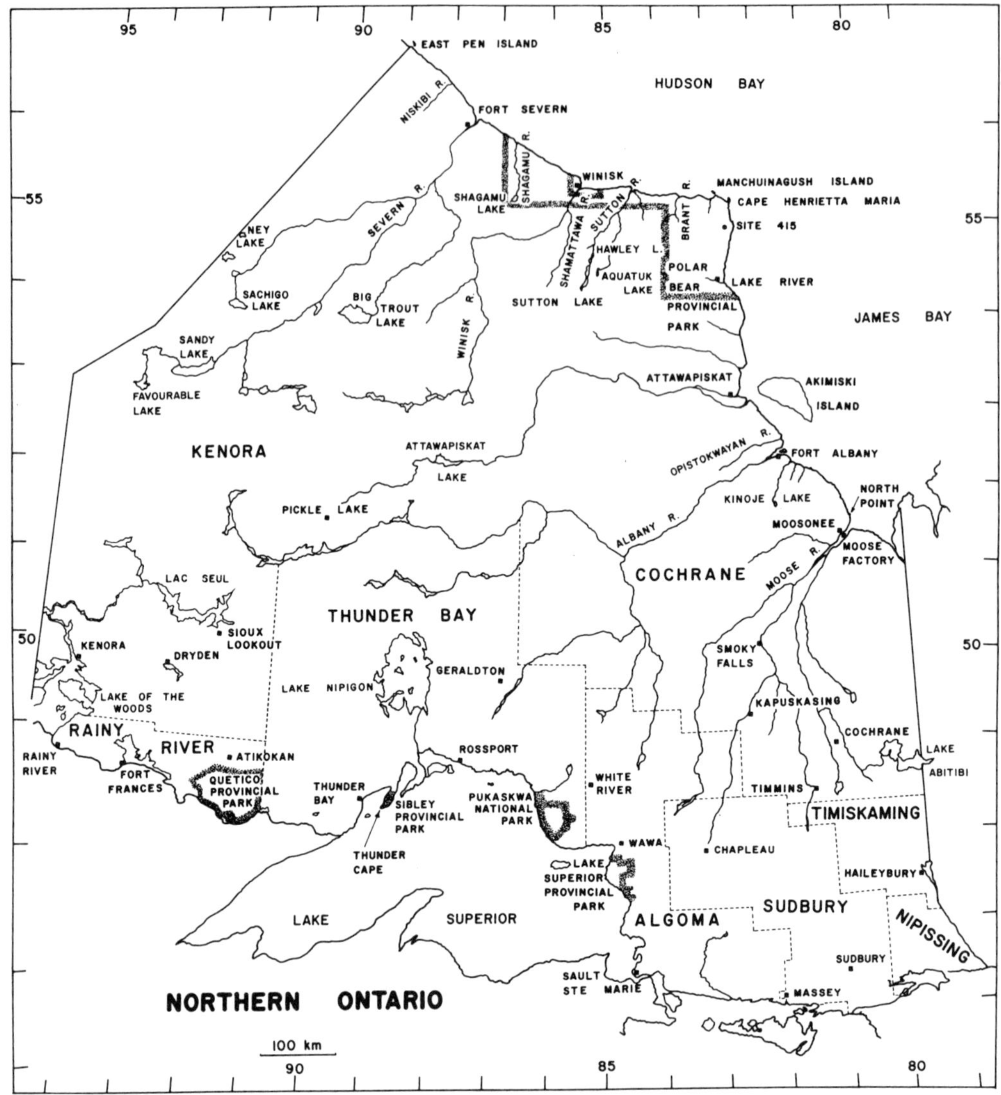

Fig. 2 Map of northern Ontario showing provincial district boundaries and localities mentioned in the text.

a shaded area on the map. However, this approach presented several problems. Very often the available records indicated the distribution of the observers but bore little relation to the range of the species in question. As well, the plotted symbols were often widely scattered, thus any line connecting them was very irregular or encompassed large areas where no records existed. We decided, therefore, to plot the available records individually with no shading, and to add a few words of explanation to amplify what is known or believed to be the true range of each species. We also considered using different-sized symbols to reflect the abundance of each species in different areas of the province. However, once again, the records indicated only the relative abundance of observers. Finally, we considered including records of summering birds on the breeding distribution maps of the species concerned. But because it was often uncertain whether the summer records were for late spring migrants, early fall migrants, or nonbreeding birds, we decided to omit such records.

Although much smaller in area, southern Ontario has far more observers than northern Ontario. We decided, therefore, to treat the two parts of the province differently. In southern Ontario (south of Lake Nipissing and including Manitoulin Island), only one symbol was plotted per county, district, or regional municipality. In the north (including northern Nipissing District), all records were plotted individually, unless numerous records existed for the same general area.

The following four symbols appear on the breeding distribution maps:

● Nesting or breeding documented by photograph or specimen.
▲ Nesting or breeding as a sight record (mostly ONRS records).
○ Historical nesting or breeding documented by photograph or specimen.
△ Historical nesting or breeding as a sight record (mostly ONRS records).

It seemed important to indicate that specimen or photographic evidence was available to substantiate nesting or breeding in the province, particularly in the case of rarer species or those for which we have but a single record. In fact, the existence of such documentation of nesting or breeding is of interest for all species, and we indicate it on each breeding distribution map.

For quite a few species our only breeding records were obtained years ago, and yet we know or strongly suspect that the species still breed in the same areas as were documented earlier. Therefore, records have been plotted regardless of their date. Where we know that a species' range has diminished, the records are indicated to be historical.

Most of the symbols on the maps represent nest records taken from the ONRS cards and include all nest and egg specimens housed in the ROM, details of which have been transferred to the ONRS cards. However, many symbols represent breeding records, that is, reports from a variety of sources of downy young or very recently fledged young. For this reason, the number of regions with symbols plotted on the breeding distribution maps may be greater than the number of regions represented by ONRS nest cards (quoted at the beginning of each species account). All available nest and breeding records up to and including 1980 have been considered for mapping.

The Ontario breeding distributions presented include records of several species at the limits of their known range anywhere. Among these, red-throated loon, snow goose, and American golden plover are nesting at the most southerly limits of their ranges, and marbled godwit at the most easterly known limit of its range.

Photographic records were not considered as material evidence if such documentation remains in private hands. Past experience has shown that such records may be lost and never added to the collections of a public institution for preservation. Most of the information used to compile breeding distribution is housed in the ROM. Many additional records used are to be found in the National Museum of Natural Sciences, National Museums of Canada (NMC), in Ottawa, and a few of the records used are scattered in several other North American museums and institutions.

Details regarding the distribution of 210 bird species known to breed in the province were presented by Baillie and Harrington (1936, 1937), and we have updated information on the ranges of these species here. For species discovered breeding in Ontario since that time, we refer interested readers to published accounts, if available, of the circumstances surrounding their discovery. In most cases the first published references are used, but if they do

Fig. 3 Map of southern Ontario showing provincial counties, districts, regional and district municipalities, and other localities mentioned in the text.

not provide adequate information, a later, more detailed reference is used. If no detailed information is published about a new Ontario breeding species, we include available information here. If the nest-record location stated in a species account is not referenced, it may be assumed that this information is on an ONRS card. Of course, within the range of breeding distribution outlined in the accounts, the species will only be found where suitable habitat is available.

Although many years of work and the examination of a large number of records have been necessary to produce a work of this nature, a perusal of some maps and species accounts will reveal that large gaps in our knowledge of Ontario breeding birds still exist. From an ornithological standpoint, most of northern Ontario and even some regions of southern Ontario, especially counties in the extreme southeast, remain relatively unexplored. For many species, multiple nest visits are needed to calculate their incubation periods. Indeed, multiple nest visits together with nest-content examinations improve the scientific value of any nest record. In most instances such examinations can be carried out with a minimum of disturbance to the birds, provided that care and imagination are used. As well, better descriptions of habitat, nest construction, and nest material are necessary. We have recently redesigned the ONRS nest card in the hope that contributors will provide more detail about each nest found. For some species we are at a stage where quality of observation is as important or more so than number of observations.

Undoubtedly there are a number of records that we have unintentionally omitted from this work. More important, there are no doubt hundreds of records by numerous individual observers that have been omitted because they were not made available to us. Such a summary is only as good as we can make it with the information we have. It is through the smaller and even seemingly insignificant observations of many that collectively we can compile useful and meaningful data on our birdlife. An understanding of the distribution and breeding biology of these birds may contribute directly to their future survival. We are concerned that future generations benefit from as complete a compilation as possible. We urge any and all who can extend in some way our knowledge of the breeding birds of Ontario to make their contributions available.

Breeding Bird Species

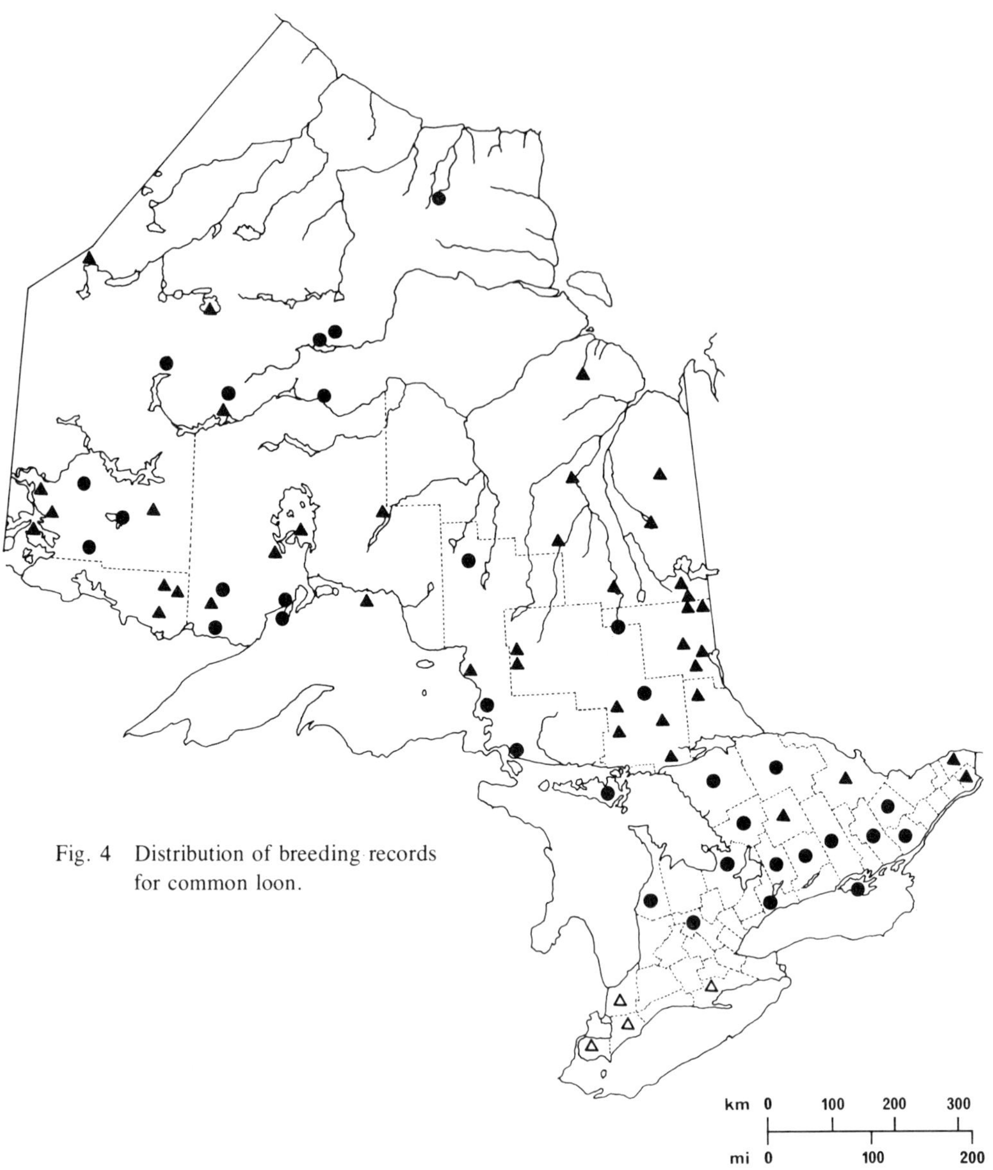

Fig. 4 Distribution of breeding records for common loon.

Common Loon, *Gavia immer* (Brünnich)

Nidiology

RECORDS 234 nests representing 26 provincial regions.
Breeds in forested and cleared regions on freshwater lakes, ponds, and occasionally rivers.

Nests were invariably situated close to water on the edges of islands and islets and on main shores of lakes and rivers, and in water on piled-up vegetation, rafts of floating vegetation, and muskrat houses.

Nests were positioned in standing cattails, bulrushes, sedges, and grasses; under trees and other vegetation such as Labrador tea, blueberry, and sweet gale; on bare ground, rock, or gravelled rocky areas; and once on top of a rotted log. Within 9 to 12 m (30 to 40 ft) of 1 island nest site there were 2 active herring gull nests.

Nest structures ranged from large, heaped-up masses of material, variously lined with cattails, grasses, mud, twigs, rootlets, mosses, pine needles, cedar sprigs, bark, and leaves, to shallow depressions in existing vegetation. At times there was no nest at all, and eggs were laid in unlined depressions in bare ground or rock. Some nests were reused in successive years.

EGGS 125 nests, 1 to 2 eggs; **1E** (33N), **2E** (91N), 4E (1N).
Average clutch range 2 eggs (91 nests).
One nest containing 4 eggs was the product of 2 females, and the eggs did not hatch. Renesting sometimes occurred after desertion or predation.

INCUBATION PERIOD 2 nests, 29 days.

EGG DATES 113 nests, 11 May to 25 August (130 dates); 57 nests, 2 June to 23 June.

Breeding Distribution

While summer sightings indicate that the common loon occurs as far north as the Hudson Bay coast, there is only one actual report of nesting in the far northern portions of the province. Once found breeding throughout the south, the species apparently no longer nests in the Deciduous Forest region and almost all recent records are from the Canadian Shield (Fig. 156B).

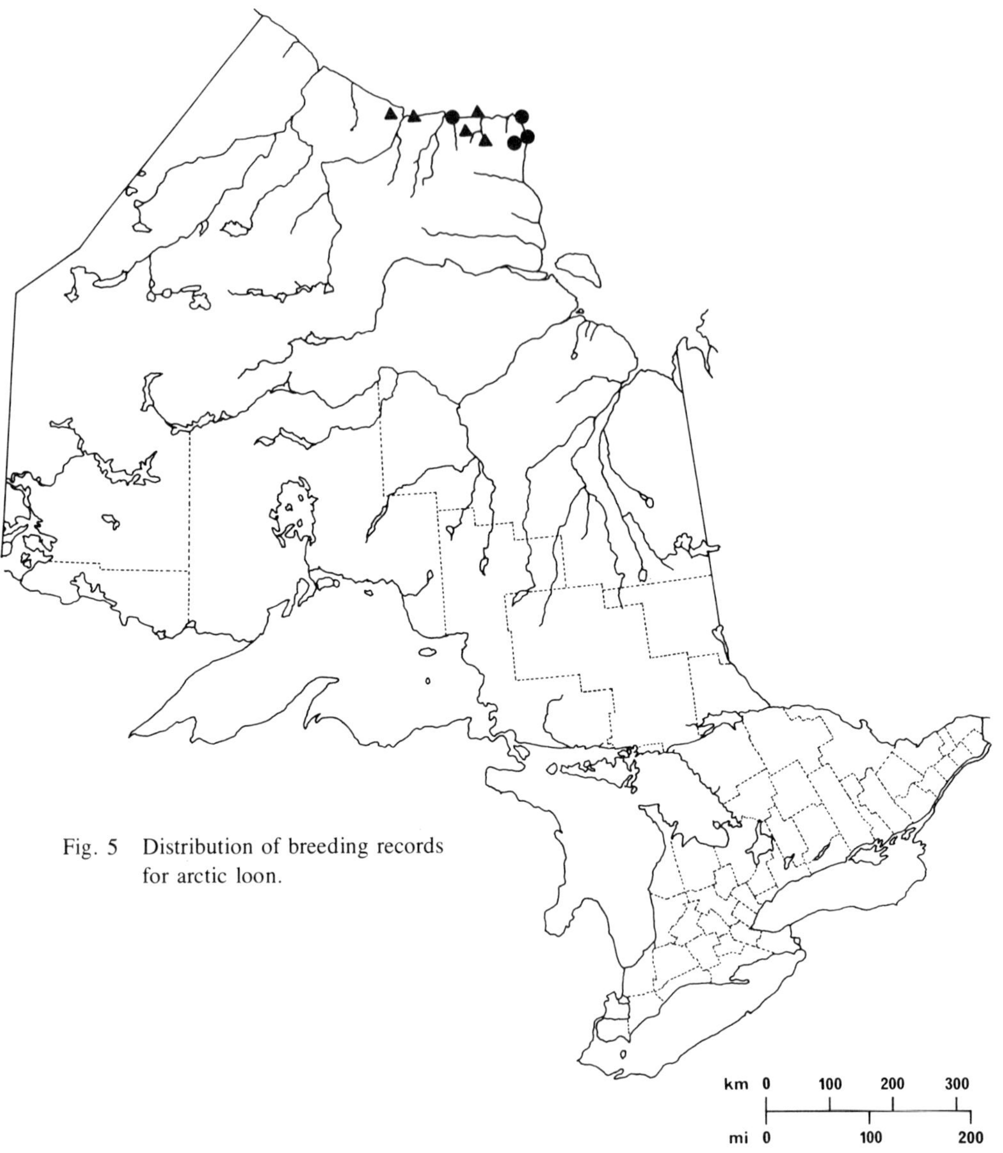

Fig. 5 Distribution of breeding records for arctic loon.

Arctic Loon, *Gavia arctica* (Linnaeus)

Nidiology

RECORDS 7 nests representing 1 provincial region.

Breeds on freshwater lakes and sloughs from 145 to 550 m (450 to 1800 ft) in length.

Nests were situated on the shore close to the water's edge (4 nests) or in the water in clumps or islets of sedge and willow 14 to 18 m (45 to 60 ft) from shore (2 nests). Water depth at 1 nest was 20 to 30 cm (8 to 12 inches). Another nest was found in the midst of a small herring gull colony on a sedge islet in a large slough, and still another was seen near a solitary herring gull nest on a small island. Two nests in adjoining sloughs were ca 550 m (1800 ft) apart.

Nests were usually in the open on solid ground, or were built-up mounds of vegetation in the water among standing growths of sedge, grass, and arctic willow. One nest on a willow islet was beneath a gnarled horizontal willow branch.

Nest structures ranged from large, heaped-up mounds of sedge and/or grass and decaying vegetation in the water to sparsely lined depressions in the ground. Nests were lined with sedge and grass stalks, mud, rootlets, and decaying vegetation. Outside diameters ranged from 46 to 61 cm (18 to 24 inches).

EGGS 7 nests, each with 2 eggs.

INCUBATION PERIOD No information.

EGG DATES 7 nests, 23 June to 4 July.

Breeding Distribution

A bird of Canada's arctic regions, the arctic loon (Fig. 166B) breeds in Ontario only in the tundra areas along the north coast. Most records are from the Cape Henrietta Maria area, and although records are lacking west of the Winisk region, this species probably breeds at least sparingly along the coast where tundra conditions prevail. Many adults, some with young, have been noted on tundra pools, even though only a few nests have been reported thus far (Peck, 1970).

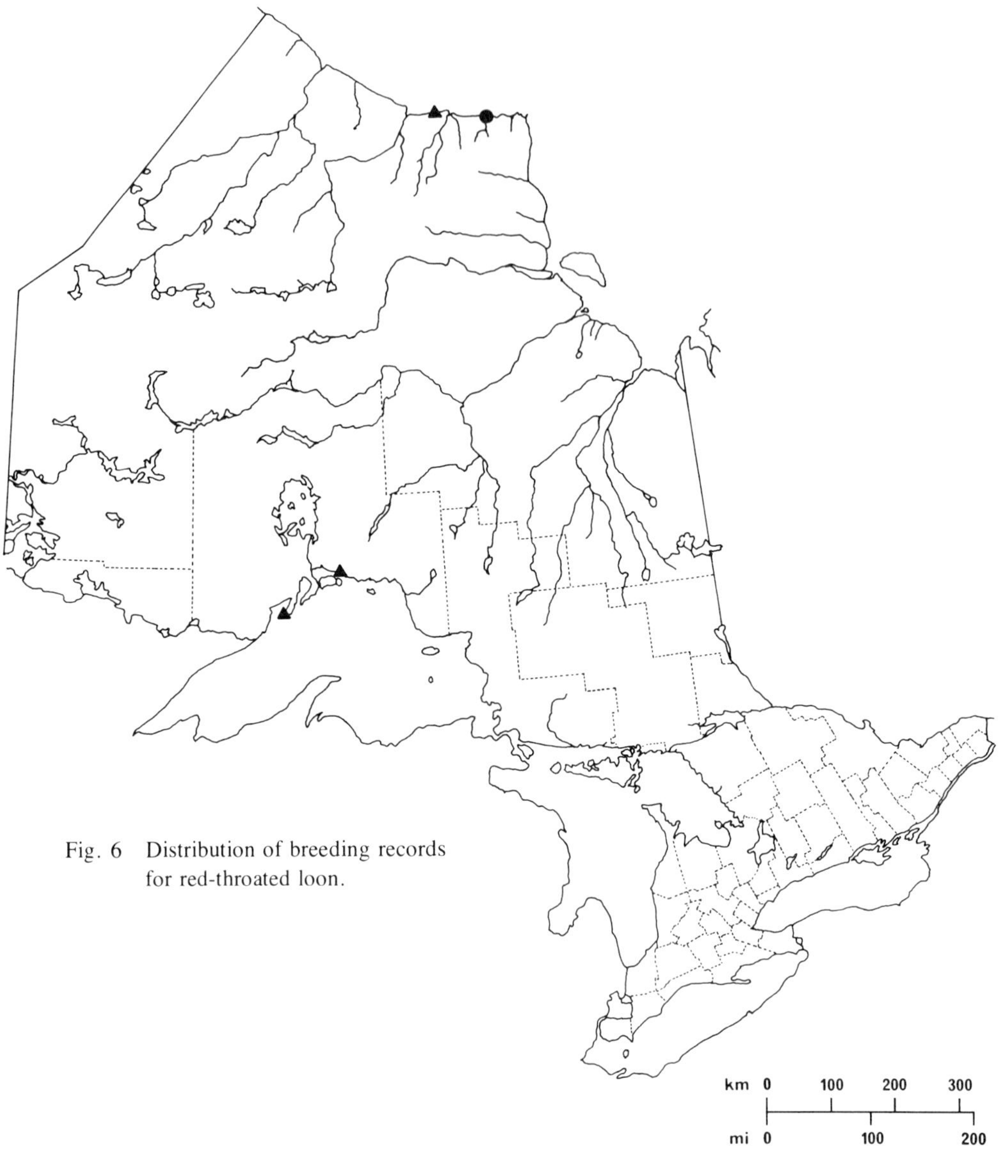

Fig. 6 Distribution of breeding records for red-throated loon.

Red-throated Loon, *Gavia stellata* (Pontoppidan)

Nidiology

RECORDS 2 nests representing 2 provincial regions.
This species usually breeds on small freshwater ponds on coastal tundra. One nest was found on a small island in Lake Superior.

One nest, with a luxuriant growth of sedge around its edge, was in a small tundra slough between 0.6 and 1.5 m (2 and 5 ft) deep. The other, in Lake Superior, was on a small, rocky island whose centre was covered with low bushes.

EGGS 2 nests, each with 2 eggs.

INCUBATION PERIOD No information.

EGG DATES 2 nests, 1 July and 3 August.

Breeding Distribution

The first evidence of breeding of the red-throated loon in Ontario came from the north shore of Lake Superior in 1912, when a nest was reported at Thunder Cape (Baillie and Harrington, 1936). This nest, together with a later breeding record at Rossport in 1941 (Baillie and Hope, 1943), went undocumented. These reports are now known to be of isolated nestings far south of the normal breeding range of this loon. The first documented breeding of the species in the province was closer to its normal breeding range: adults with downy young were photographed on the tundra area west of Cape Henrietta Maria in 1966 (Simkin, 1968). This record, together with one other nest record from the north coast, indicates a very limited breeding distribution in northeastern Ontario.

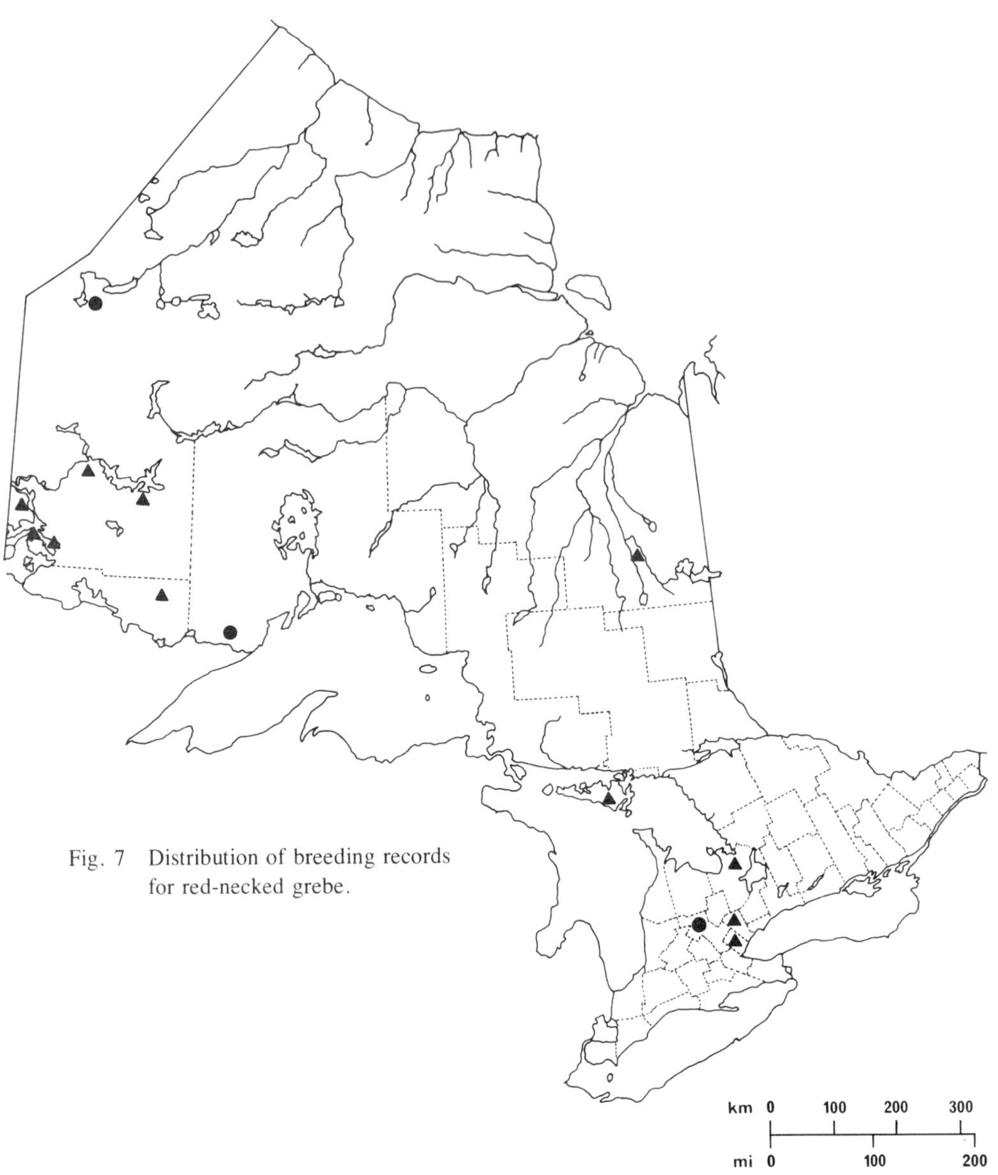

Fig. 7 Distribution of breeding records for red-necked grebe.

Red-necked Grebe, *Podiceps grisegena* (Boddaert)

Nidiology

RECORDS 71 nests representing 8 provincial regions.
This mainly western-breeding grebe nests in Ontario in marshy areas of freshwater lakes. It was once found nesting in a boat harbour in a lake-side town.

Nests were floating structures attached to standing growths of bulrush, cattail, and similar emergent vegetation, and at 1 location they were fastened to submerged willow branches. In a boat harbour nests were built on manmade wooden floats and rafts. Nest-site water depths ranged from 25 to more than 150 cm (10 to 60 inches). Nest positions ranged from close to shore (ca 1 m [3.5 ft] away) to as far as 18 m (60 ft) from it, in open water.

Nests were built-up mounds of mud, decaying vegetation, and sticks, and were variously lined with dead and living stems of sedge, bulrush, and cattail, as well as with pondweed, water milfoil, rootlets, and leaves. As in most grebe nests, eggs were covered with decaying vegetation when the nest was vacated by the adult. Outside diameters of 3 nests ranged from 36 to 46 cm (14 to 18 inches); inside diameter of 1 nest was 18 cm (7 inches); inside depth of 1 nest was 4 cm (1.5 inches).

EGGS 52 nests with 1 to 7 eggs; 1E (11N), **2E** (11N), **3E** (7N), **4E** (14N), **5E** (5N), **6E** (3N), **7E** (1N).
Average clutch range 3 to 5 eggs (26 nests).

INCUBATION PERIOD 1 nest, 23 days.

EGG DATES 60 nests, 15 May to 17 September (72 dates); 30 nests, 12 June to 19 June.

Breeding Distribution

The red-necked grebe apparently breeds regularly only in the portion of the province lying west of Thunder Bay in the south and Favourable Lake in the north. However, an isolated colony recorded as far east as Cochrane indicates that the species possibly breeds locally over a more widespread area throughout the less accessible regions of central Ontario.
Isolated nestings have occasionally occurred at several locations in southern Ontario. The irregular and isolated nature of occurrences in the east and south makes documentation of the species' breeding distribution difficult.

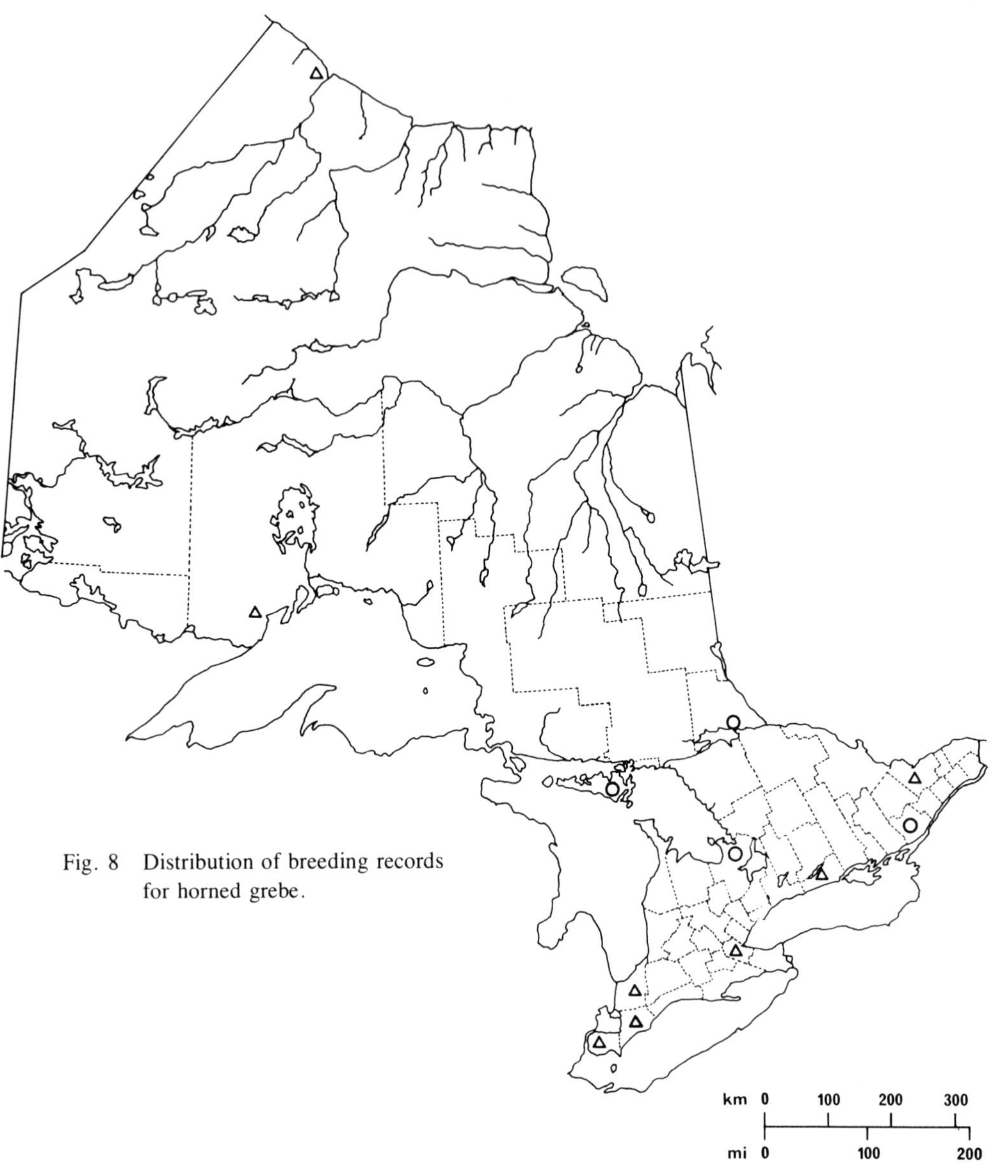

Fig. 8 Distribution of breeding records for horned grebe.

Horned Grebe, *Podiceps auritus* (Linnaeus)

Nidiology

RECORDS 12 nests representing 5 provincial regions.
This species formerly nested sporadically in a few Ontario locations. Because of the similarity of its nest and eggs to those of the pied-billed grebe, a degree of uncertainty exists about the authenticity of most of the few nest records available. The only egg specimens were from nests amid bulrushes at the edge of a large lake.

Nests were floating structures or, in shallow water (ca 30 cm [1 ft] in depth), were built up from the bottom. Nests were often placed in exposed situations and attached to standing reeds and other vegetation. One nest was 15 m (50 ft) from shore. At another nest the water depth was 60 cm (2 ft).

Nests were made of wet rushes and decaying vegetation. One nest was built up above water to a height of 10 cm (4 inches).

EGGS 10 nests with 1 to 8 eggs; 1E (1N), 2E (1N), **3E** (1N), **4E** (2N), **5E** (2N), **6E** (1N), 7E (1N), 8E (1N).
Average clutch range 3 to 6 eggs (6 nests).

INCUBATION PERIOD No information.

EGG DATES 10 nests, 16 May to 1 July; 5 nests, 29 May to 12 June.

Breeding Distribution

A species of widespread distribution in western Canada, the horned grebe was believed to breed right across central Ontario (Godfrey, 1966) and earlier in this century across southern Ontario (Palmer, 1962); there are virtually no substantiated nesting or breeding records for northern Ontario. The documented records from southern Ontario are based on eggs that are indistinguishable from those of the pied-billed grebe. Given the very secretive nature of both species' behaviour near the nest, these eggs could have easily been misidentified. In fact, there have been no verified nests of this species in Ontario since 1938 and it may no longer breed in the province.

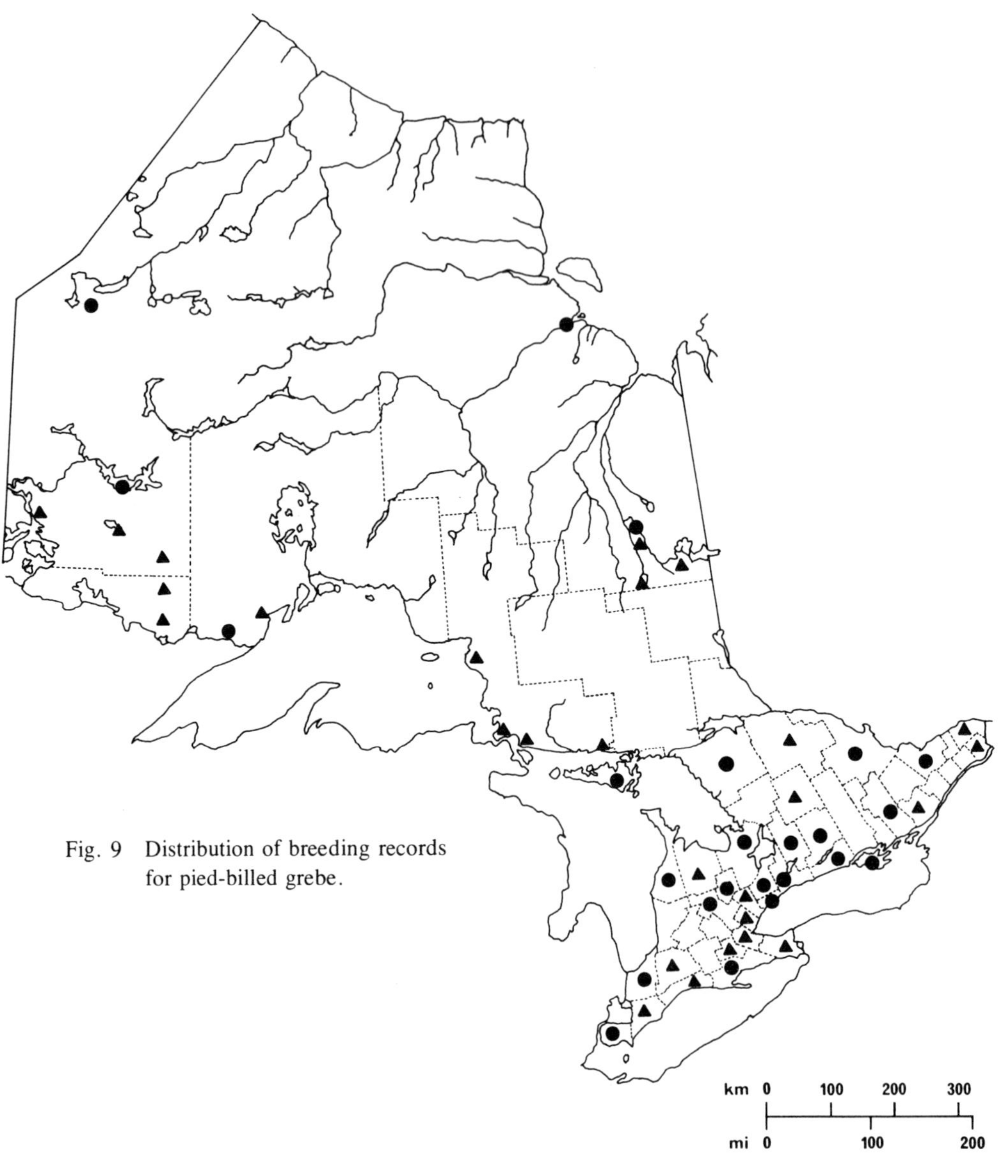

Fig. 9 Distribution of breeding records for pied-billed grebe.

Pied-billed Grebe, *Podilymbus podiceps* (Linnaeus)

Nidiology

RECORDS 368 (387 nests) representing 28 provincial regions.
Breeds in marshes and in marshy areas of large and small freshwater lakes, rivers, ponds, and bogs.

Nests were almost invariably floating structures that were situated in standing marsh vegetation (cattails, bulrushes, sedges, bur-reeds), anchored to dead trees, or attached to logs. Occasionally they were placed on a more solid foundation such as the middle of a floating, marshy islet. Water depths at nest sites ranged from a few centimetres to more than a metre. Nests were found as close as 9 m (30 ft) to shore. They were usually situated in vegetation from 2.5 to 15 m (9 to 50 ft) from open water but were occasionally found in open water.

The nest mounds were occasionally as high as 38 cm (15 inches); outside diameters ranged from 20 to 46 cm (8 to 18 inches). The base, composed of mud and rotting vegetation, was covered over and variously lined with coarse marsh grasses or with sedges, mosses, aquatic plants, and dead and living stalks of cattail and bulrush.

EGGS 127 nests with 1 to 10 eggs; 1E (10N), 2E (5N), 3E (5N), **4E** (12N), **5E** (15N), **6E** (18N), **7E** (32N), **8E** (23N), 9E (4N), 10E (3N).
Average clutch range 5 to 8 eggs (88 nests).

INCUBATION PERIOD No information.

EGG DATES 131 nests, 3 May to 22 August (144 dates); 65 nests, 20 May to 9 June.

Breeding Distribution

The pied-billed grebe breeds throughout the agricultural part of southern Ontario, but appears to be very sparsely distributed in the Canadian Shield. It may be absent from most of the Boreal Forest region and seldom reaches the northern limits plotted on the breeding-range map on the facing page.

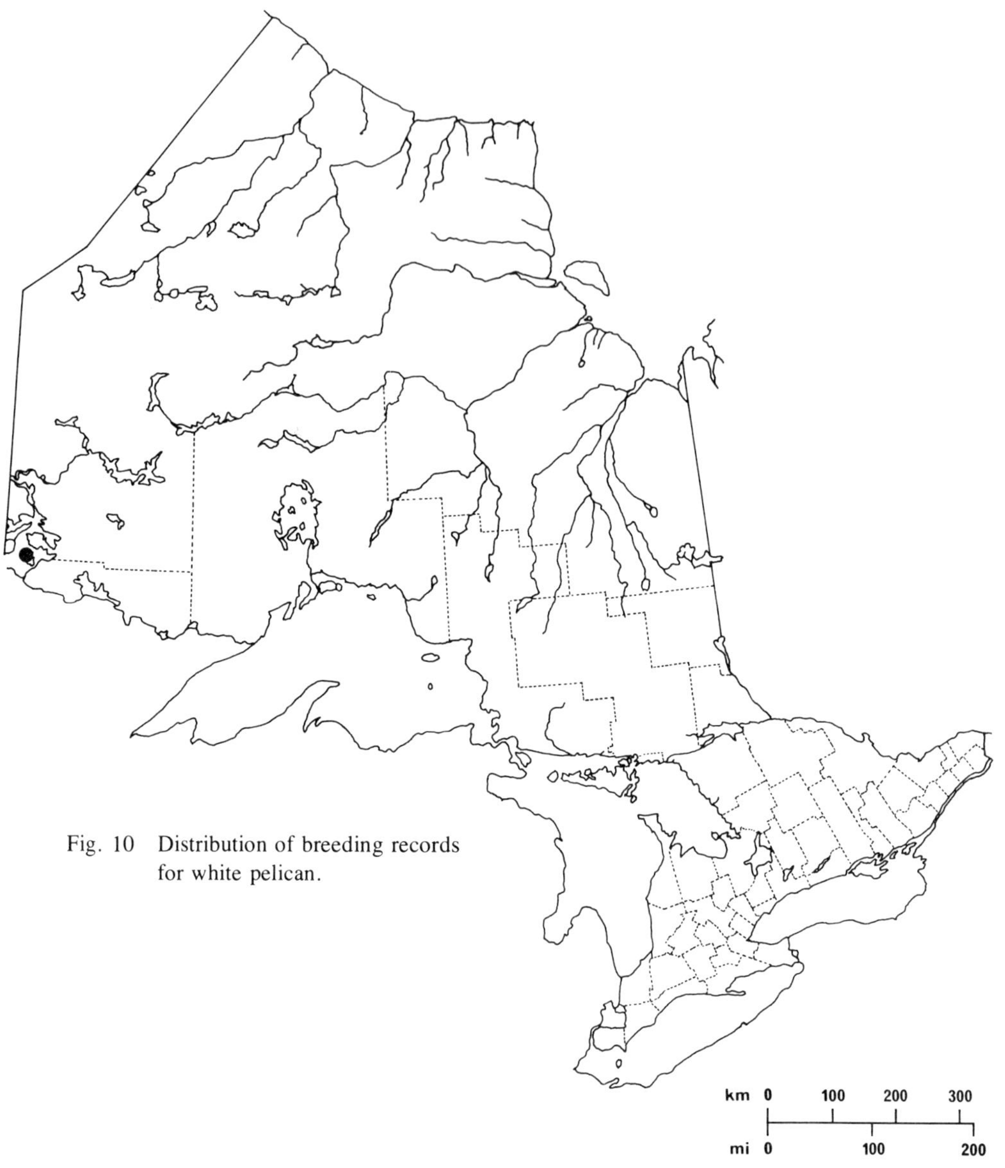

Fig. 10 Distribution of breeding records for white pelican.

White Pelican, *Pelecanus erythrorhynchos* Gmelin

Nidiology

RECORDS 18 (33 colonies, ca 3513 nests) representing 1 provincial region.
The white pelican is a colonial species and nests in homogeneous colonies or in company with double-crested cormorant and herring gull. The average size of 33 colonies was 110 nests.

Colonies were located on small, low, bedrock islands from 0.4 to 1.2 ha (1 to 3 acres) in size. These islands were sparsely vegetated with grasses, nettles, and smartweed, or more thickly covered with high shrubs and trees such as elder and bur oak.

Nests were on the ground, in depressions in the thin soil or sand, and were composed of soil, sticks, and vegetation piled up on the outside. Nests were lined with finer sticks, feathers, leaves, and stems of nearby vegetation.

EGGS 162 nests with 1 to 3 eggs; **1E** (45N), **2E** (113N), 3E (2N).
Average clutch range 2 eggs (113 nests).

INCUBATION PERIOD No information.

EGG DATES 16 records, 8 May to 27 July (21 dates); 8 records, 4 June to 30 June.

Breeding Distribution

A summer resident of the western provinces, the white pelican is known to breed at only one location in Ontario, in the southern part of Lake of the Woods (Fig. 144B). A colony of these pelicans was first reported there in 1938, when about eight pairs were found on Dream Island (Baillie, 1960). Since 1938, subsequent colonies have been reported at the same location. There are now two colonies located on two of the Three Sister Islands and they have grown to about 3000 birds (1500 pairs).

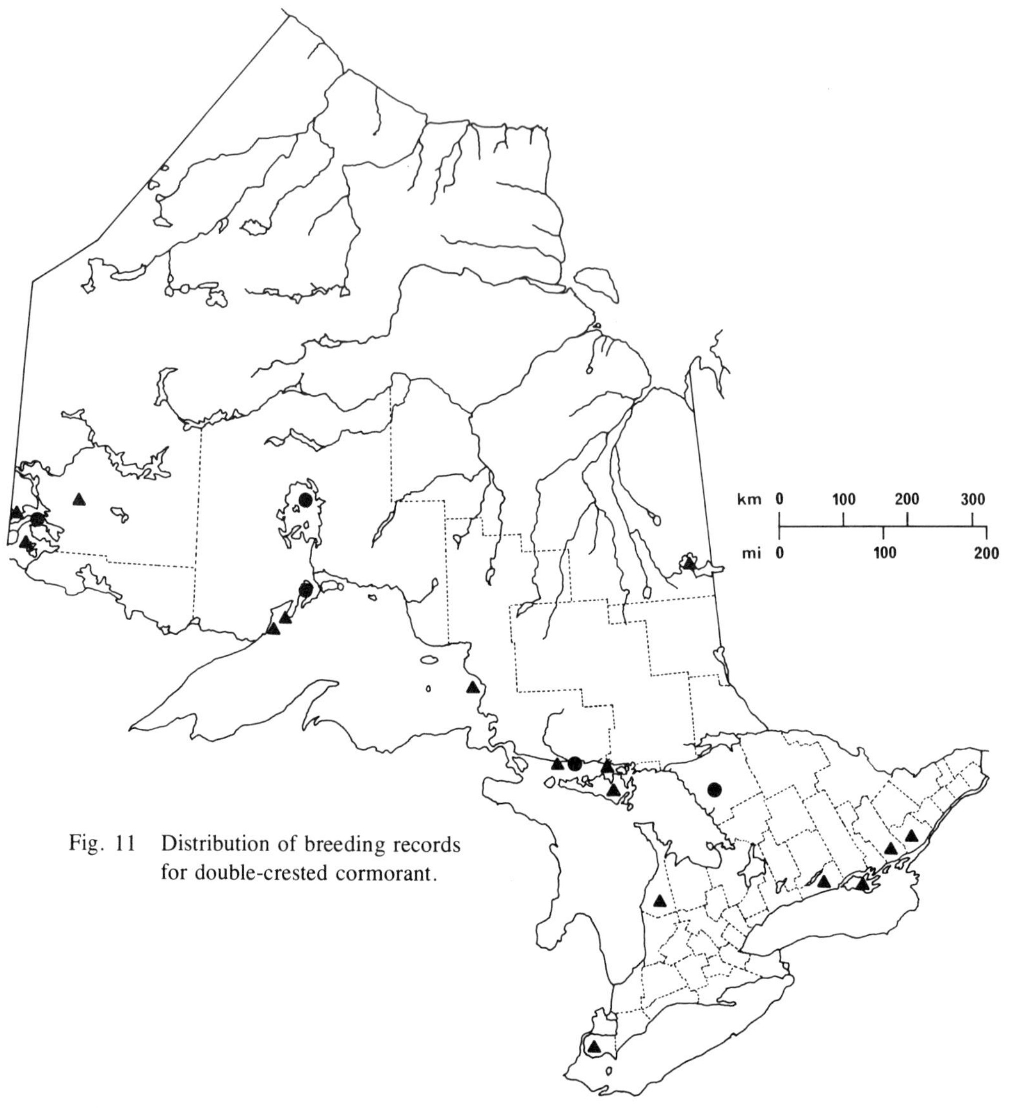

Fig. 11 Distribution of breeding records for double-crested cormorant.

Double-crested Cormorant, *Phalacrocorax auritus* (Lesson)

Nidiology

RECORDS 200 (85 colonies, 7 isolated nestings, ca 8006 nests) representing 11 provincial regions.

For almost 60 years this colonial species has been recorded as breeding in Ontario. Colonies are rarely alone, and are usually in company with herring gull and occasionally with white pelican. The average size of 83 colonies was 96 nests.

Colonies were situated on small (ca 1.2 ha [3 acres] or less) rocky islands in large lakes and rivers. The nesting islands were of bare rock or were sparsely to densely covered with grasses, goldenrod, nettles, smartweed, and other ground vegetation; with shrubs (chokecherry, elder, and hackberry); or with trees (willow spp., cedar, oak spp., poplar spp., and others).

Nests were variously positioned on bare rock; on thin pockets of soil, sand, and gravel; in rank vegetation and shrubs; at the bases of or in trees; and on stumps. Nests were usually on the ground; however, when dead or living trees were present on the island, a few to all the nests were placed in them, at heights ranging from 0.2 to 7.5 m (0.5 to 25 ft). As many as 20 nests were noted in a single tree.

Nests were bulky mounds composed of twigs and sticks that were up to 1.3 cm (0.5 inches) in diameter, and at times nests contained twine, rope, netting, stones, or skeletons of gulls and pelicans. They were variously lined with straw, coarse grasses, rootlets, and feathers, and often with green aquatic plants and other green and dead vegetation. Nests often became heavily covered with excrement as the nesting season progressed. Outside depths of 89 ground nests ranged from 3.8 to 74 cm (1.5 to 29 inches), with 45 averaging 14 to 31 cm (5.5 to 12 inches); outside diameters of 42 nests ranged from 33 to 56 cm (13 to 22 inches), with 21 averaging 42 to 48 cm (16.5 to 19 inches); inside diameters of the same 42 nests ranged from 19 to 33 cm (7.5 to 13 inches), with 21 averaging 25 to 28 cm (10 to 11 inches).

EGGS 1022 nests with 1 to 7 eggs; 1E (125N), 2E (139N), **3E**(307N), **4E** (406N), **5E** (36N), 6E (8N), 7E (1N).
Average clutch range 3 to 4 eggs (713 nests).
In a colony with 44 nests, there were 13 records of renesting after egg or chick loss, and at 1 nest renesting was attempted twice. In a second colony with 27 nests renesting was observed once.

INCUBATION PERIOD 12 nests, ca 24 to 27 days: 1 of ca 24 days, 6 of ca 25 days, 1 of ca 26 days, 4 of ca 27 days. Because incubation appeared to commence before the clutch was complete, hatching was not simultaneous and exact incubation periods were difficult to determine.

EGG DATES 80 colonies, 27 April to 30 August (94 dates); 40 colonies, 5 June to 5 July.

Breeding Distribution

The double-crested cormorant has undoubtedly nested in Ontario for several centuries at least because Tanner (*in* James, 1956) reports it at Lake of the Woods in the 1790s. However, the first documented breeding in the province was in 1924 on an island in Lake Nipigon. Other colonies were probably established some years before that date on Lake Superior (Baillie, 1947); the species was at Agawa Rocks on Lake Superior possibly as early as 1913 (F. Novy pers. comm. to S. Postupalsky), but may have spread from Lake of the Woods to western Lake Superior even earlier. By 1938 the cormorant had spread throughout the Great Lakes and was nesting on a few islands in each lake. It has also nested on Lake Abitibi at least since 1953 and more probably since about 1935 (Smith, 1957). A colony was found in southern James Bay in 1912 (Todd, 1963), but no evidence of nesting in Ontario areas of the bay was ever secured.

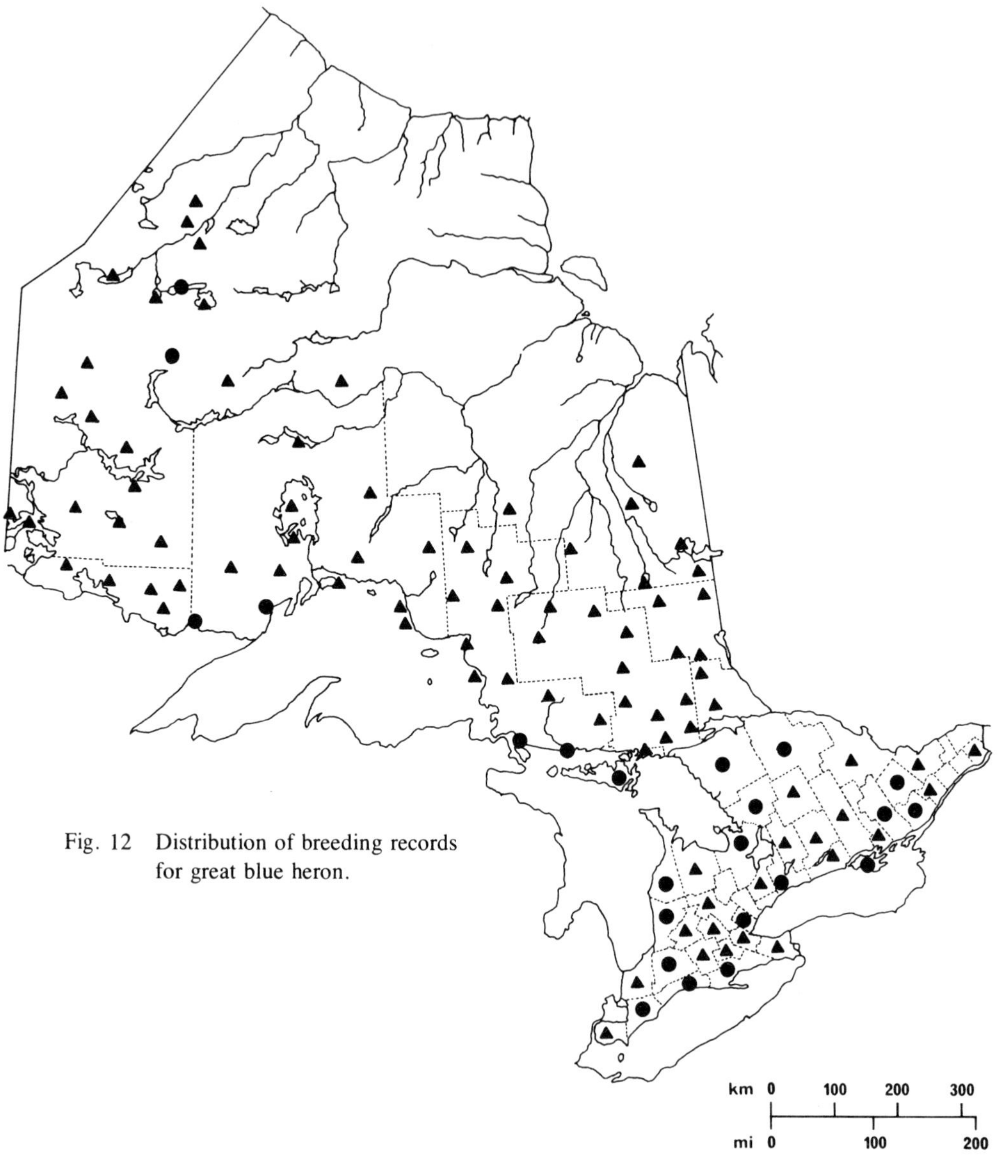

Fig. 12 Distribution of breeding records for great blue heron.

Great Blue Heron, *Ardea herodias* Linnaeus

Nidiology

RECORDS 337 (248 colonies, 34 isolated nestings, ca 12 211 nests) representing 38 provincial regions.

The great blue heron is a colonial species and usually nests in homogeneous or mixed heronries, but occasionally nests singly. In mixed heronries it is found with green heron, cattle egret, great egret, and black-crowned night heron. Its preferred habitats are wet or dry wooded areas, sparsely treed islands, beaver ponds, and marshes. The average size of 248 colonies was ca 49 nests.

Nests were almost invariably situated in trees, and were over water, on islands, or in mainland wood lots at some distance from water. In Thunder Bay District 1 ground nest in a colony of 19 nests, 5 ground nests in a colony of 11, and 1 single ground nest were recorded. Tree nests were in living or dead trees, and both deciduous (19 spp.) and coniferous (6 spp.) trees were used. Nests were most often in upper limbs at some distance from the trunk and sometimes in forks or crotches against the trunk. There were as many as 18 nests in a single tree. Heights of 785 nests ranged from 2 to 30 m (6 to 100 ft), with 384 averaging 11 to 15 m (35 to 50 ft).

Ospreys were recorded nesting in both active and deserted heronries of great blue heron; in 1 active heronry a herring gull nested in a tree nest of this species; in another heronry there were 2 nests of great horned owl, apparently both active in the same season as the heron nests.

Nests were bulky platforms of twigs and branches with outside nest diameters of 45 to 91 cm (18 to 36 inches). They were usually lined with finer sticks, and at least 1 nest was lined with green pine needles.

EGGS 124 nests with 1 to 8 eggs; 1E (13N), 2E (22N), **3E** (21N), **4E** (27N), **5E** (33N), 6E (7N), 8E (1N).

Average clutch range 3 to 5 eggs (81 nests).

INCUBATION PERIOD No information.

EGG DATES 33 colonies, 24 April to 30 June (35 dates); 16 colonies, 3 May to 23 May.

Breeding Distribution

The great blue heron nests throughout Ontario, at least as far north as Big Trout Lake in the west, and in the east, as far north as Cochrane and probably considerably farther.

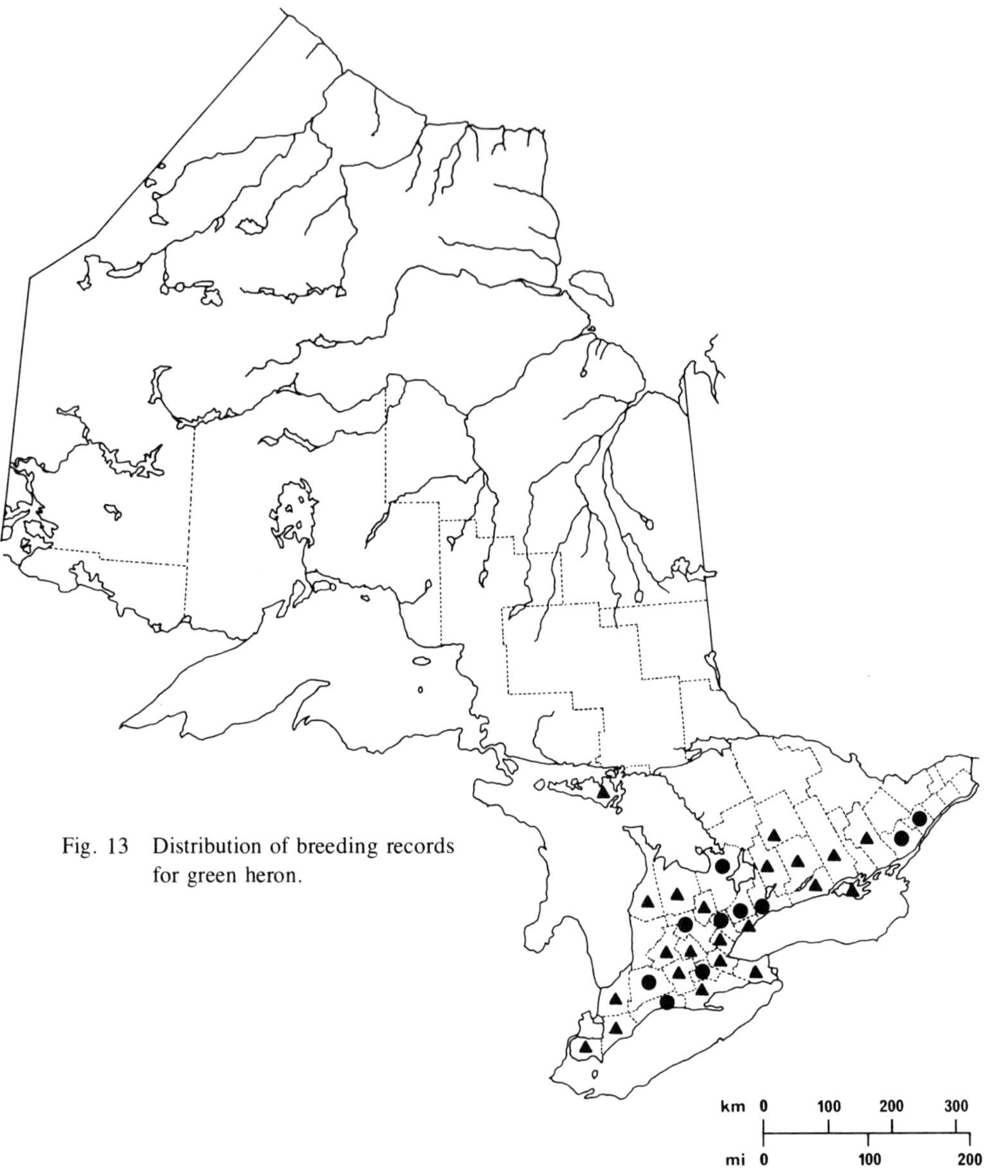

Fig. 13 Distribution of breeding records for green heron.

Green Heron, *Butorides striatus* (Linnaeus)

Nidiology

RECORDS 161 (ca 233 nests) representing 30 provincial regions.
This species often breeds singly and also in small, loose homogeneous colonies or in mixed heronries with great blue heron, cattle egret, great egret, and black-crowned night heron.
Nests were found in both wet and dry and both open and wooded areas: in borders of drainage ditches, creeks, rivers, ponds, lakes, bogs, and marshes; in fields with hawthorn groves and in shrubby tangles; in deciduous, mixed, and coniferous woods and plantations; and in cedar, tamarack, and alder swales and bogs.

Nests were situated either over water or as far as 274 m (900 ft) from the nearest water, in deciduous trees and shrubs (15 spp., 81 nests) and in coniferous trees (9 spp., 65 nests).
One tree nest was built on an old squirrel drey. One nest was placed in a wood duck nesting box. Tree nests were in forks and crotches and on horizontal limbs, either against the trunk or up to 2.4 m (8 ft) distant from it. Preferred nest trees were willow spp. (25 nests), cedar and juniper (23 nests), pine spp. (21 nests), hawthorn spp. (15 nests), and tamarack (13 nests). Heights of 49 nests ranged from 0.2 to 15 m (0.7 to 50 ft), with 23 averaging 3 to 6 m (9 to 20 ft).

Nests were usually unlined and were frail, homogeneous platforms of twigs and sticks; some nests contained roots and coarse grasses. They were usually flat but sometimes well hollowed. Outside diameters of 8 nests ranged from 25.5 to 45.5 cm (10 to 18 inches); outside depth of 1 nest was 10 cm (4 inches); inside diameters of 2 nests were 25.5 and 30.5 cm (10 and 12 inches); inside depth of 1 nest was 7.5 cm (3 inches).

EGGS 96 nests with 1 to 6 eggs; 1E (8N), 2E (6N), **3E** (14N), **4E** (30N), **5E** (33N), **6E** (5N).
Average clutch range 3 to 5 eggs (77 nests).

INCUBATION PERIOD 5 nests: 1 of 19 days, 1 of ca 20 days, 2 of ca 21 days, 1 of ca 22 days.
Incubation commenced before the clutch was complete.

EGG DATES 42 records, 4 May to 25 July (48 dates); 21 records, 23 May to 14 June.

Breeding Distribution

The green heron breeds widely throughout the agricultural areas of southern Ontario but becomes very scarce in southern Canadian Shield country and does not extend its range into northern provincial areas. Although specific records are lacking, the species apparently breeds regularly in eastern Ontario as far north as the Ottawa area.

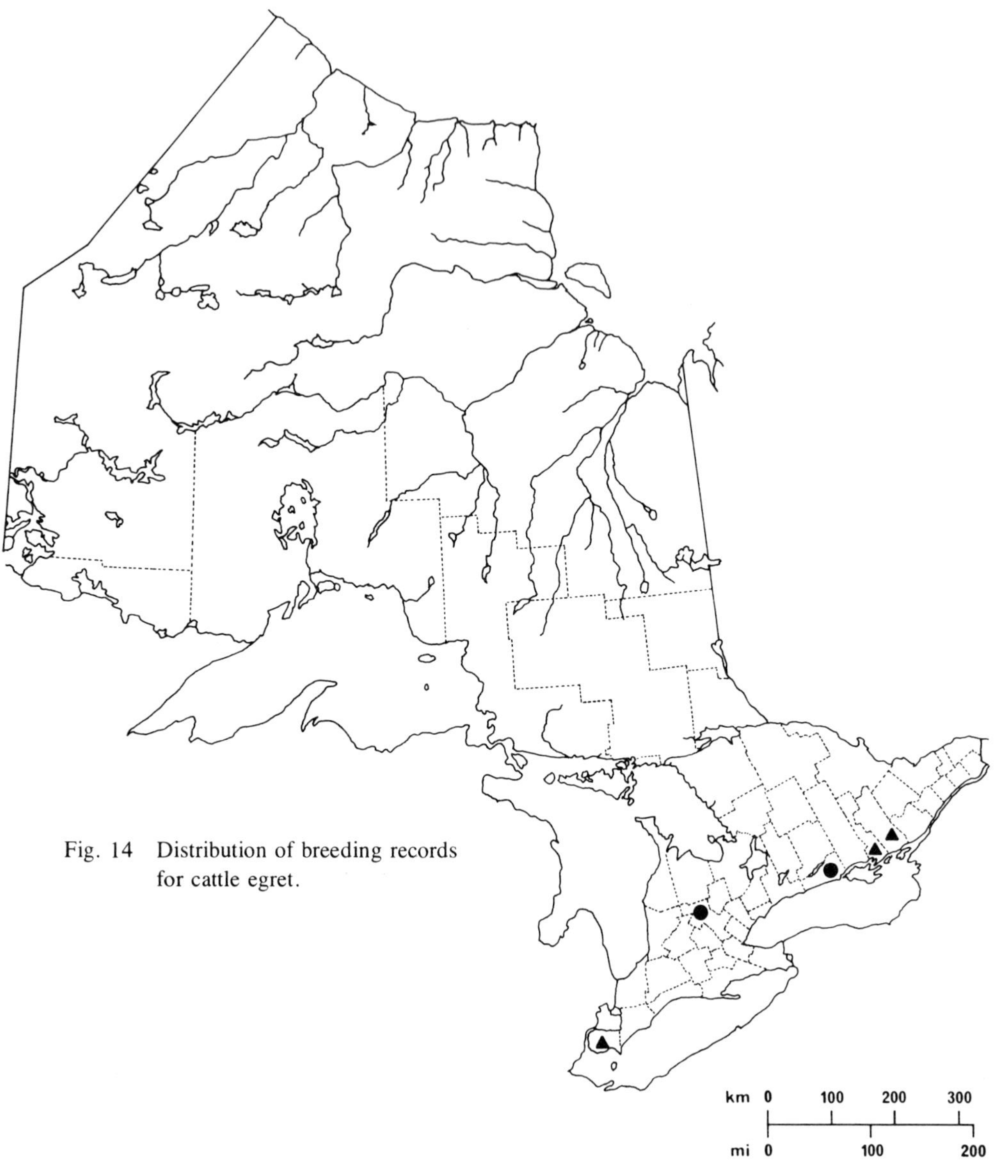

Fig. 14 Distribution of breeding records for cattle egret.

Cattle Egret, *Bubulcus ibis* (Linnaeus)

Nidiology

RECORDS 30 (13 colonies, 50 nests) representing 5 provincial regions.
This species usually breeds in mixed colonies, most often with black-crowned night heron and less often with great blue heron, green heron, and great egret. It also breeds in small homogeneous colonies and occasionally breeds singly. This species often coexists with colonies of gulls and terns. Nests are in sparsely or thickly wooded areas that often border marshy areas on or near the borders of islands in lakes (Fig. 143).

Nests were situated over water, or near it, in trees and shrubs (ash spp., buttonbush, cedar, sumac, and lilac), and heights of 25 nests ranged from 0.8 to 4.5 m (2.5 to 15 ft), with 13 averaging 1.2 to 2.1 m (4 to 7 ft). Nests were in dead or living shrubs and trees, and were variously positioned at the trunk and towards the outer edge of the tree.

Nests were flimsy, loosely built platforms of sticks, with relatively flat tops, and were saddled on limbs or positioned in crotches. One old nest of black-crowned night heron was used, and fresh material, including a green sprig of sumac with attached leaves, was placed in the lining. Other nests were made mostly of deciduous twigs with linings of smaller twigs, herbaceous stalks, and other vegetation; many nests were unlined. One nest measured 30.5 by 61 cm (12 by 24 inches) and another was square with sides of 30.5 cm (12 inches).

EGGS 21 nests with 1 to 4 eggs; 1E (1N), 2E (1N), **3E** (12N), **4E** (7N).
Average clutch range 3 eggs (12 nests).

INCUBATION PERIOD No information.

EGG DATES 11 nests, 4 June to 14 July (15 dates); 6 nests, 9 June to 12 June.

Breeding Distribution

The cattle egret (Fig. 144A) rapidly expanded its range in eastern North America and its first occurrence in Ontario was reported in 1956. The first nests were reported in 1962 at Luther Marsh in Wellington County (Buerkle and Mansell, 1963) and, a few weeks later, at Presqu'ile Provincial Park in Northumberland County (Baillie, 1963). Southern Ontario seems to be the northern limit of this species' range. Since 1962 only three new nesting sites have been found, none significantly farther north. At best, the breeding population is represented by a very few pairs.

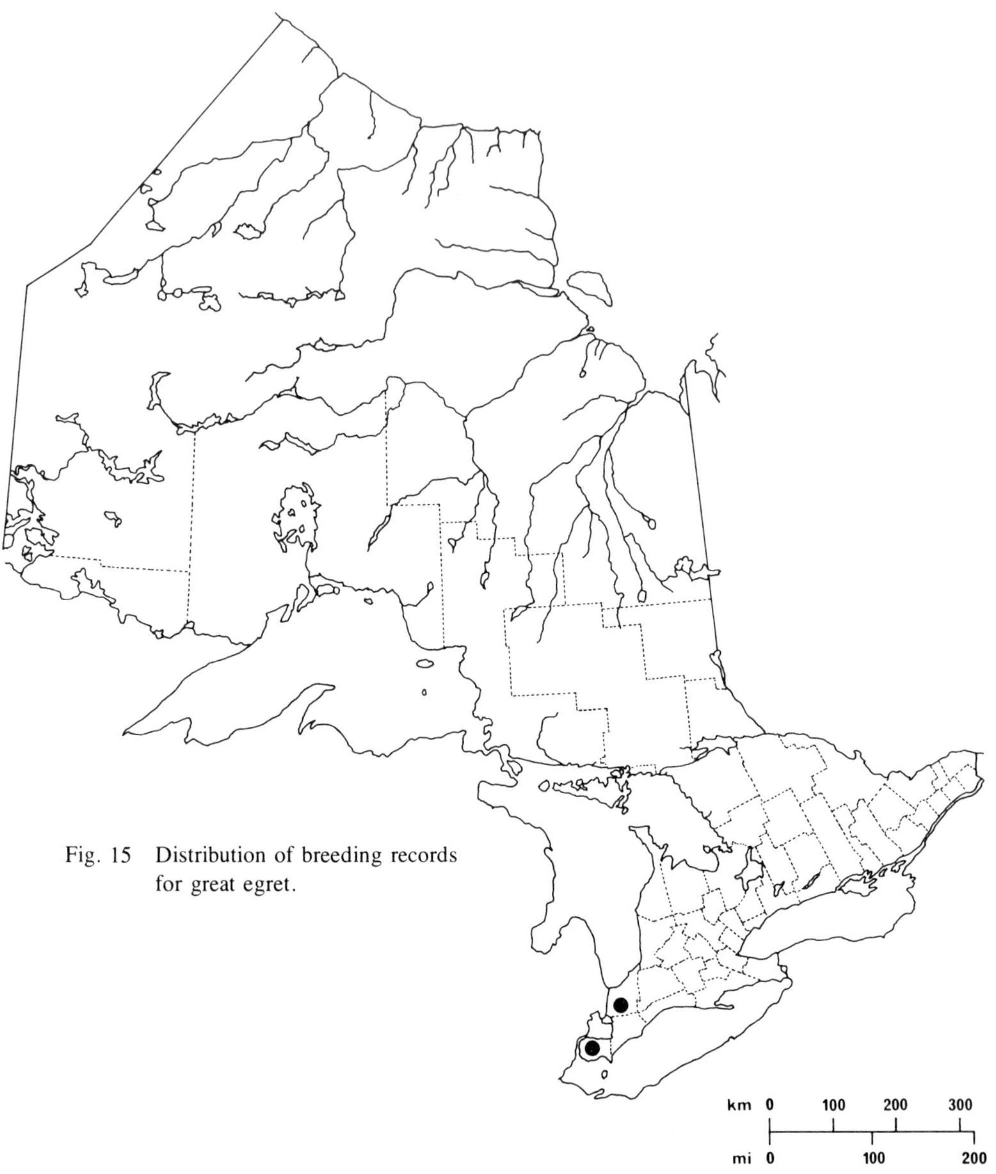

Fig. 15 Distribution of breeding records for great egret.

Great Egret, *Casmerodius albus* (Linnaeus)

Nidiology

RECORDS 18 (13 colonies, 1 isolated nesting, 88 nests) representing 2 provincial regions. This species usually breeds on islands in large lakes in mixed colonies with herons or other egrets. A few pairs nested in large colonies of great blue heron and black-crowned night heron, occasionally with green heron, and since 1974, with cattle egret. The average size of 13 colonies was ca 7 nests. Nests were reported in the shallow, depressed, central portion of a 14 ha (35 acre) wooded island; in a mainland hardwood grove bordered on 3 sides by a marsh; in the wooded border of a marshy lagoon on a large island; and in willow and alder shrubs in an island cattail marsh.

Nests were situated either over or near water in shrubs, trees, and vines (grape, alder, willow, ash, elm, and hackberry), both living and dead. Heights of 16 nests ranged from 0.6 to 11 m (2 to 35 ft), with 8 averaging 4.5 to 9 m (15 to 30 ft).

Nests were bulky platforms of sticks and twigs placed in the branches of shrubs and in tree crotches, either against the trunk or at some distance from it.

EGGS 13 nests, 1 to 4 eggs; 1E (2N), 2E (2N), **3E** (1N), **4E** (8N).
Average clutch range 4 eggs (8 nests).

INCUBATION PERIOD No information.

EGG DATES 4 colonies, 11 May to 24 June (14 nests, 6 dates); 2 colonies, 7 June to 8 June.

Breeding Distribution

The great egret apparently nested in the province for the first time in 1953, when a single nest was found on East Sister Island in Lake Erie (Fig. 143) (Baillie, 1963). Since then the species has continued to increase slowly in numbers and to expand into several other colonies, although it is still confined to a very limited area in the extreme south of Ontario. There are now three island colonies in Lake Erie; in addition, an adjacent mainland heronry in Essex County was reported in 1959 and 1960, and another island colony was discovered in 1977 in Lake St. Clair, Lambton County.

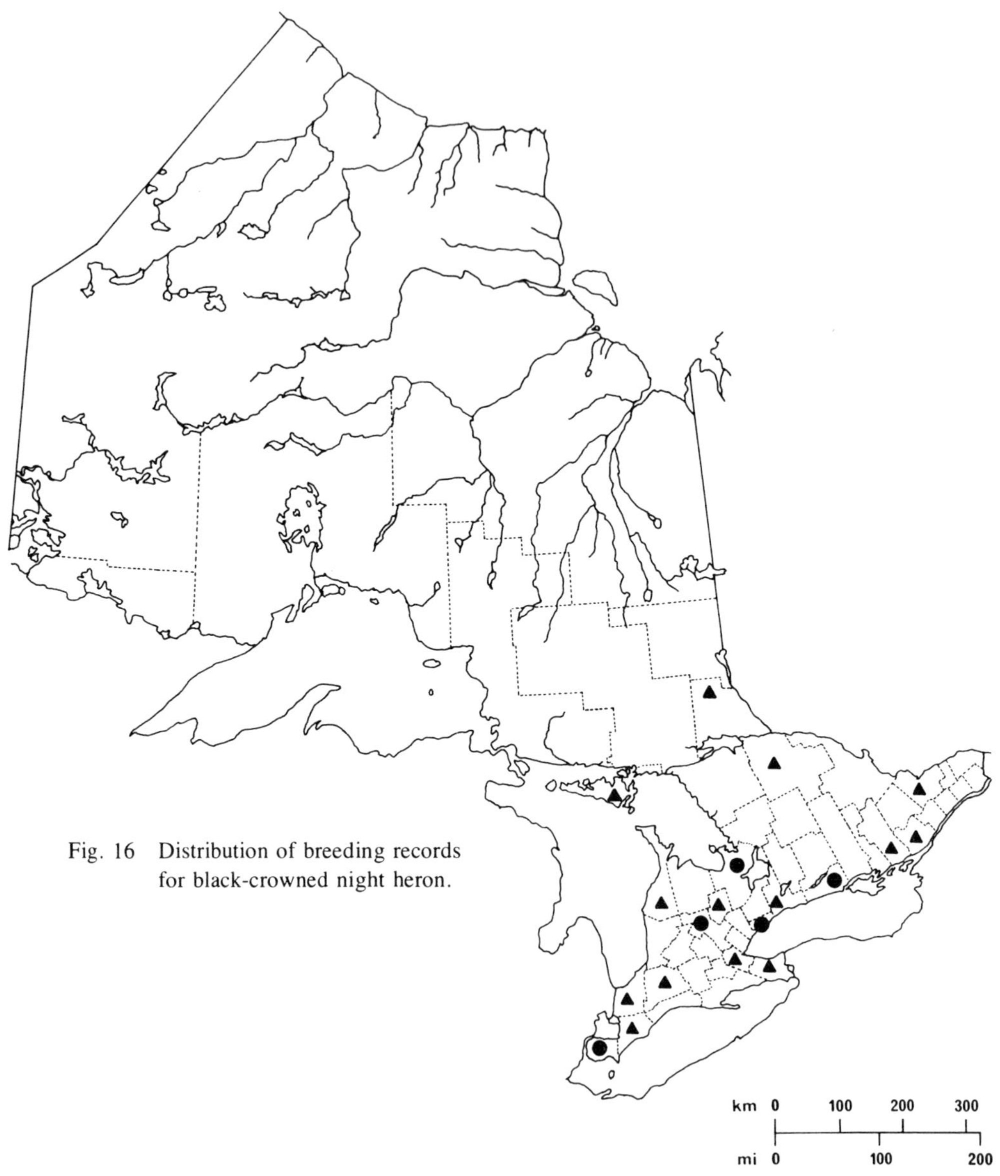

Fig. 16 Distribution of breeding records for black-crowned night heron.

Black-crowned Night Heron, *Nycticorax nycticorax* (Linnaeus)

Nidiology

RECORDS 198 (68 colonies, 4 isolated nestings, ca 8340 nests) representing 13 provincial regions.
This species nests most often in homogeneous colonies or in mixed colonies with great blue and green herons and cattle and great egrets; occasionally it nests singly. Nesting habitats are wet or dry wooded areas, such as sparsely or heavily treed islands, wooded river banks, cattail marshes, and swamps. Although undocumented colonies of 10 000 pairs have been reported, the average size of 68 colonies was ca 123 nests.

Nests were usually located in shrubs and trees, either over land or over water. Nest trees were most often deciduous (14 spp.) and occasionally coniferous (2 spp.). For several seasons 1 colony was located in a marsh of narrow-leaved cattail, with nests on matted vegetation just above water level. Heights of 544 nests in trees and shrubs ranged from 0.3 to 12 m (1 to 40 ft), with 272 averaging 1.8 to 3.7 m (6 to 12 ft). As many as 5 nests were noted in a single tree.

Most nests were platforms of twigs and sticks, with outside diameters ranging from 30 to 60 cm (12 to 24 inches). These nests were sometimes lined with fine twigs, dry grasses, and leaves. The cattail-marsh nests were constructed of broken leaves and stems of dead cattail bent inward to form the nest platform that was situated at water level.

EGGS 568 nests with 1 to 7 eggs; 1E (54N), 2E (108N), **3E** (156N), **4E** (204N), **5E** (40N), 6E (4N), 7E (2N).
Average clutch range 3 to 4 eggs (360 nests).
A number of 1- and 2-egg clutches were incomplete.

INCUBATION PERIOD 17 nests: 3 of at least 21 days, 11 of 21 to 22 days, 3 of 22 to 23 days. Since incubation commences with the first egg, exact incubation periods are difficult to determine unless eggs are marked.

EGG DATES 46 colonies, 6 May to 9 September (61 dates); 23 colonies, 1 June to 17 June.

Breeding Distribution

The black-crowned night heron breeds almost exclusively in southern Ontario. Baillie and Harrington (1936) reported colonies from only four counties along the province's southern border. Since then the species has expanded its breeding range and now occupies a few widely scattered heronries in the agricultural part of southern Ontario and possibly two others that have been reported in the Canadian Shield, in Nipissing District.

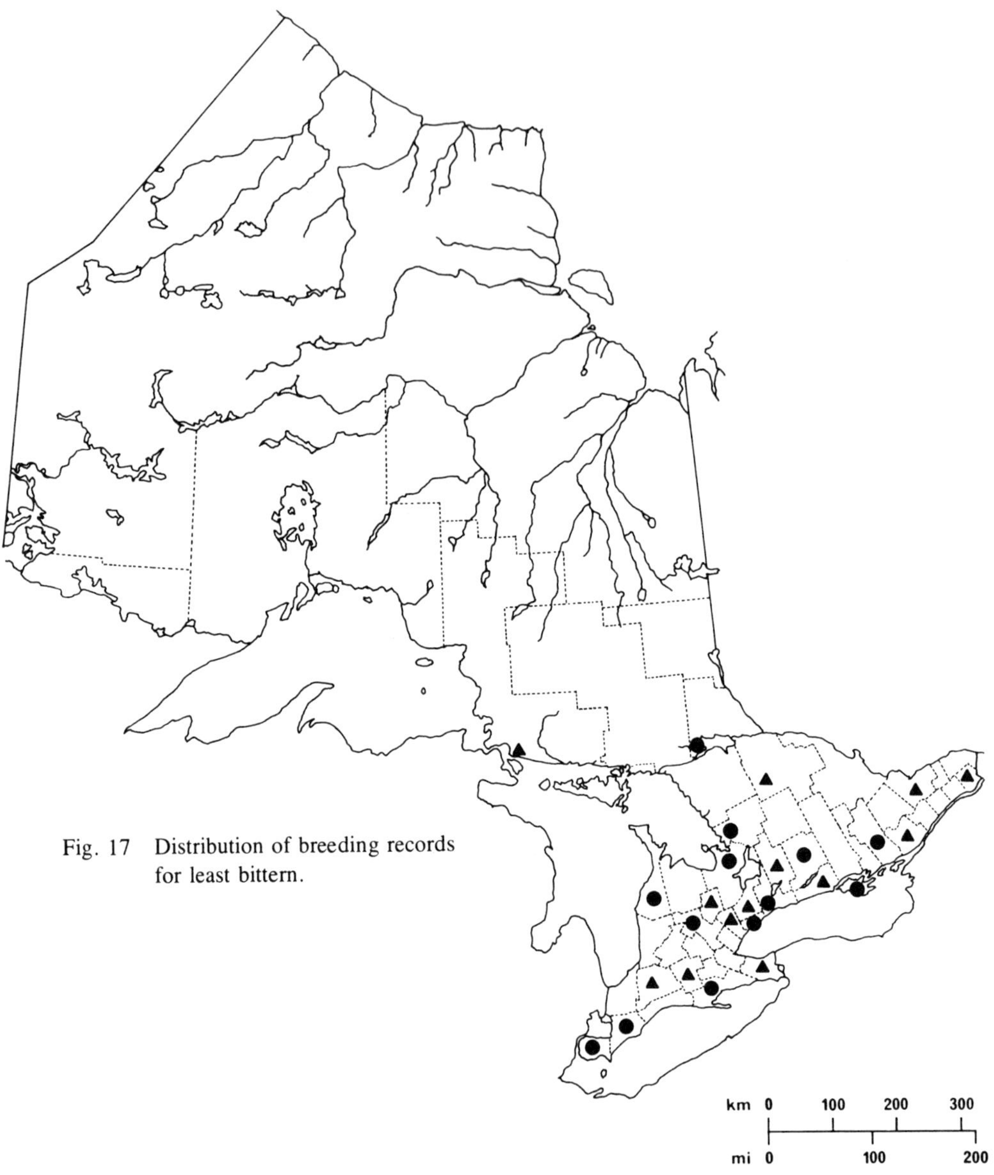

Fig. 17 Distribution of breeding records for least bittern.

Least Bittern, *Ixobrychus exilis* (Gmelin)

Nidiology

RECORDS 247 (248 nests) representing 25 provincial regions.
Breeds in freshwater marshes; in marshy areas of drainage ditches, creeks, rivers, lakes, and occasionally bogs.

Nests were over water, which ranged in depth from a few centimetres to more than 91.5 cm (36 inches). They were situated in standing vegetation (cattails, bulrushes, common reed grasses, horsetails, sedges, grasses, willows, dogwood, and heaths) as far as 200 m (660 ft) from shore; some nests were deep in marsh vegetation as far as 45 m (150 ft) from open water.

Nests were flat or shallow-cupped platforms and were woven from pieces of dead cattail and other marsh reeds, and from grasses, small twigs, and sticks. Some nests were lined with leaves, grasses, fine rush stems, and small twigs. They were supported by living and dead standing rushes, by grasses, shrubs, and small trees. One nest was built on top of the occupied nest of a long-billed marsh wren. Heights of 68 nests ranged from 5 to 122 cm (2 to 48 inches), with 34 averaging 30 to 60 cm (12 to 24 inches). The bottoms of some nests just touched the surface of the water. Diameters of 3 nests ranged from 15 to 30.5 cm (6 to 12 inches); outside depths of 4 nests ranged from 10 to 25.5 cm (4 to 10 inches); inside depth of 1 nest was 5 cm (2 inches).

EGGS 196 nests with 1 to 9 eggs; 1E (9N), 2E (23N), **3E** (33N), **4E** (65N), **5E** (54N), **6E** (11N), 9E (1N).
Average clutch range 4 to 5 eggs (119 nests).

INCUBATION PERIOD No information.

EGG DATES 198 nests, 15 May to 2 August (221 dates); 99 nests, 6 June to 21 June.

Breeding Distribution

The least bittern breeds throughout southern Ontario, usually outside the Canadian Shield, and occasionally as far north as Sault Ste Marie and Lake Nipissing.

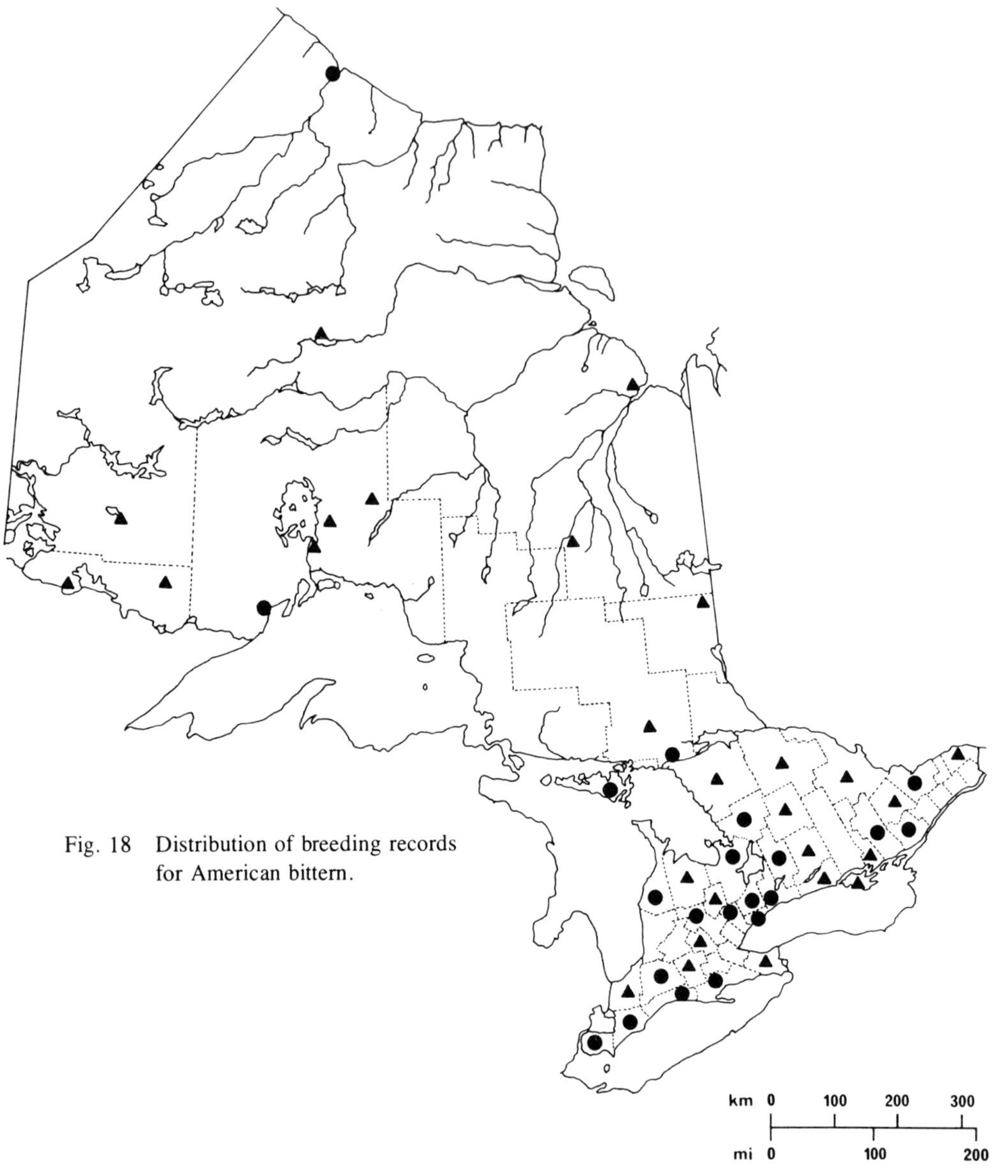

Fig. 18 Distribution of breeding records for American bittern.

American Bittern, *Botaurus lentiginosus* (Rackett)

Nidiology

RECORDS 199 nests representing 33 provincial regions.
Breeds typically in wet areas such as freshwater marshes, swamps, and bogs, and occasionally in dry areas such as grassy fields, pastures, and corn fields, up to 90 m (300 ft) from the nearest water. It possibly breeds in brackish and saltwater marshes along the seacoasts in northern Ontario, where it occurs during the breeding season.

Nests were often situated over water as deep as 45.5 cm (18 inches), and were in standing cattails, bulrushes, and sedges; on mounds of sphagnum moss; and among bushes of leatherleaf, cranberry, Labrador tea, and willow. Nests were less often situated on dry ground in fields of grasses and weeds, in crop fields, in bushes, in wild grape vines, and once in a raspberry patch. Two nests were only 7.5 m (25 ft) apart and another nest was within 90 m (300 ft) of an active nest of marsh hawk. Ground nests were sometimes elevated to heights of 15 cm (6 inches); heights above water of 19 nests ranged from 10 to 122 cm (4 to 48 inches), with 10 averaging 15 to 30 cm (6 to 12 inches).

Nests were flat or shallow-cupped platforms, and were homogeneously constructed of reeds, sedges, grasses, weedstalks, sticks, and twigs. They were sometimes lined with finer material. On rare occasions nest platforms were floating masses of vegetation. Sometimes eggs were laid on the ground with little or no nest evident. Paths leading through vegetation to the nest were sometimes reported. Diameters of 2 nests were 30 and 35 cm (12 and 14 inches); inside diameters of 3 nests ranged from 20.5 to 30.5 cm (8 to 12 inches).

EGGS 177 nests with 1 to 9 eggs; 1E (7N), 2E (10N), **3E** (27N), **4E** (87N), **5E** (40N), **6E** (5N), 9E (1N).
Average clutch range 3 to 5 eggs (154 nests).
Two nests, each with 5 eggs of American bittern, also contained 1 egg of ruddy duck.

INCUBATION PERIOD 1 nest, ca 23 days. Incubation commenced before the clutch was complete.

EGG DATES 146 nests, 5 May to 8 July (160 dates); 73 nests, 30 May to 11 June.

Breeding Distribution

Although records are few and widely scattered in the north, the American bittern breeds throughout the province in suitable habitat.

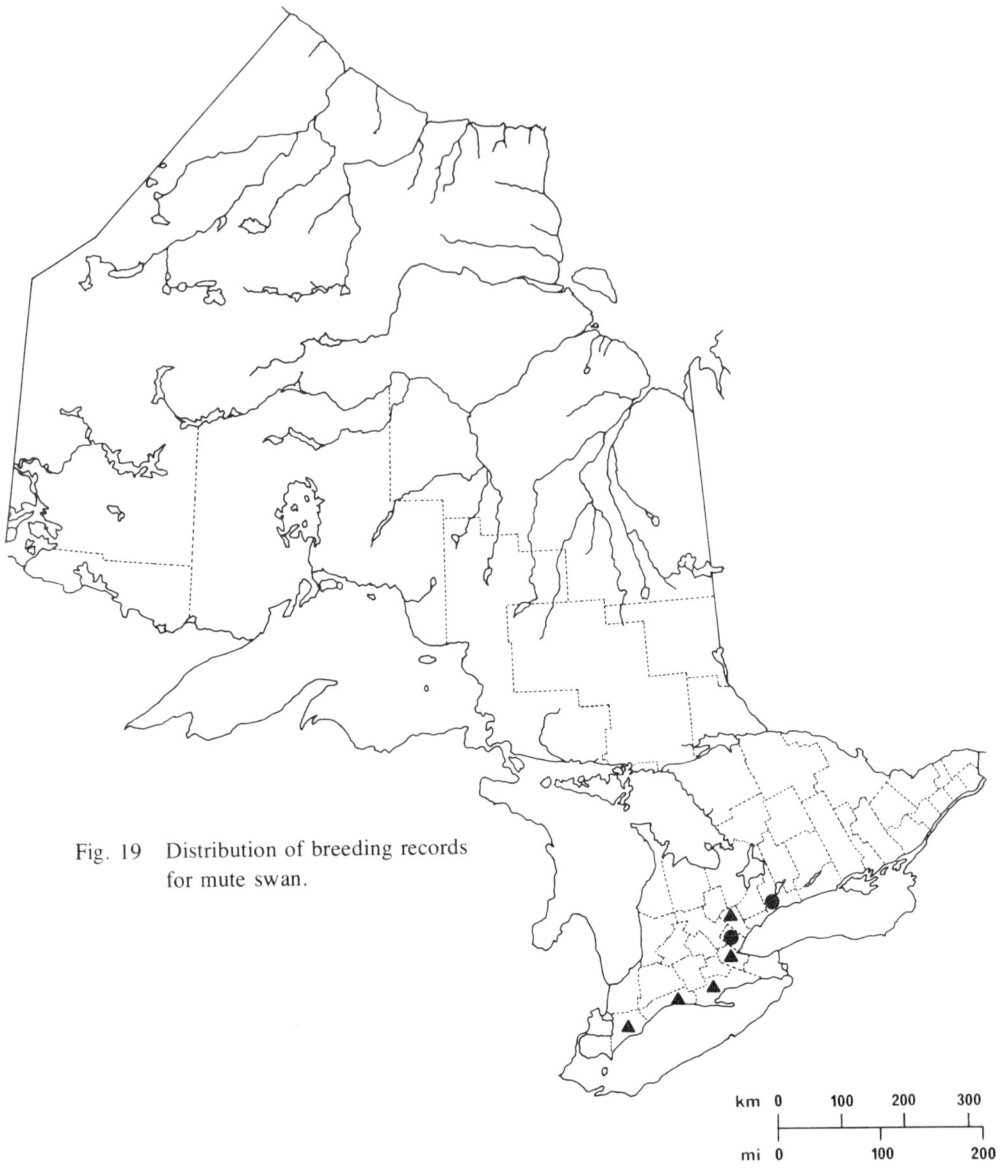

Fig. 19 Distribution of breeding records for mute swan.

Mute Swan, *Cygnus olor* (Gmelin)

Nidiology

RECORDS 10 nests representing 5 provincial regions.
This species has been established as a feral breeding bird in Ontario for more than 20 years. Most of the reported nests were in cattail marshes, although 2 were on the shore of a small pond.

The marsh nests were situated in cattails from 13.5 to 90 m (45 to 300 ft) from shore and were placed on mats of floating vegetation, anchored by standing cattails, or on muskrat houses.

Nest structures were large mounds of dead cattails. One nest was lined with some down and feathers. Nest diameters were 2 m (6 ft) and nest depths ranged from 20.5 to 61 cm (8 to 24 inches).

EGGS 8 nests with 1 to 9 eggs; 1E (1N), **4E** (1N), **5E** (3N), **8E** (2N), 9E (1N).
Average clutch range 4 to 8 eggs (6 nests).

INCUBATION PERIOD No information.

EGG DATES 8 nests, 4 April to 29 June (9 dates); 4 nests, 9 May to 22 May.

Breeding Distribution

The mute swan is kept in captivity in many parts of the province. It also lives in a semi-wild state in parks, where it can be fed and maintained through the winter months. However, feral birds have succeeded in maintaining themselves throughout the winter and such birds began breeding in the wild as early as 1958 (Peck, 1966). Because this species is nonmigratory, it is confined to areas of suitable habitat near the southern Great Lakes where open water is found all winter.

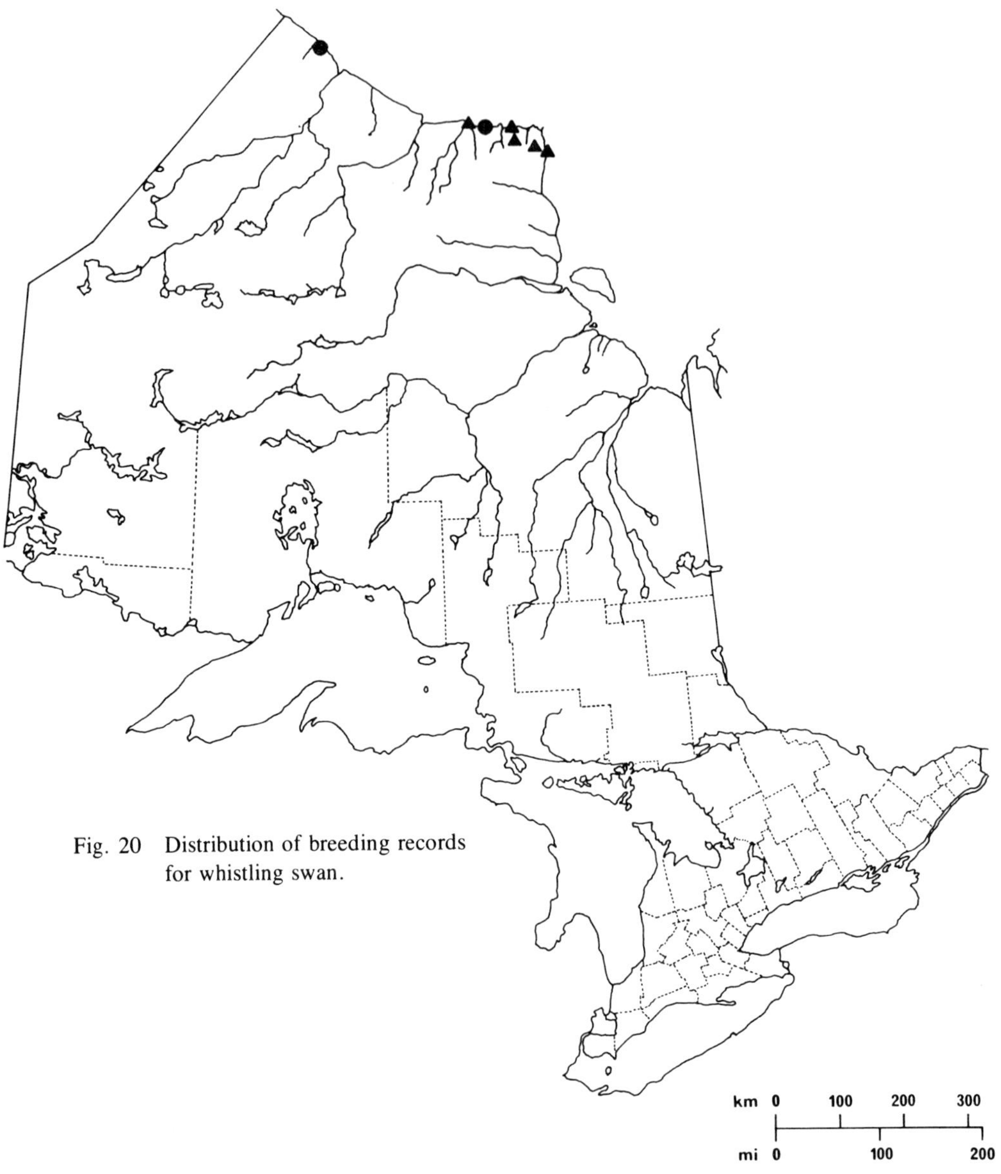

Fig. 20 Distribution of breeding records for whistling swan.

Whistling Swan, *Olor columbianus* (Ord)

Nidiology

A rare summer resident of the narrow strip of subarctic tundra bordering the Hudson Bay and northern James Bay coasts, this species was first documented as a breeding bird in Ontario in August 1974, when photographs were secured of a captured, flightless cygnet (ROM PR 566–568) at the mouth of the Niskibi River on Hudson Bay, west of Fort Severn, Kenora District. Earlier, on 23 July 1973, near Cape Henrietta Maria, 32 km (20 mi) east of the Sutton River on Hudson Bay, a sighting was made and aerial photographs were secured of a pair of adults with 2 young that were almost certainly of this species (ROM PR 68–69).

More recently, a pair of adults with 3 cygnets was recorded on a tundra lake 14.5 km (9 mi) east of the forks of the Brant River on 5 July 1978, and still another pair with 3 cygnets was noted on 6 July 1979, on a tundra pool 16 km (10 mi) west of abandoned radar site 415 in the Cape Henrietta Maria region, Kenora District.

Breeding Distribution

Historical references indicate that the whistling swan probably bred along the Hudson Bay coast during the early days of fur trading in North America (Lumsden, 1975). However, its populations dwindled as a result of over-exploitation by fur traders. It had apparently disappeared as a breeding species in Ontario by the end of the 19th century.

Subsequently, protection of the species has permitted it to increase its population size and to expand its breeding range in the north. In 1964 a pair of swans was observed from an aircraft, one of them sitting on what appeared to be a nest. This sighting was near the mouth of the Sutton River and provided the first recent indication that whistling swans were breeding again in the province (Lumsden, 1975). The species' breeding status in Ontario was verified only in 1973 and 1974 when cygnets were photographed (Peck, 1976).

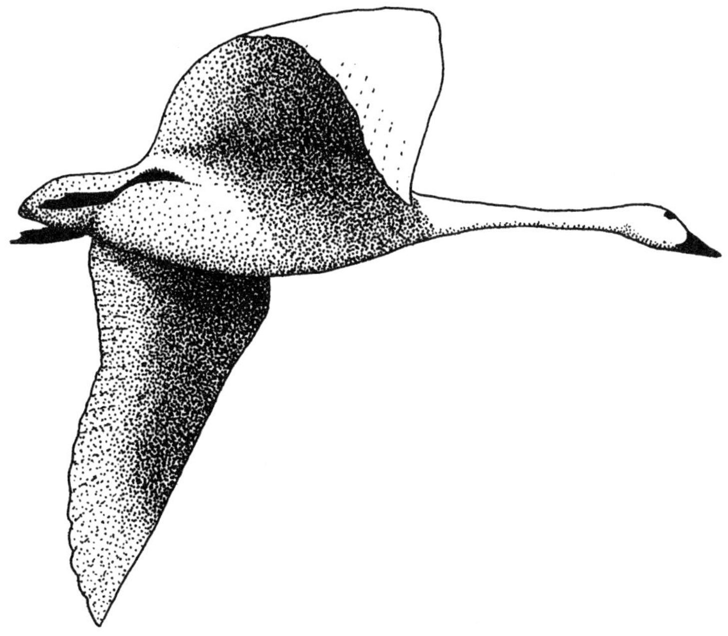

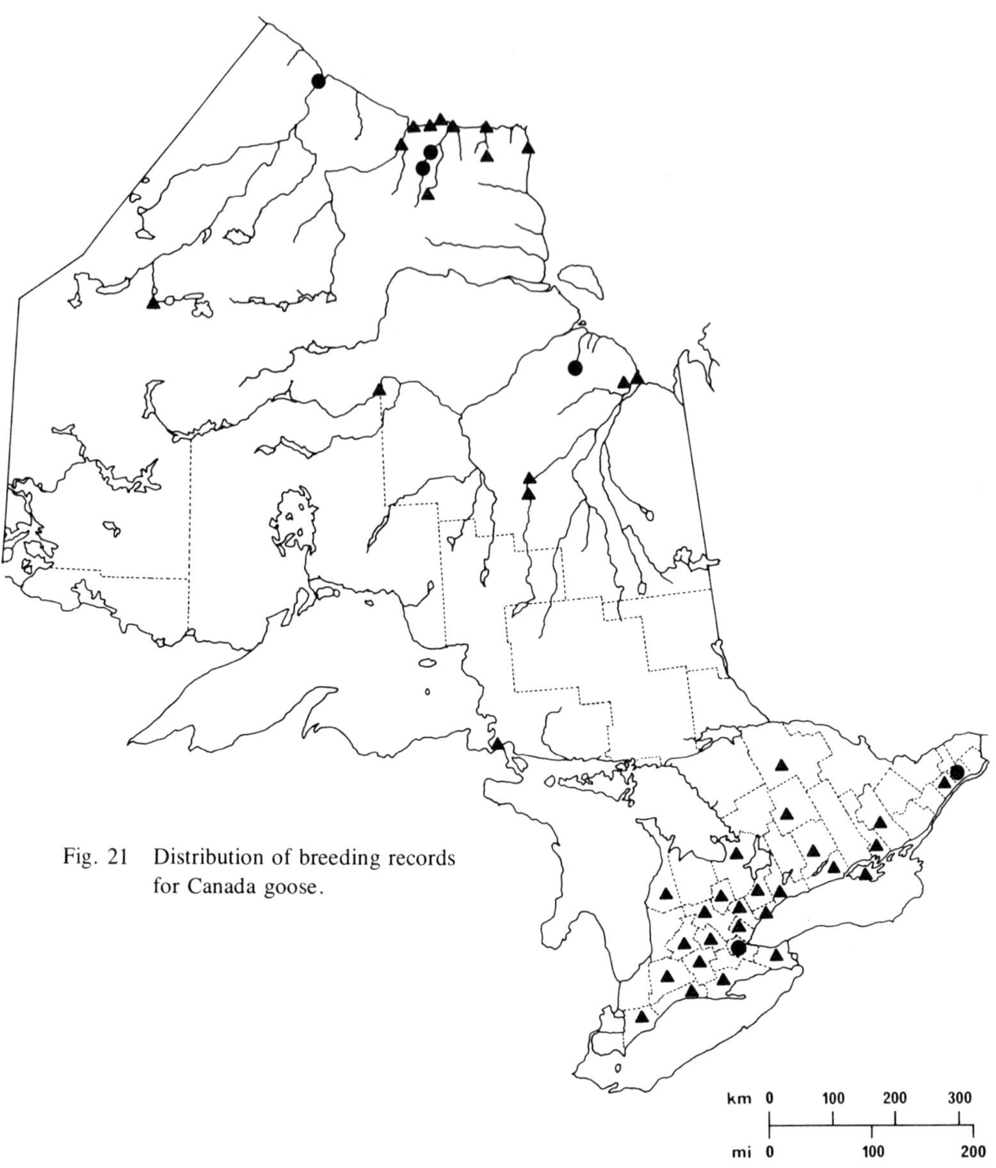

Fig. 21 Distribution of breeding records for Canada goose.

Canada Goose, *Branta canadensis* (Linnaeus)

Nidiology

RECORDS 541 (556 nests) representing 22 provincial regions.
During the last half-century captive birds have established a feral breeding population throughout southern Ontario and these largely nonmigratory geese remain quite separate from the migratory geese breeding in natural populations farther north in the province.

The species breeds in diverse locations, usually not more than 45 m (150 ft) from water: on tundra and in muskeg; on shores and small islands in lakes, rivers, ponds, and marshes; in open, grassy fields; in shrubby areas; and in flooded gravel pits.

Nests were usually placed on hummocks but were occasionally in depressions. They were positioned as follows: in grasses, sedges, rushes, and bushes; on muskrat houses and marsh vegetation; on or among rock piles and on various artificial sites (oil drums, tubs, tires, the deck of an old barge); and in tree crotches from 0.5 to 2.5 m (1.5 to 8 ft) in height. Many southern nests were placed at the bases of trees, bushes, logs, and rocks.

Most nests were mounds of living and dead grasses and rushes, with leaves, branches, twigs, bark, and straw. They were lined with down, feathers, twigs, and fine grasses. Outside depths ranged from 15 to 61 cm (6 to 24 inches); outside diameters ranged from 76 to 91.5 cm (30 to 36 inches); inside diameters ranged from 30.5 to 40.5 cm (12 to 16 inches); 1 inside depth was 7.5 cm (3 inches).

EGGS 140 nests with 1 to 9 eggs; 1E (3N), 2E (8N), 3E (12N), **4E** (26N), **5E** (36N), **6E** (36N), **7E** (16N), 8E (2N), 9E (1N).
Average clutch range 4 to 6 eggs (98 nests).

INCUBATION PERIOD No information.

EGG DATES 161 nests, 29 March to 30 June (205 dates); 80 nests, 28 April to 19 May. On Akimiski Island, James Bay, 12 nests all hatched between 11 June and 17 June.

Breeding Distribution

Natural populations breed mainly in the Hudson Bay Lowland (Fig. 163) in the north. A few early records of breeding in more remote regions of southern Ontario were probably of injured birds from these natural populations that were incapable of migrating (MacLulich, 1938), but most of the records scattered throughout the south are from reintroduced geese. Giant Canada geese (*Branta c. maxima*) probably formerly bred naturally in only a few regions in the extreme south of Ontario and in western Ontario near Lake of the Woods (Palmer, 1976), but reintroduced populations of these geese are also now scattered throughout the southern part of the province.

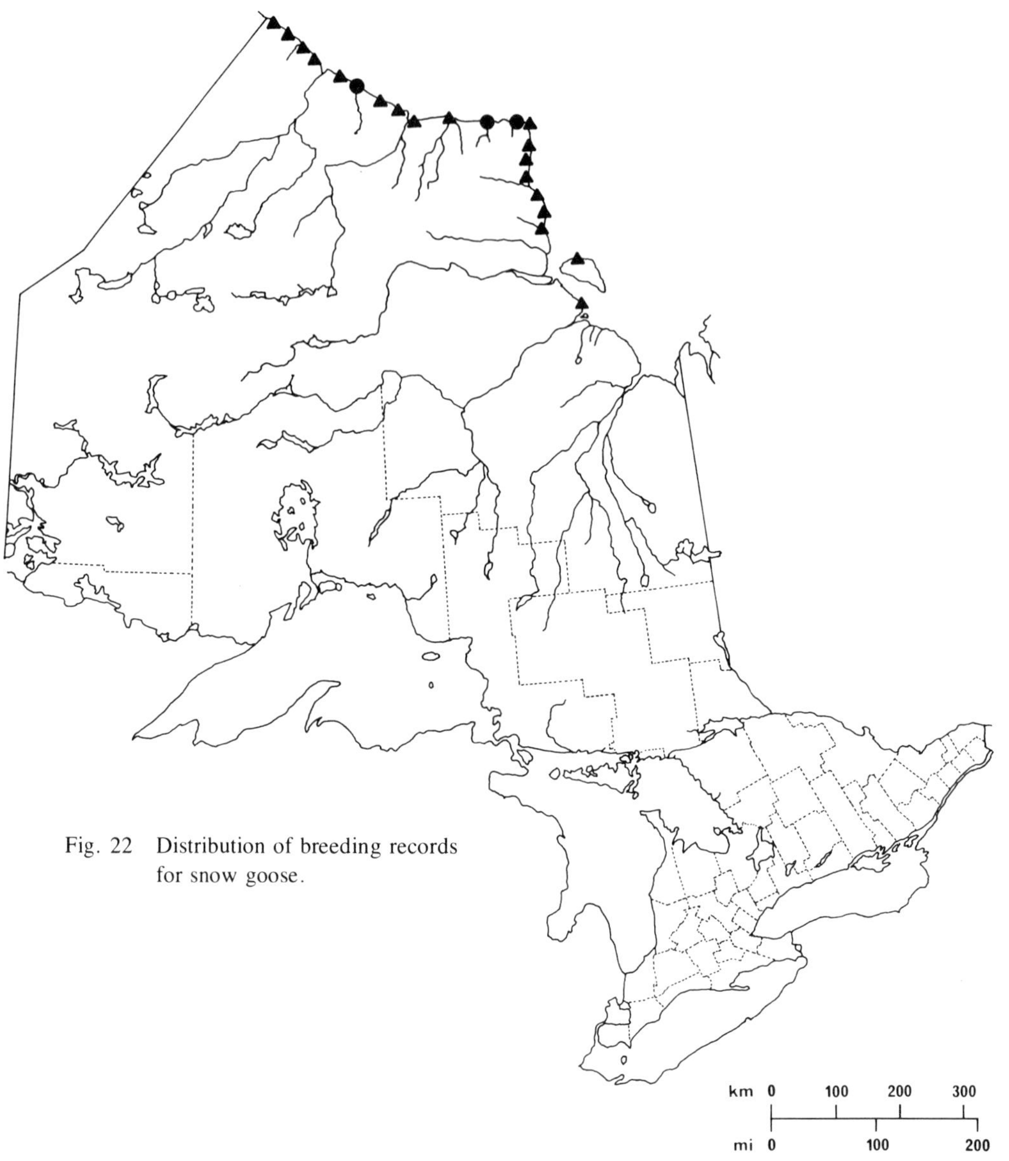

Fig. 22 Distribution of breeding records for snow goose.

Snow Goose, *Chen caerulescens* (Linnaeus)

Nidiology

RECORDS 6 (481 nests) representing 1 provincial region and an adjacent island in James Bay (NWT).

This colonial species breeds mainly in several locations on the narrow strip of tundra bordering the Hudson Bay and James Bay coasts. The principal colony straggles over a large area, and nests are situated on pond tundra and at a braided river mouth.

Nests were positioned among arctic willow and also on the open tundra near ponds and creeks.

Nests were built-up mounds of mosses, sedges, grasses, and twigs, and were lined with down.

EGGS 481 nests with 1 to 9 eggs; 1E (17N), 2E (29N), **3E** (124N), **4E** (233N), **5E** (61N), **6E** (15N), 7E (1N), 9E (1N).
Average clutch range 3 to 5 eggs (418 nests).

INCUBATION PERIOD No information.

EGG DATES 2 colonies, 2 June to 22 June (3 dates).

Breeding Distribution

The first report of breeding in Ontario by the snow goose was in 1947, when a few birds with young were seen from the air near Cape Henrietta Maria (Lumsden, 1959a). The colony in the Cape region expanded rapidly and by 1975 comprised more than 50 000 adults (Dzubin et al., 1975). At the time when this colony was established blue-phase birds made up only about one-third of the population (Cooch, 1963), but by 1975 about 75 per cent of the birds were blue phase (Dzubin et al., 1975). By 1979 this colony contained 55 000 nests (H. G. Lumsden, pers. comm.).

In addition to the large colony near Cape Henrietta Maria, recent reports of breeding include such locations as an island in James Bay (NWT), the mouth of the Opistokwayan River at James Bay, other locations on the James Bay coast, and scattered locations on the Hudson Bay coast west to the Manitoba border (Hanson et al., 1972).

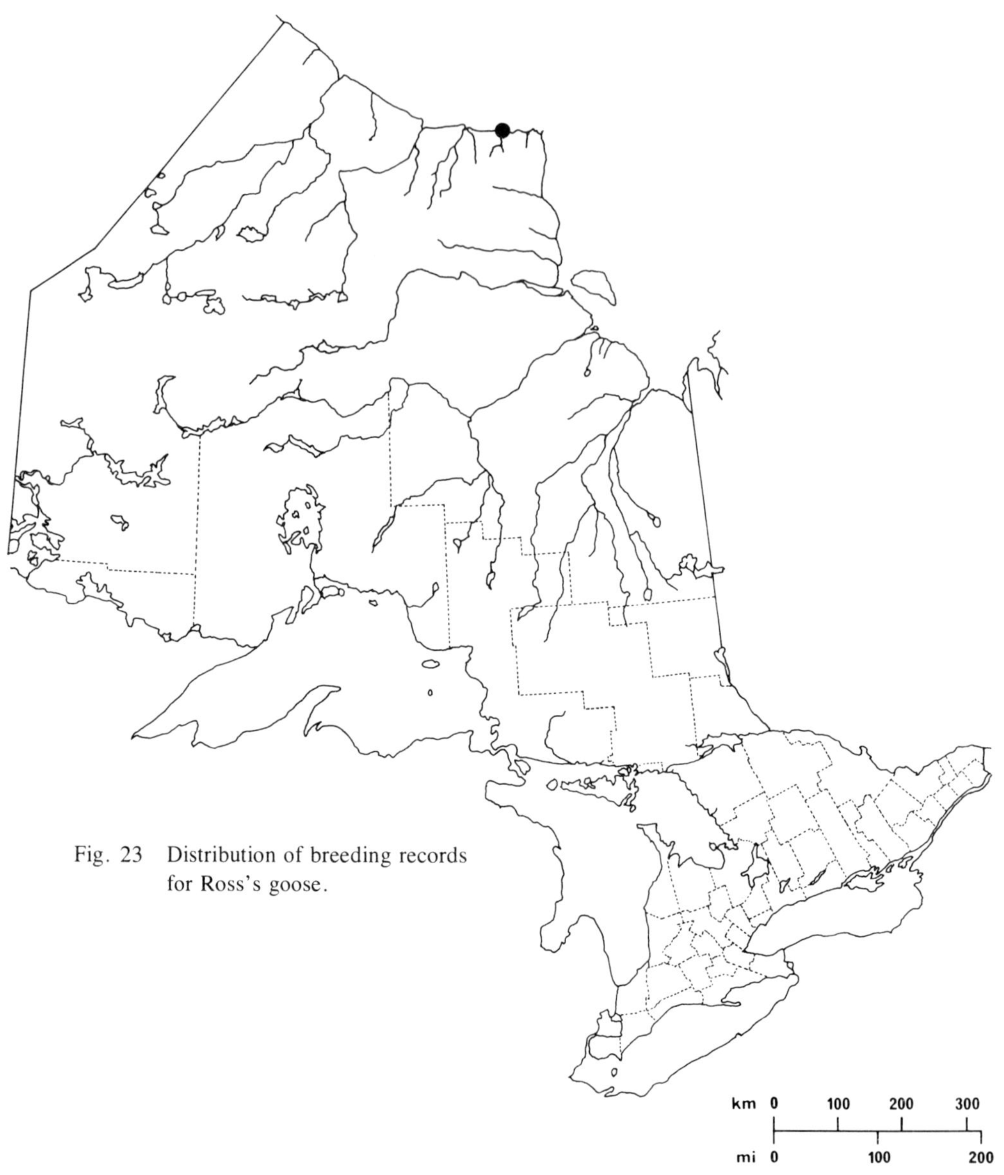

Fig. 23 Distribution of breeding records for Ross's goose.

Ross's Goose, *Chen rossii* (Cassin)

Nidiology

On 29 July 1975 a family of 2 adults and 3 prefledging juveniles was captured, measured, banded, and photographed near the mouth of the Brant River on the Hudson Bay coast, Kenora District, at a snow goose breeding colony (Fig. 165). Copies of these confirming photographs, on file in the Royal Ontario Museum (ROM PR 78–81), depict a male adult Ross's goose, a larger adult female, which appears to be a hybrid between a Ross's goose and a snow goose, and flightless young birds (Prevett and Johnson, 1977).

Breeding Distribution

Although the record described above is the first and only verified evidence of breeding of the Ross's goose in the province, the hybrid female may have been the progeny of a previous nesting by a Ross's goose in Ontario. There is also an unconfirmed report of breeding as early as 1967 near the Shagamu River mouth (Prevett and Johnson, 1977). Since numbers of this species have been increasing in other snow goose colonies along the western Hudson Bay coast north of Ontario, an increase is to be expected in the province as well.

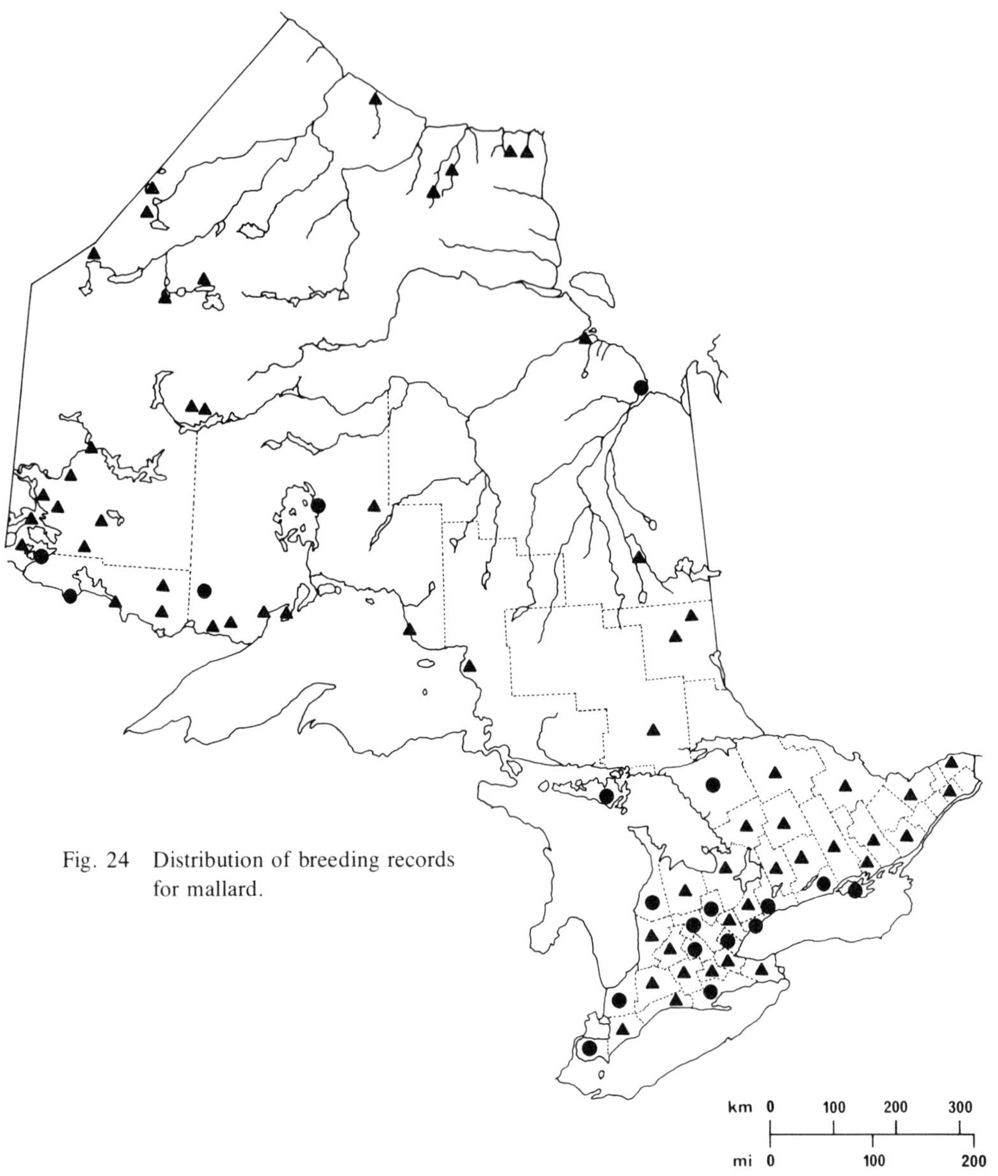

Fig. 24 Distribution of breeding records for mallard.

Mallard, *Anas platyrhynchos* Linnaeus

Nidiology

RECORDS 1032 nests representing 44 provincial regions.
Breeds in freshwater marshes and bogs; in meadows, crop fields, and pastures; in woodlands and coniferous plantations; on islands and islets in lakes, rivers, and ponds; on sand and gravel beaches; in dykes and ditches; on road and railway shoulders; on golf courses; in garbage dumps; at sewage disposal ponds; in gravel pits; and on lawns and in residential gardens. Some nests were noted in ring-billed gull colonies. Nest locations ranged from just over water to more than 1.6 km (1 mi) distant from it.

Nests were in hollows on earth, sand, and rock; on floating mats of vegetation; on muskrat houses and bog hummocks; in clumps of grass, sedge, rushes, weeds, fern, moss, and horsetail; under vines, bushes, and trees; in tree roots; on tops of stumps and in tree crotches (once in an old nest of common crow) from 0.3 to 6.5 m (1 to 22 ft) in height (24 nests); in boxes on nesting rafts; in baled hay; under logs and brush piles; on rock piles; and once on steel dock girders. Two nests were found 0.6 m (2 ft) from each other; 1 nest was 3 m (10 ft) from a nest of black duck; 1 nest was 0.6 m (2 ft) from a nest of ring-billed gull.

Nests were hollowed-out bowls of dead and living grasses, rushes, and weeds, with corn stalks, leaves, conifer needles, mosses, small sticks, and wood chips. They were lined with down and feathers, both increasing in amount with clutch completion, and also with plant down, mosses, leaves, and pine needles. One nest contained cellophane. Outside diameters ranged from 25.5 to 30.5 cm (10 to 12 inches); outside depths ranged from 7.5 to 18 cm (3 to 7 inches). Inside depths of 6.5 cm (2.5 inches) were reported. At times there was a dome of surrounding vegetation over the nest.

EGGS 318 nests with 4 to 24 eggs; 4E (2N), 5E (7N), **6E** (25N), **7E** (23N), **8E** (41N), **9E** (63N), **10E** (48N), **11E** (46N), **12E** (30N), **13E** (12N), 14E (6N), 15E (5N), 17E (3N), 18E (3N), 20E (2N), 21E (1N), 24E (1N).
Average clutch range 7 to 11 eggs (221 nests).
Large sets were probably the product of more than 1 female; 1 20-egg set was incubated simultaneously by 2 females. Three nests contained runt eggs; 2 nests with mallard eggs also contained eggs of ring-necked pheasant.

INCUBATION PERIOD 16 nests, 24 to 31 days, with 11 averaging 25 to 28 days.

EGG DATES 324 nests, 2 April to 20 July (504 dates); 162 nests, 16 May to 3 June.

Breeding Distribution

The mallard breeds throughout the province.

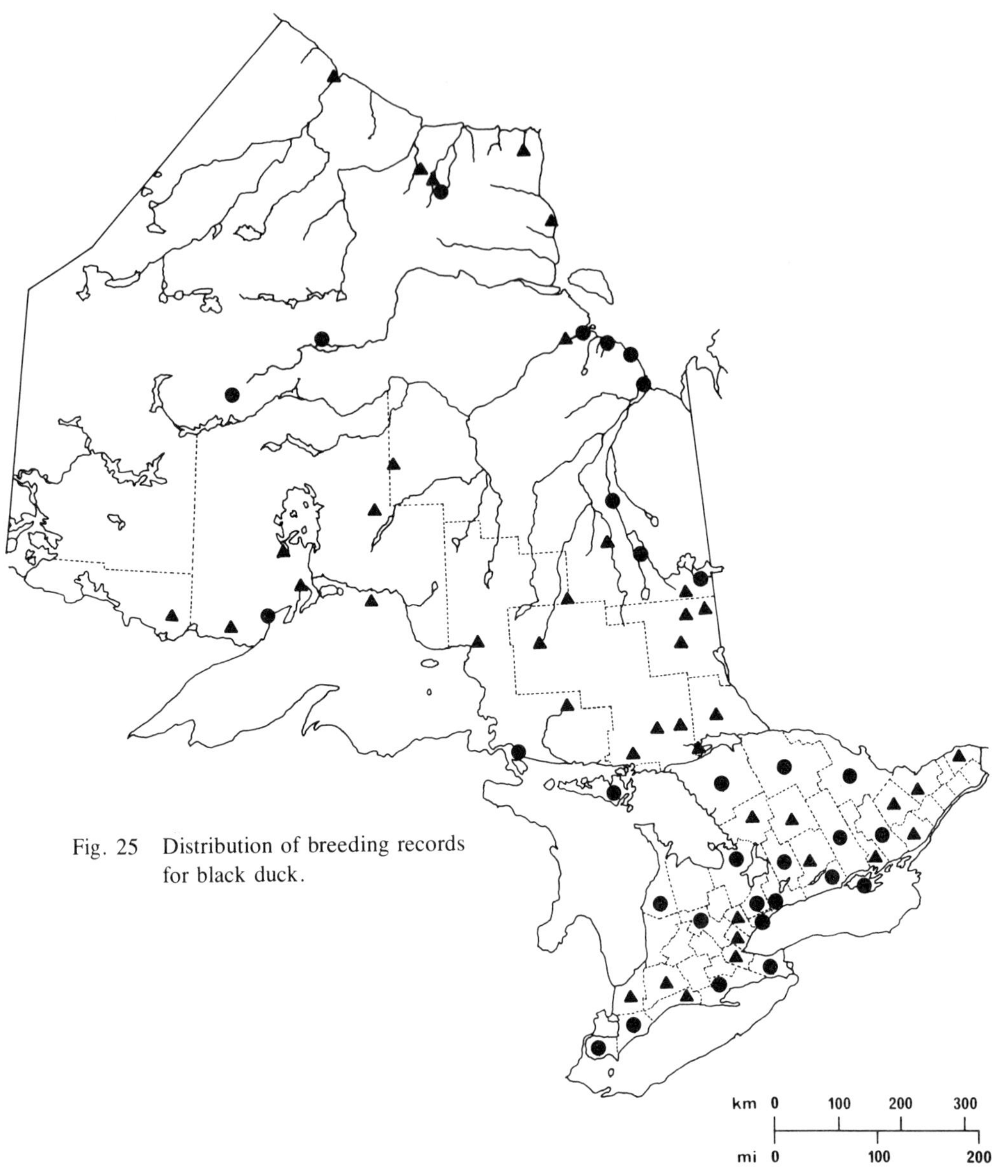

Fig. 25 Distribution of breeding records for black duck.

Black Duck, *Anas rubripes* Brewster

Nidiology

RECORDS 284 nests representing 36 provincial regions.
Breeds in freshwater marshes of rivers, lakes, and ponds; in heath and sphagnum bogs and alder swamps; along ditches; in woods and coniferous plantations; in scrubby, grassy fields; on barren, rocky, or wooded islands; and on open sand beaches. Some nests were noted in the midst of a herring gull colony. Locations of nests ranged from directly over water to as far as 0.8 km (0.5 mi) distant from it.

Most nests were situated as follows: in slight hollows in the ground, rock, or sand, surrounded by grasses, weeds, mosses, and dense shrubs; at the bases of trees; or under bushes and logs. Some nests were in marsh sites, situated in rushes and on muskrat houses. Occasionally nests were in hollow stumps, tree crotches, and in crow and hawk nests in trees, at heights of 1 to 12 m (3 to 40 ft). Nest boxes and nesting rafts were sometimes used and 1 nest was found under an overturned boat.

Nests were composed of grasses, weeds, rushes, leaves, bark, twigs, and conifer needles, and were lined with down and feathers. Outside diameter of 1 nest was 23 cm (9 inches); outside depths of 2 nests were 20.5 and 30.5 cm (8 and 12 inches), but most nests were not as deep; inside depths of 2 nests were 7.5 and 10 cm (3 and 4 inches).

EGGS 243 nests with 1 to 14 eggs; 1E (9N), 2E (8N), 3E (3N), 4E (7N), 5E (8N), **6E** (13N), **7E** (24N), **8E** (34N), **9E** (47N), **10E** (41N), **11E** (33N), **12E** (10N), 13E (1N), 14E (5N).
Average clutch range 8 to 10 eggs (122 nests).

INCUBATION PERIOD 10 nests, 23 to 29 days, with 5 averaging 24 to 26 days.

EGG DATES 242 nests, 1 April to 18 July (290 dates); 121 nests, 12 May to 2 June.

Breeding Distribution

The black duck probably breeds throughout Ontario, but reports are lacking for the central-western part of the province.

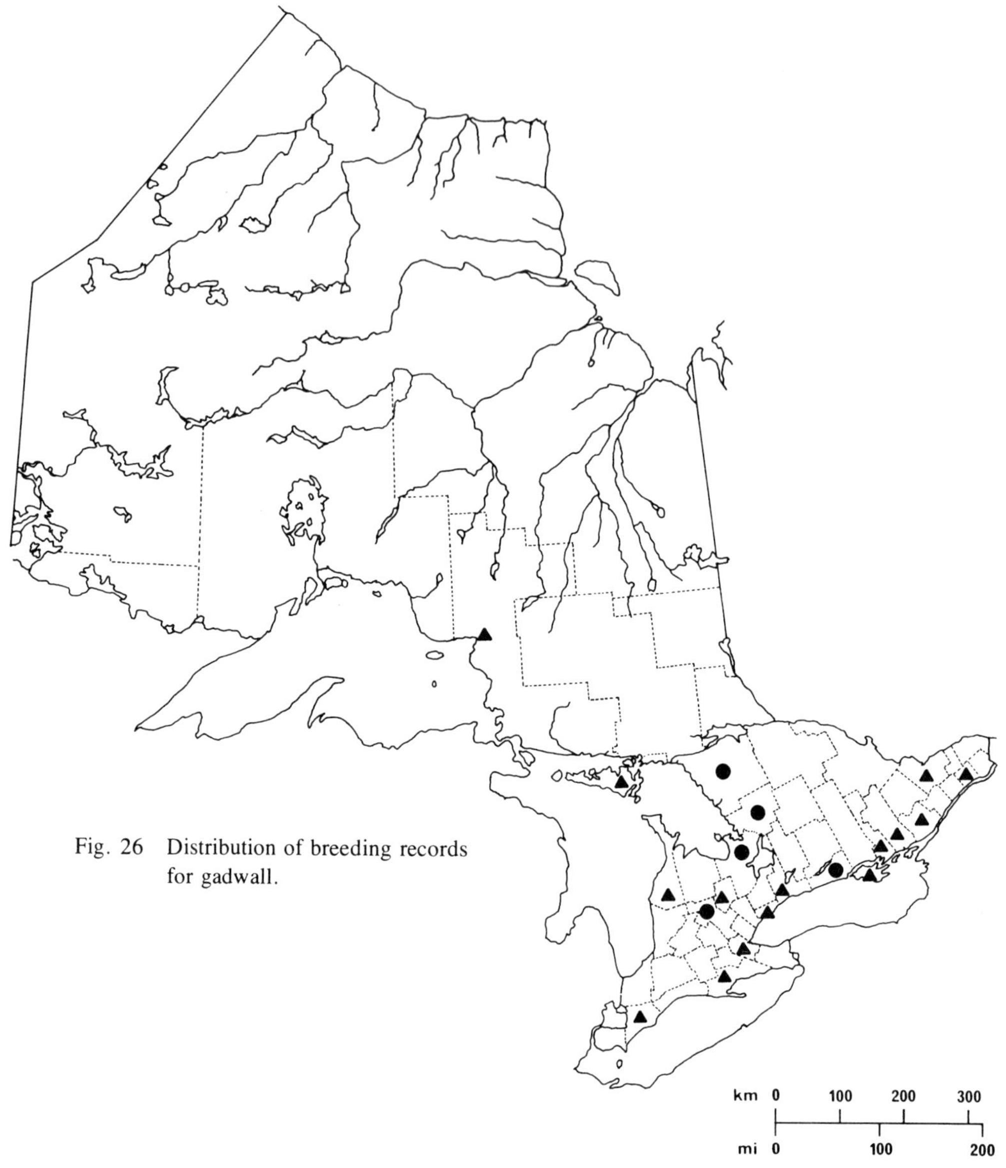

Fig. 26 Distribution of breeding records for gadwall.

Gadwall, *Anas strepera* Linnaeus

Nidiology

RECORDS 137 nests representing 15 provincial regions.
Breeds in grassy fields and clearings, bogs, and marshy borders of lakes and rivers; and on grass-covered or rocky islands, occasionally in or near herring gull, ring-billed gull, and common tern colonies.

Nests were shallow depressions in the ground, usually in clumps of grass, weeds, or nettles, and occasionally in cattails and leatherleaf. They were in the open or at the bases of small trees or bushes, situated from a few metres to 180 m (600 ft) from water.

Nests were constructed of grasses, weed stalks, mosses, leaves, rootlets, and branches of leatherleaf, and were lined with down, feathers, grasses, and twigs.

EGGS 127 nests with 1 to 18 eggs; 1E (2N), 3E (5N), 4E (2N), 5E (2N), **6E** (6N), **7E** (14N), **8E** (21N), **9E** (17N), **10E** (28N), **11E** (24N), 12E (4N), 14E (1N), 18E (1N).
Average clutch range 8 to 10 eggs (66 nests).
One nest with 6 gadwall eggs also contained 1 egg of redhead.

INCUBATION PERIOD 15 nests, 23 to 27 days, with 13 averaging 25 to 26 days.

EGG DATES 128 nests, 15 April to 25 July (208 dates); 64 nests, 19 June to 24 June.

Breeding Distribution

Although the gadwall was mentioned as breeding in Ontario as early as 1906 (Cooke, 1906), no detailed information on nesting was ever obtained until 1955, when the first nest was reported from the Lake St. Clair area (Baillie, 1963). In the past 10 to 15 years the gadwall has expanded its range rapidly throughout most of southern Ontario. A single breeding record in the Algoma District, plus a number of summer sightings as far north as James Bay, indicate that a further expansion of its range is to be expected.

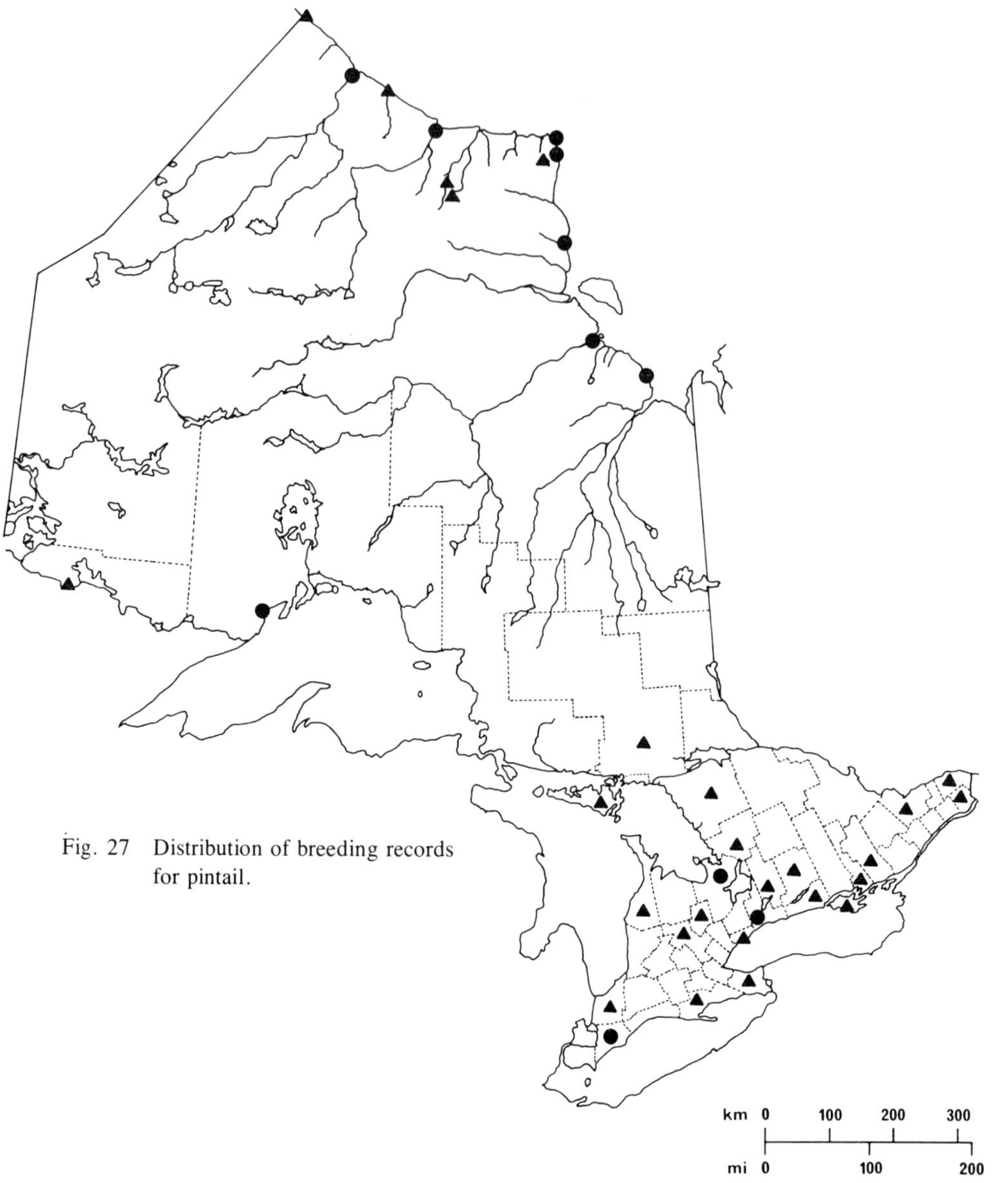

Fig. 27 Distribution of breeding records for pintail.

Pintail, *Anas acuta* Linnaeus

Nidiology

RECORDS 33 nests representing 18 provincial regions.
Breeding habitats recorded included grassy and shrubby areas in fields and orchards, burnt-over areas, marshes and heath bogs, grassy and rocky islands, wet tundra, willow borders of tidal flats, and a dense spruce-willow-alder area near a small pond. Nests were placed from a few metres to more than 90 m (300 ft) from open water.

Nests were shallow depressions, commonly located beneath low trees and shrubs, but also in sedge clumps, tall grasses, cattails, and leatherleaf. A beaten path through grass to 1 nest was noted. Nests were constructed of dead rushes, fine grasses, and grass stems, and were lined with down.

EGGS 28 nests with 1 to 12 eggs; 1E (1N), 2E (1N), 3E (1N), **7E** (7N), **8E** (5N), **9E** (7N), **10E** (1N), **11E** (3N), 12E (2N).
Average clutch range 7 to 9 eggs (19 nests).

INCUBATION PERIOD 1 nest, 23 to 24 days.

EGG DATES 26 nests, 10 April to 30 June (32 dates), 13 nests, 28 May to 22 June.

Breeding Distribution

The pintail is known to be a common breeding species along the Hudson Bay and James Bay coasts and for some distance inland, despite the paucity of nest records. It is also more thinly scattered throughout the south as far as the northern limits of the Great Lakes-St. Lawrence Forest region. However, records are lacking for the central part of Ontario in the Boreal Forest region.

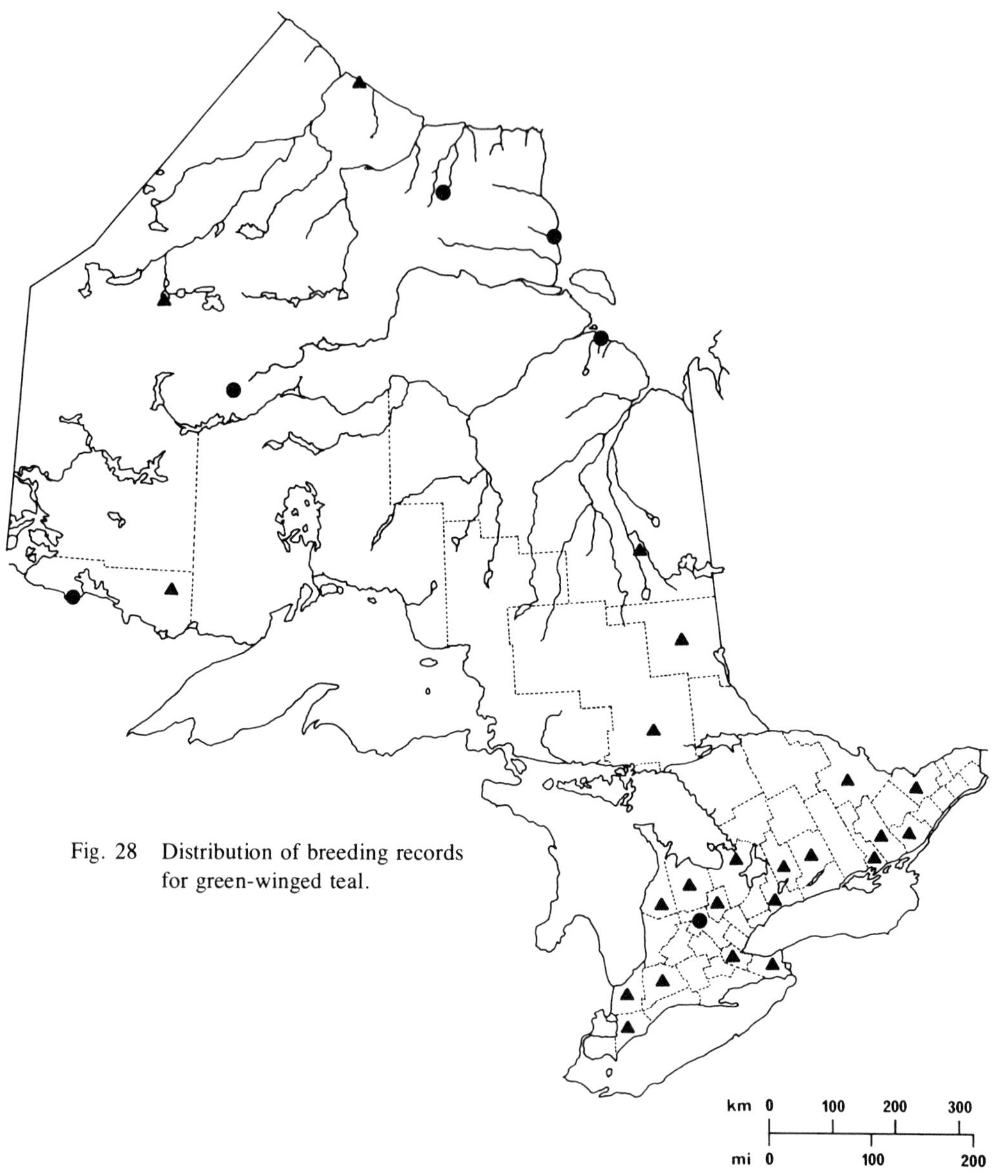

Fig. 28 Distribution of breeding records for green-winged teal.

Green-winged Teal, *Anas crecca* Linnaeus

Nidiology

RECORDS 19 nests representing 9 provincial regions.
Breeds in marshes and bogs; in grassy beach areas and rocky areas near water; and in upland sites such as fields, slashes, and coniferous plantations. Nests were at times close to water but usually from 36.5 to 55 m (120 to 180 ft) distant from it.

Most nests were shallow depressions on hummocks and in clumps of grass, rushes, heath, and bracken. Some were at the bases of trees or beneath fallen trees. One nest was elevated ca 25 cm (10 inches) on an old rotted stump. Nests were neatly woven cups of grasses, rushes, ferns, pine needles, and poplar leaves, and were lined with down and feathers as clutches became complete. Outside diameter of 1 nest was 15 cm (6 inches).

EGGS 18 nests with 3 to 11 eggs; 3E (1N), 4E (1N), 6E (1N), **7E** (1N), **8E** (3N), **9E** (5N), **10E** (3N), **11E** (3N).
Average clutch range 8 to 11 eggs (14 nests).

INCUBATION PERIOD No information.

EGG DATES 17 nests, 11 May to 6 July (25 dates); 9 nests, 27 May to 4 June.

Breeding Distribution

The highest concentrations of breeding green-winged teal appear to be along the north coast of the province, but the species is to be found sparingly throughout Ontario in summer. The first breeding record was established in 1947 (Manning, 1952), but not until 1957 was a nest with eggs located (Tozer and Richards, 1974).

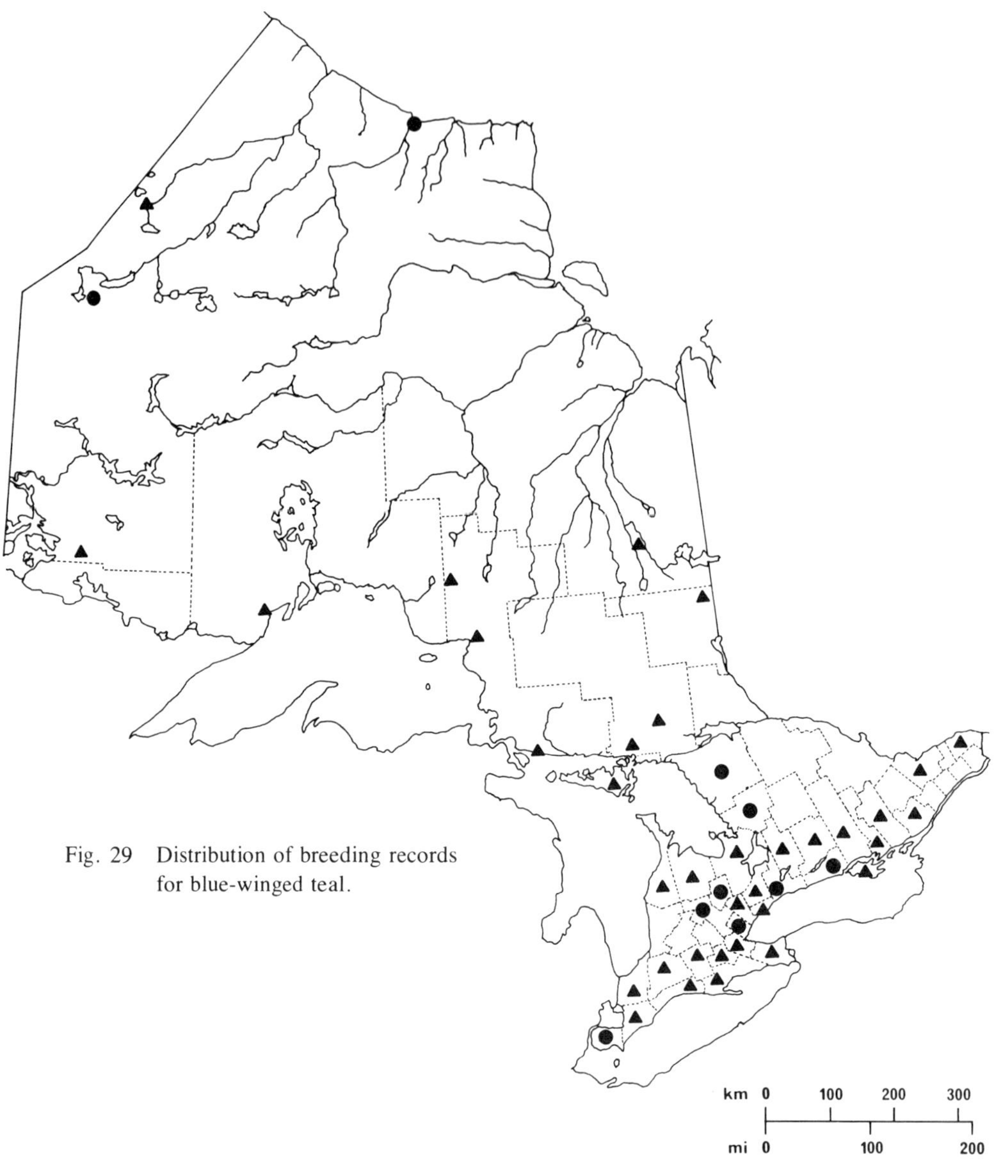

Fig. 29 Distribution of breeding records for blue-winged teal.

Blue-winged Teal, *Anas discors* Linnaeus

Nidiology

RECORDS 262 (270 nests) representing 30 provincial regions.
Breeds in diverse habitats such as grassy fields and meadows, wet and dry pastures, orchards, coniferous plantations, deciduous wood lots, swamps, marshes, bogs, beaver ponds, sand dunes, and edges of ditches and railroad tracks. Some nests were beside or over water, but more often they were from 4.5 to 230 m (15 to 750 ft) distant from it.

Nests were variously situated in clumps or tussocks of living and dead grasses up to 0.6 m (2 ft) in height; in weeds, cattails, sedges, bulrushes, heaths, and bushes (raspberry, juniper); and in small shoots of willow and alder. Some nests were placed under ferns, small trees, or fallen logs, and a few upon muskrat houses.

Nests were well-formed, tightly woven bowls of dead grass stems, straw, sedges, cattails, and bulrushes. They were lined with dead leaves and mosses, and with down and feathers, which increased in amount as the clutch neared completion. One nest had a runway, 0.8 m (2.5 ft) in length, leading to it.

EGGS 128 nests with 1 to 13 eggs; 1E (7N), 2E (2N), 3E (3N), 4E (8N), 5E (6N), 6E (5N), **7E** (7N), **8E** (12N), **9E** (17N), **10E** (14N), **11E** (18N), **12E** (28N), 13E (1N).
Average clutch range 8 to 12 eggs (89 nests).
One nest contained 8 eggs of blue-winged teal and 2 eggs of ring-necked pheasant; another held 9 eggs of blue-winged teal and 1 egg of redhead.

INCUBATION PERIOD 6 nests, 23 to 25 days.

EGG DATES 132 nests, 4 May to 29 July (150 dates); 66 nests, 24 May to 16 June.

Breeding Distribution

The blue-winged teal was formerly believed to breed only as far north as Lake Nipissing or Thunder Bay (Baillie and Harrington, 1936). However, recent records indicate that it is at least thinly scattered throughout Ontario as far north as Hudson Bay.

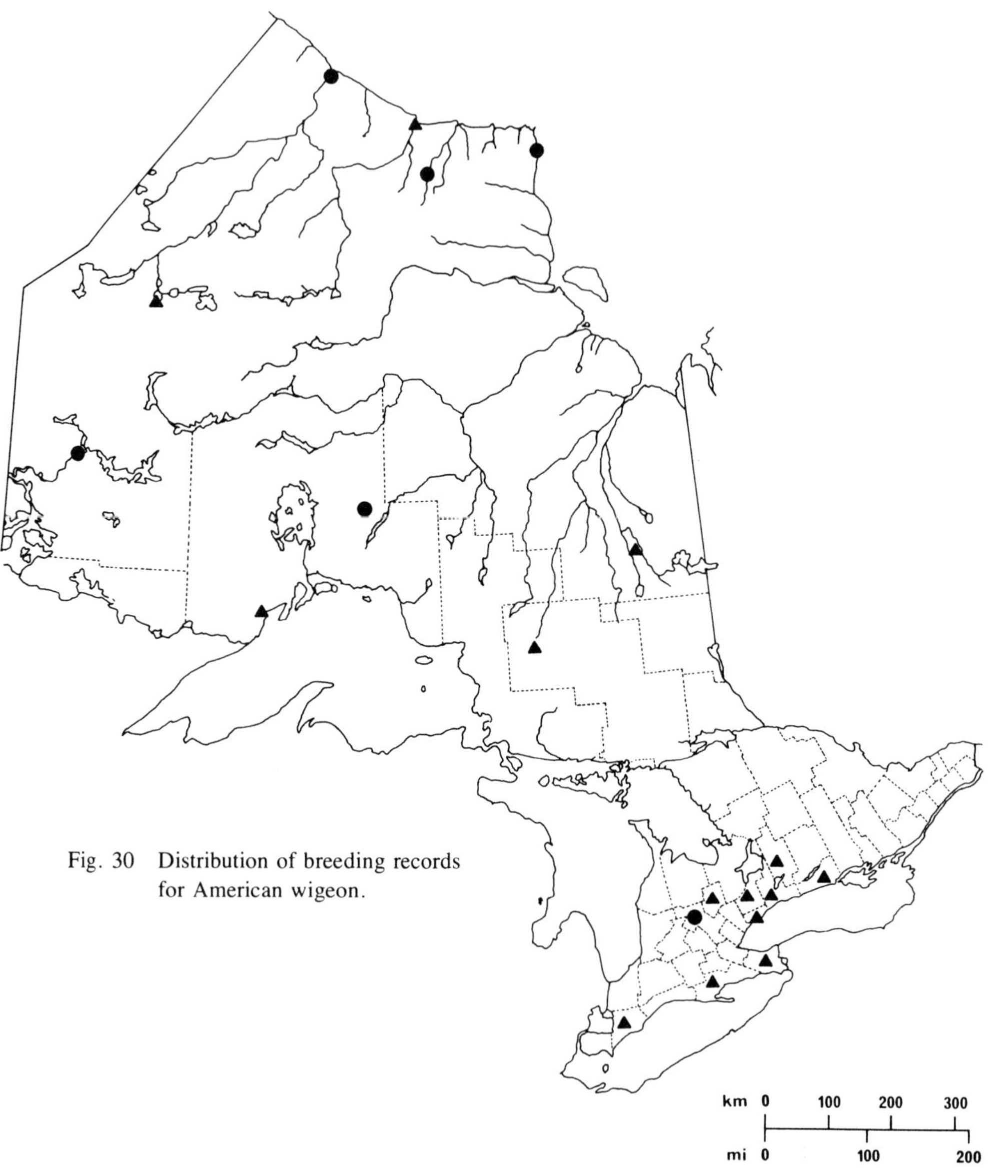

Fig. 30 Distribution of breeding records for American wigeon.

American Wigeon, *Anas americana* Gmelin

Nidiology

RECORDS 22 nests representing 7 provincial regions.
Breeds in cattail marshes, heath bogs, open coniferous woods, and grassy meadows; on islands and shores of lakes and rivers; and on sand beaches (1 record). Nests were positioned from just over water to as far as ca 180 m (600 ft) distant from it.

Nests were usually shallow depressions in clumps of grass, moss, weeds, cattail, Labrador tea, leatherleaf, and nettles; they were sometimes placed at the bases of alder and dogwood bushes; 1 nest was beside a fallen tree and another under an uprooted stump.

Nests were composed of grasses and cattails and were lined with down and feathers.

EGGS 21 nests with 3 to 11 eggs; 3E (2N), 5E (1N), 6E (1N), **7E** (2N), **8E** (3N), **9E** (7N), **10E** (4N), **11E** (1N).
Average clutch range 7 to 10 eggs (16 nests).

INCUBATION PERIOD No information.

EGG DATES 20 nests, 9 May to 19 August (24 dates); 10 nests, 17 June to 26 June.

Breeding Distribution

The American wigeon is widely distributed in western Canada but is much scarcer in Ontario. The fact that the first report of a nest in Ontario was in 1934 at Toronto (Baillie, 1960) would seem to indicate that the species has only recently expanded its breeding range into the province. However, a relatively large population on the Hudson Bay coast that was reported only in 1952 (Manning, 1952) actually dates back to the 1850s. This indicates a much longer residency in the province that went unnoticed for many years. Other than on the Hudson Bay coast, the species is sparingly distributed, but it probably nests in any part of Ontario.

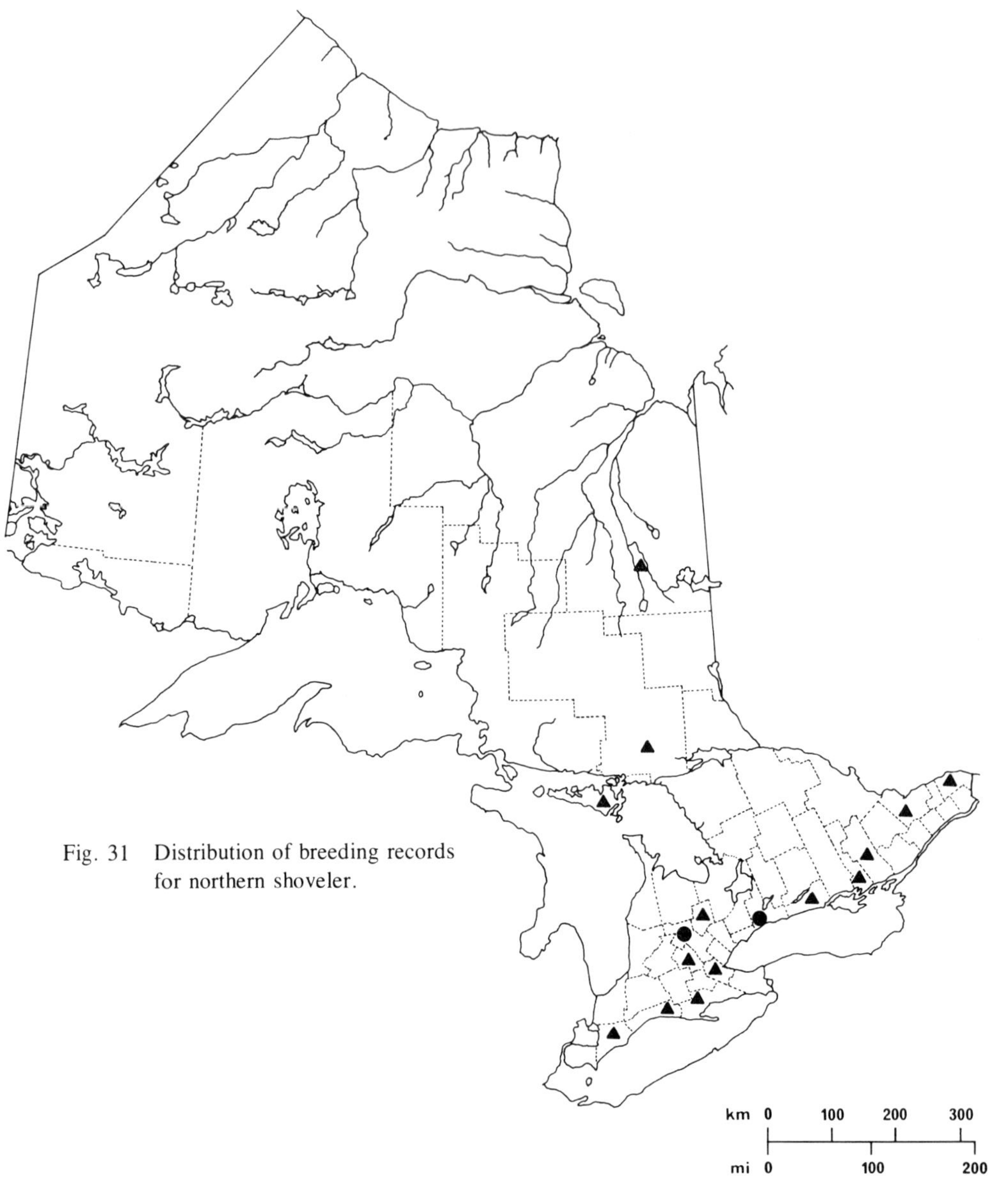

Fig. 31 Distribution of breeding records for northern shoveler.

Northern Shoveler, *Anas clypeata* Linnaeus

Nidiology

RECORDS 5 nests representing 5 provincial regions.
Breeds in large and small cattail marshes, grassy fields, and abandoned pastures. Nests were situated over water and on dry land as far as 90 m (300 ft) distant from water.

Nests were slight depressions in the ground, once at the base of a small tree, or were semi-floating on flattened cattails. They were lined with cattails, grasses, and down. Measurements of 1 nest were as follows: outside diameter, 43 by 48.5 cm (17 by 19 inches); outside depth, 20.5 cm (8 inches); inside diameter, 18 to 20.5 cm (7 to 8 inches); and inside depth, 8.5 cm (3.25 inches).

EGGS 5 nests, with 8 to 13 eggs; 8E (1N), 9E (1N), 10E (2N), 13E (1N).

INCUBATION PERIOD No information.

EGG DATES 5 nests, 22 May to 23 June (7 dates); 3 nests, 31 May to 18 June.

Breeding Distribution

There are only a few nesting records for the northern shoveler. Early breeding evidence of this western duck was based on sight records of adults with young—in 1886 and 1887 in Haldimand County (Haldimand-Norfolk RM) (Baillie and Harrington, 1936) and from 1951 to 1954 in Kent County—and on a nest record in 1959, also in Kent County. The first material evidence of breeding was that of a nest with nine eggs collected in Ontario County (Durham RM) on 30 May 1967 (ROM 11933) (Tozer and Richards, 1974).

Today there are still only a few scattered reports of nesting. Almost all records are from southern Ontario but regular summer occurrences along the northern coasts indicate that the species may also breed in that area.

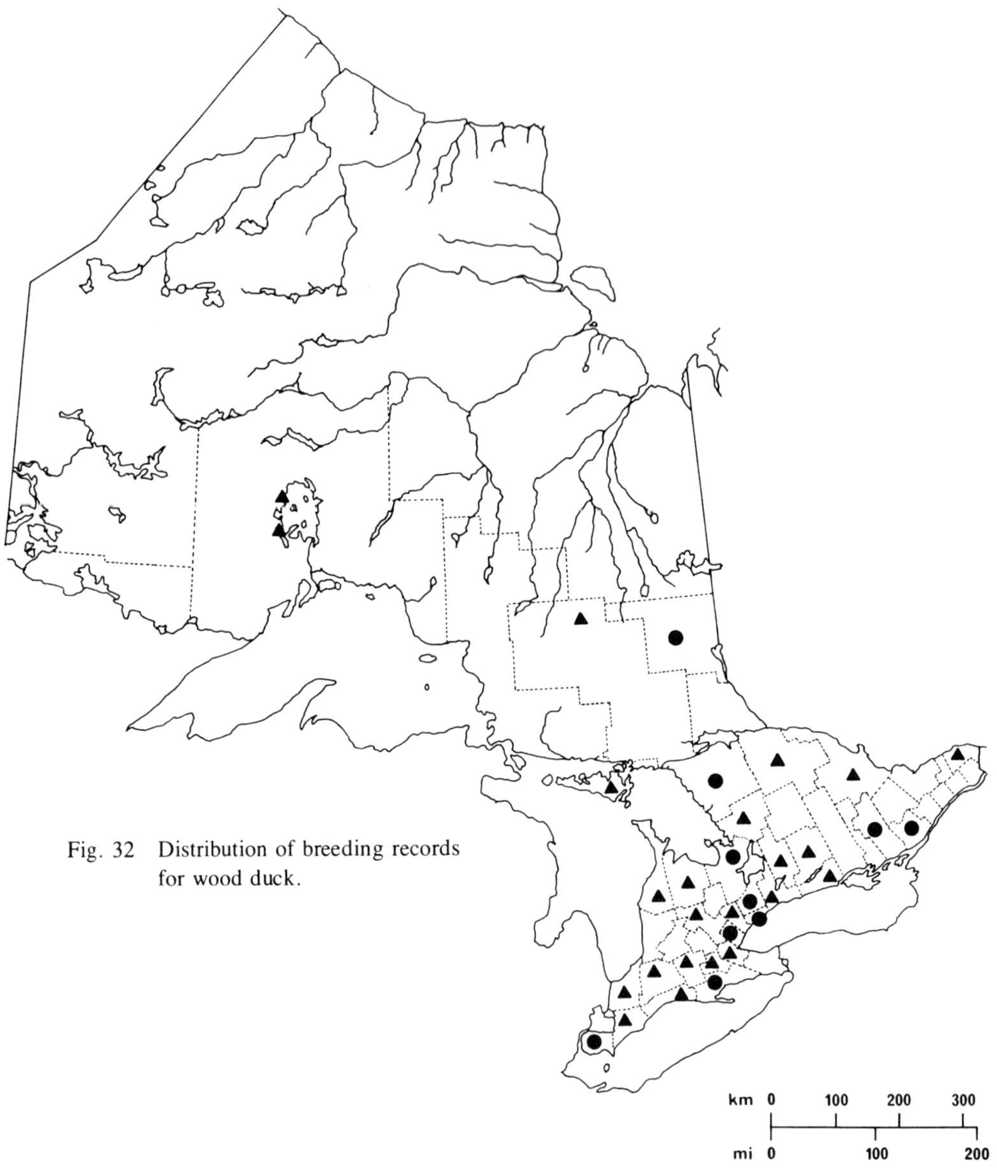

Fig. 32 Distribution of breeding records for wood duck.

Wood Duck, *Aix sponsa* (Linnaeus)

Nidiology

RECORDS 105 nests representing 21 provincial regions.
Breeding habitats recorded were cattail marshes and marshy borders of lakes, rivers, and ponds where trees were nearby to serve as nest sites; wooded swamps and beaver ponds; both wet and dry deciduous and mixed woodlands; sewage lagoons; and islands (there is 1 record of breeding in a residential backyard). Nests were usually over or near water, but were sometimes as far as 200 m (660 ft) distant from it.

Nests were usually in natural cavities or old woodpecker cavities (20 nests), in dead or partly dead deciduous (16 nests) or coniferous (3 nests) trees (1 nest tree unspecified). Where available, nest boxes (72 nests) were often used; an old, open nest of common crow was once used. Heights of 20 nests in natural sites ranged from 2 to 14 m (6 to 45 ft), with 10 averaging 4.5 to 9 m (15 to 30 ft).

Nest cavities contained existing materials such as wood chips, bark fibre, and, occasionally, grasses. Nests were lined with feathers and down, which increased in amount as the clutch neared completion. One cavity opening measured 10 cm (4 inches) in diameter, and cavity depths ranged from several centimetres to ca 1 m (a few inches to several feet).

EGGS 69 nests with 1 to 30 eggs; 1E (1N), 3E (1N), 4E (2N), 5E (2N), 6E (1N), 7E (1N), **9E** (9N), **10E** (10N), **11E** (10N), **12E** (12N), **13E** (6N), **14E** (4N), 15E (2N), 16E (2N), 17E (2N), 18E (1N), 19E (1N), 22E (1N), 30E (1N).
Average clutch range 10 to 13 eggs (38 nests).
Large sets were the result of laying by more than 1 female. One nest contained 16 eggs of wood duck and 8 eggs of hooded merganser.

INCUBATION PERIOD 5 nests: 3 of 24 to 26 days, 1 of 28 days, 1 of 29 days.

EGG DATES 75 nests, 28 March to 18 July (100 dates); 38 nests, 9 May to 25 May.

Breeding Distribution

Baillie and Harrington (1936) considered the wood duck to be a rare breeding species in southern Ontario that was greatly reduced in numbers from former years. Its decline was perhaps the result of hunting, forest cutting, and swamp drainage. However, in more recent times, perhaps partly in response to the erection of nest boxes, the species appears to be increasing somewhat and expanding its range. It is now found throughout southern Ontario and extends into the north in small numbers as far as Lake of the Woods, Lake Nipigon, and Timiskaming District.

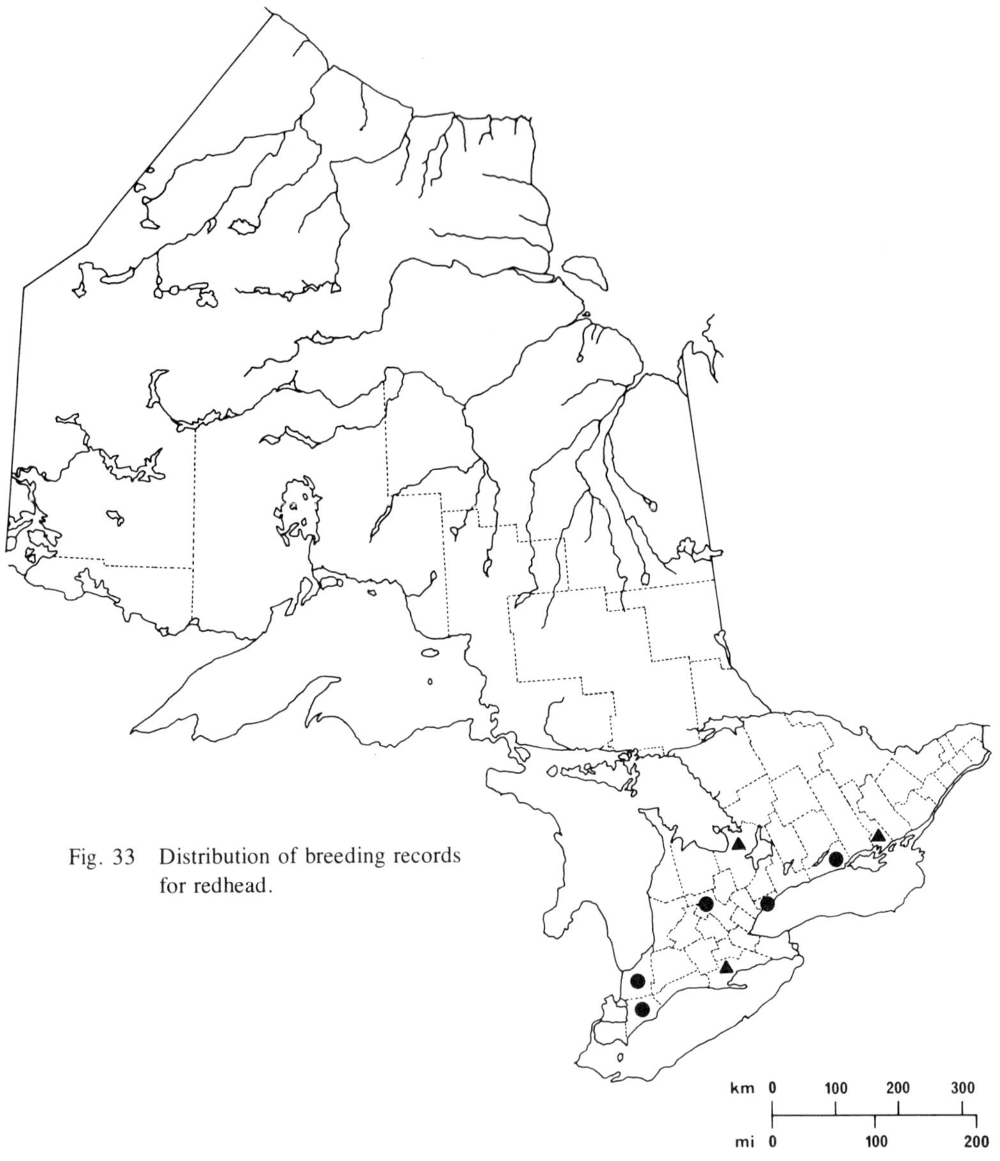

Fig. 33 Distribution of breeding records for redhead.

Redhead, *Aythya americana* (Eyton)

Nidiology

RECORDS 20 (31 nests) representing 7 provincial regions.
Breeds in cattail and bulrush marshes, leatherleaf bogs, and in wet and dry areas of islands in lakes.

Nests were situated in wet (19 nests), semi-wet (7 nests), and dry-ground (5 nests) habitats. Wet-area nests were in standing cattails and bulrushes, over water that ranged from a few centimetres to 90 cm (36 inches) in depth and up to 9 m (30 ft) from open water; 1 nest was on a floating log. Semi-wet-area nests were in leatherleaf bushes in bogs, often at considerable distances from open water. Dry-ground nests were in clumps of mustard, nettles, and dogwood bushes on a small limestone island.

Nests were constructed of dead and living cattails, bulrushes, grasses, weeds, leatherleaf, and dogwood branches, and were lined with twigs, grasses, weed stems, feathers, and down. At least 1 nest over water had a long ramp leading to it.

EGGS 19 nests with 2 to 15 eggs; 2E (1N), **6E** (5N), **7E** (2N), **8E** (1N), **9E** (1N), **10E** (5N), **11E** (1N), 12E (1N), 15E (2N).
Average clutch range 6 to 10 eggs (14 nests).
Single eggs of redhead were found in a gadwall nest with 6 eggs and in a blue-winged teal nest with 9 eggs; 2 and 5 eggs of redhead were reported in 2 nests of ring-necked duck, with 5 and 8 eggs respectively; 2 eggs of redhead were found in a lesser scaup nest with 11 eggs.

INCUBATION PERIOD No information.

EGG DATES 30 nests, 24 May to 4 July (32 dates); 15 nests, 24 May to 21 June.

Breeding Distribution

The redhead was reported to breed in good numbers in the Lake St. Clair area as long ago as 1877. However, Baillie and Harrington (1936) were unaware that specimens had been collected in that area in 1882 and in about 1900, and they did not include the species in their list of Ontario's breeding birds (Baillie, 1958). Since 1949 there have been a number of new nesting or breeding reports for a few scattered localities in southern Ontario, south of the Canadian Shield, and a good-sized population is apparently present in the dyked marshes of Walpole Island in Lake St. Clair (H. G. Lumsden, pers. comm.).

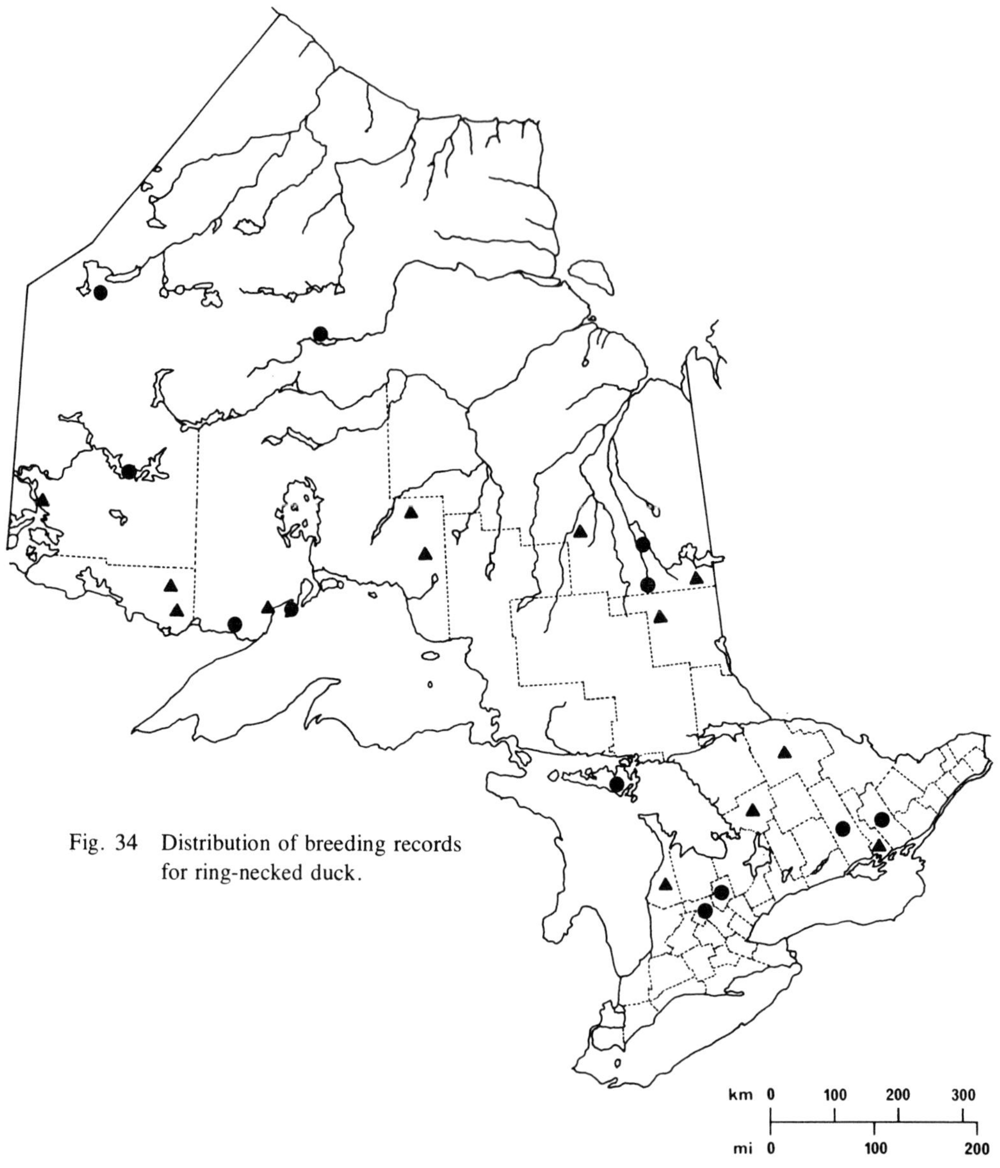

Fig. 34 Distribution of breeding records for ring-necked duck.

Ring-necked Duck, *Aythya collaris* (Donovan)

Nidiology

RECORDS 24 nests representing 6 provincial regions.
Breeds in cattail and sedge marshes, heath fens, beaver meadows, and the marshy borders of lakes and islands.

Nests were usually situated in clumps of sedge, cattail, and sweet gale, from 5 to 23 cm (2 to 9 inches) above water that ranged in depth from 15 to ca 90 cm (6 to 36 inches). A dry-ground nest was situated in grasses at the base of a thorny bush.

Nests were formed of interwoven vegetation: sedges, cattails, and grasses. They were sometimes canopied and were lined with leaves, dead stems of marsh vegetation, small sticks, and down.

EGGS 21 nests with 5 to 13 eggs; 5E (1N), 6E (1N), **7E** (2N), **8E** (6N), **9E** (5N), **10E** (4N), **11E** (1N), 13E (1N).
Average clutch range 8 to 10 eggs (15 nests).
One nest contained 8 eggs of ring-necked duck and 5 eggs of redhead; another held 5 eggs of ring-necked duck and 2 eggs of redhead; a third held 6 eggs of ring-necked duck and 1 egg of ruddy duck.

INCUBATION PERIOD 1 nest, ca 27 days.

EGG DATES 22 nests, 19 May to 28 July (27 dates); 11 nests, 10 June to 18 June.

Breeding Distribution

There is no positive evidence that the ring-necked duck bred in Ontario before 1919, when downy young were collected at Lac Seul. There is no record of its presence along the north coast during the period of the fur trade (Alison, 1975). Since 1920, however, the species has spread across the entire province except for counties in the extreme south near Lake Erie.

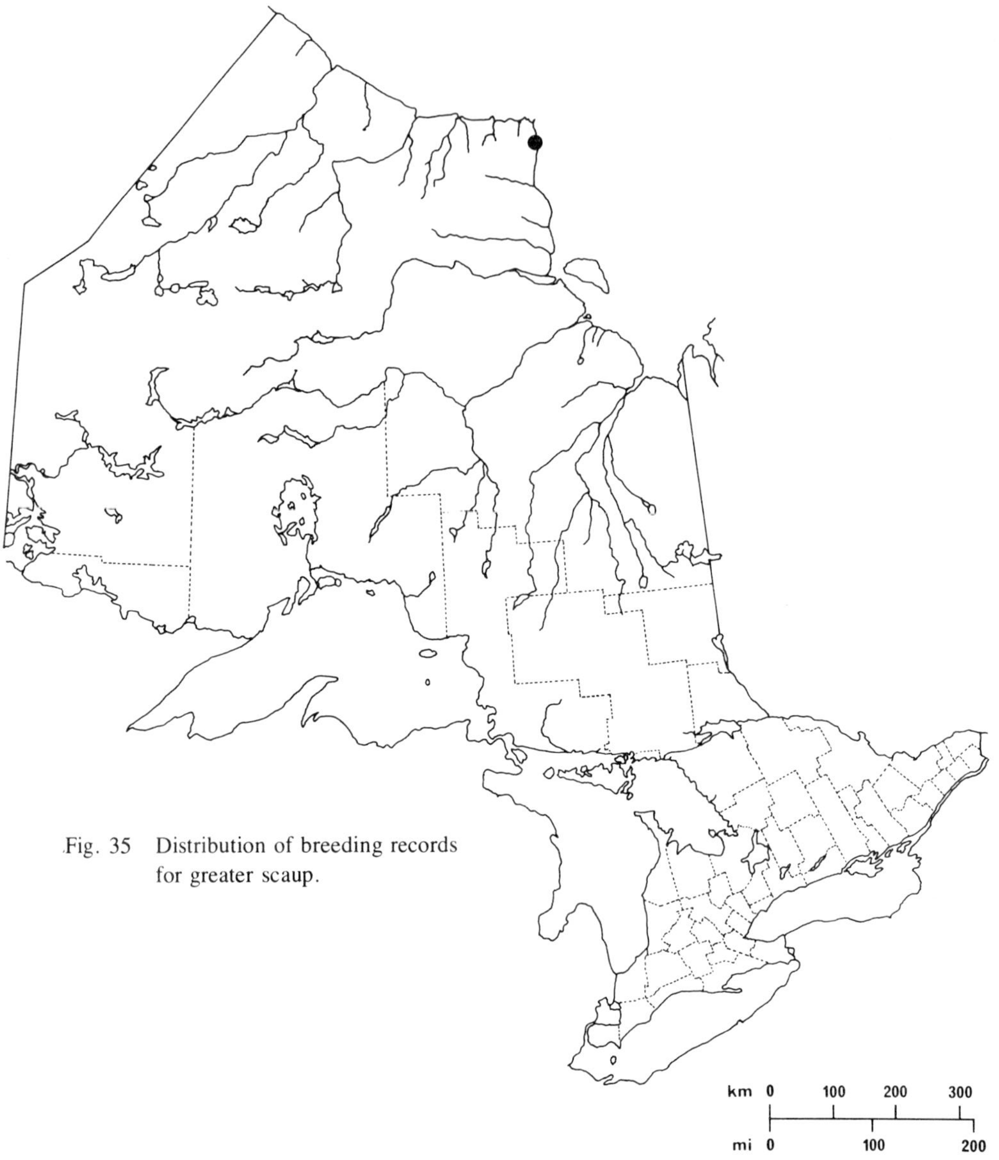

Fig. 35 Distribution of breeding records for greater scaup.

Greater Scaup, *Aythya marila* (Linnaeus)

Nidiology

Although no nests of this duck have been reported in Ontario, the probability of breeding was first established on 18 June 1940 at Fort Severn, Kenora District, by the collection of a female with well-developed ova without shells. On 28 June 1948 in the Cape Henrietta Maria region, Kenora District, another female about to lay eggs was collected. In the latter area on 16 July 1948, a female and 9 downy young were collected (ROM 75517–75525, 75536), and on 19 July, 3 more downy young were collected (ROM 75529–75531).

Breeding Distribution

Because of the difficulty in distinguishing greater scaup from lesser scaup, little information is available on the extent of the summer breeding range of the greater scaup. Only two documented breeding records for the province are available (Baillie, 1961). However, it appears that the species breeds all along the Hudson Bay coast, and probably for some distance inland along the major rivers (Lumsden, 1957).

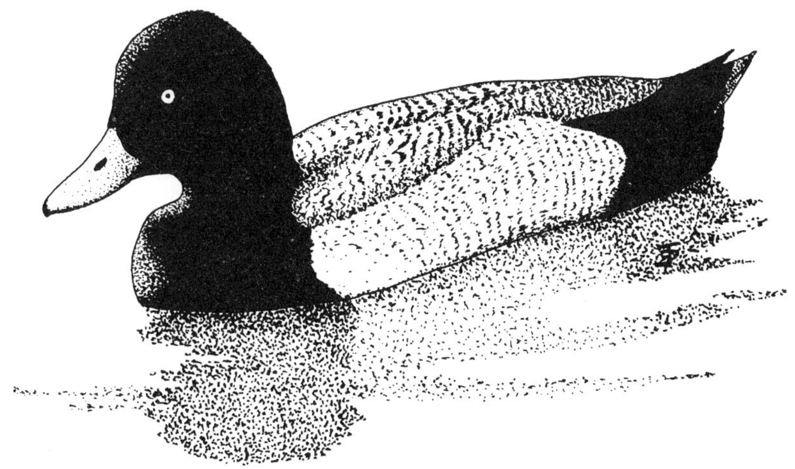

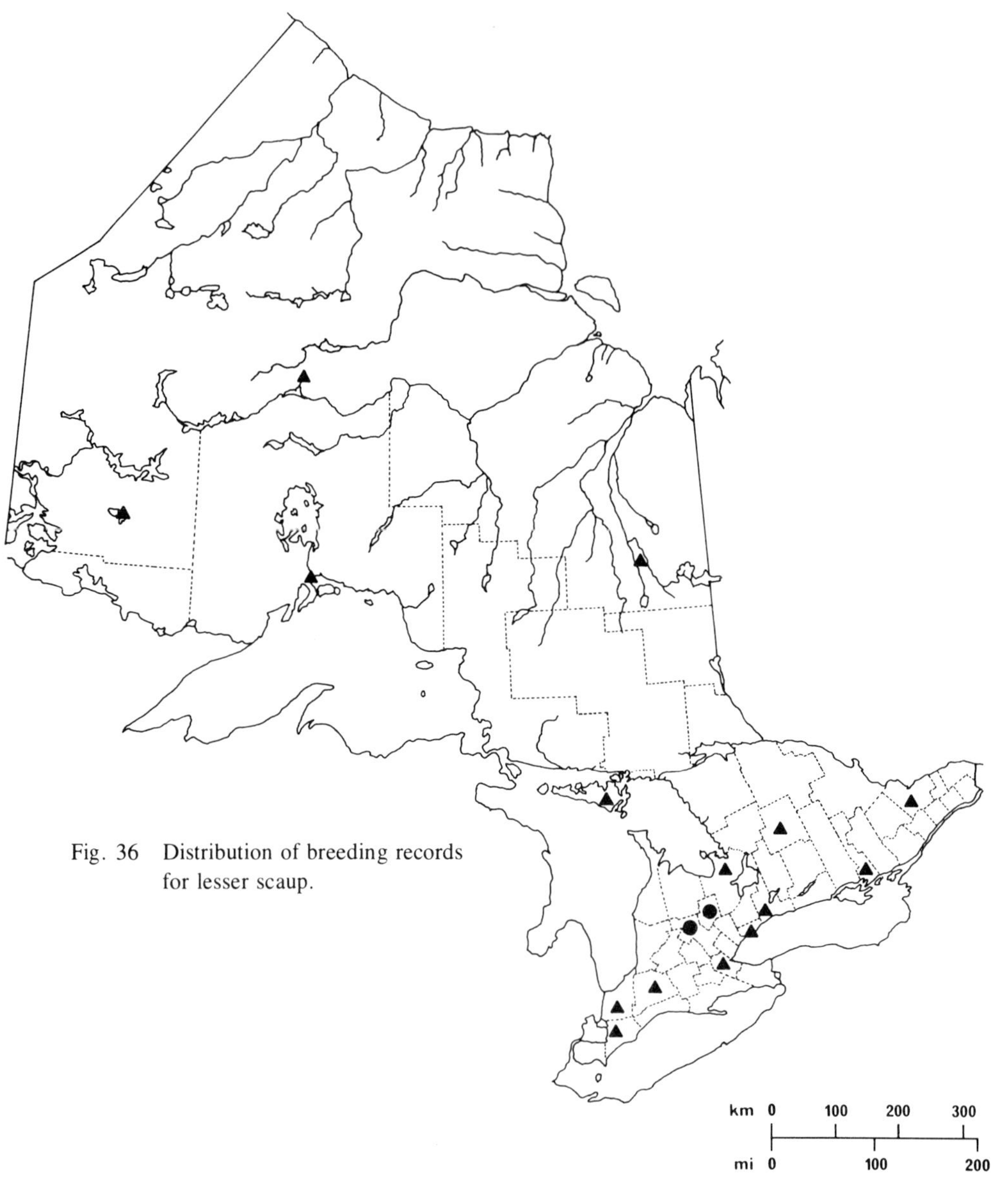

Fig. 36 Distribution of breeding records for lesser scaup.

Lesser Scaup, *Aythya affinis* (Eyton)

Nidiology

RECORDS 7 nests representing 4 provincial regions.
Breeds in sedge, cattail, and heath-bog borders of lakes and small islands in lakes.

Nests were placed 15 to 25.5 cm (6 to 10 inches) above water in leatherleaf bushes and sedge clumps that formed cover 51 to 61 cm (20 to 24 inches) in height. Water depth under 1 nest was 30.5 cm (12 inches). Nests were situated as far as 21 m (70 ft) from open water. One nest was placed 3 m (10 ft) from a nest of blue-winged teal and another was 30 m (100 ft) from a nest of redhead.

Nest structures were of interwoven dead and living sedge stems. They were lined with reeds, grasses, mosses, leaves, feathers, and down, and were sometimes canopied. Inside diameter of 1 nest was 19 by 21.5 cm (7.5 by 8.5 inches).

EGGS 7 nests with 10 to 12 eggs; 10E (3N), 11E (2N), 12E (2N).
One nest contained 11 eggs of lesser scaup and 2 eggs of redhead.

INCUBATION PERIOD 1 nest, 23 days.

EGG DATES 7 nests, 27 May to 11 July (11 dates); 4 nests, 5 June to 17 June.

Breeding Distribution

Although the lesser scaup is mainly a western breeding species, it has probably always nested in Ontario in small numbers because records exist back to the turn of the century. The species is very scarce in the province during the breeding season, and the few records available give only a poor idea of the full extent of its range. It appears, however, to be very thinly scattered as far north as Cochrane in the east and near the Hudson Bay coast in the west, and it possibly nests throughout the province.

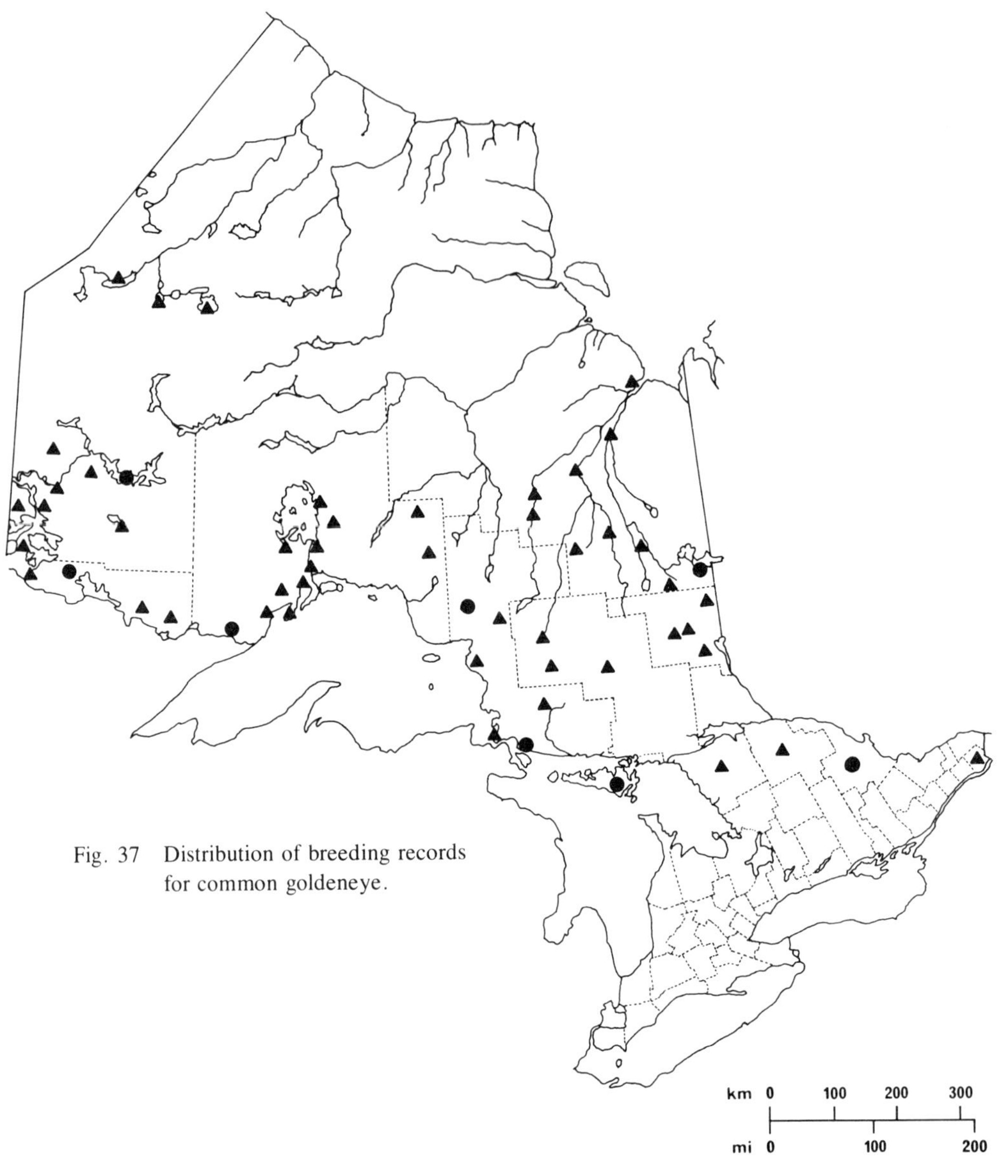

Fig. 37 Distribution of breeding records for common goldeneye.

Common Goldeneye, *Bucephala clangula* (Linnaeus)

Nidiology

RECORDS 37 nests representing 5 provincial regions.
Breeding habitats recorded for this tree-nesting duck were deciduous woods (8 nests) and mixed woods (1 nest) near lakes and rivers, and an island in a river (1 nest). Nest situations ranged from the water's edge to 90 m (300 ft) distant from it.

Nest boxes (27 nests) were readily used when available. Natural-site nests were invariably in living or dead trees, poplar (8 nests) and birch (1 nest). Nests were in either natural cavities (7 nests) or old pileated woodpecker cavities (2 nests). Diameter of 1 nest tree was 35.5 cm (14 inches). Heights of 8 nests in natural sites ranged from 4.5 to 18 m (15 to 60 ft), with 4 averaging 6 to 9 m (20 to 30 ft).

Nests cavities were lined with a few feathers and down. One cavity measured about 76 cm (30 inches) in depth.

EGGS 30 nests with 1 to 11 eggs; 1E (3N), 2E (3N), 3E (1N), 5E (2N), **6E** (3N), **7E** (8N), **8E** (5N), **9E** (3N), **10E** (1N), 11E (1N).
Average clutch range 6 to 8 eggs (16 nests).
One nest containing 5 eggs of common goldeneye also contained 5 eggs of hooded merganser, and another nest with 9 eggs of hooded merganser also contained 1 egg of common goldeneye.

INCUBATION PERIOD 3 nests: 1 nest of 28 days, 1 nest of 29 days, 1 nest of ca 30 days.

EGG DATES 36 nests, 11 May to 21 July (46 dates); 18 nests, 19 May to 4 June.

Breeding Distribution

The common goldeneye breeds across central Ontario. However, it is absent as a breeding bird in most of southern Ontario and scarce where it does occur. Although it is commonly seen as far north as Hudson Bay in summer, these reports may be sightings of nonbreeding birds because no nest records have been reported north of Sandy Lake and Moosonee.

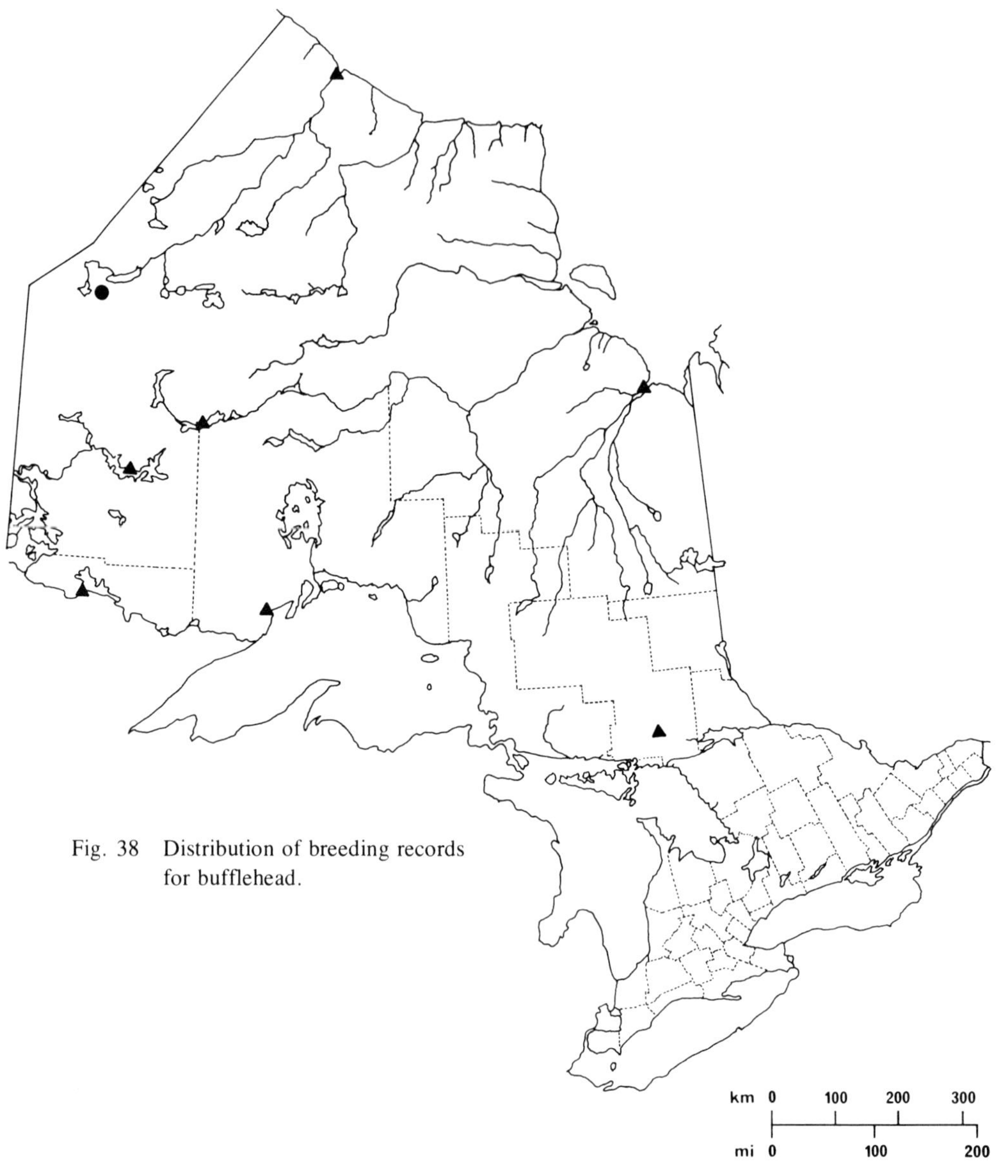

Fig. 38 Distribution of breeding records for bufflehead.

Bufflehead, *Bucephala albeola* (Linnaeus)

Nidiology

RECORDS 1 nest representing 1 provincial region.

This species owes its place on the Ontario breeding-bird list to the collection of 5 young birds with an adult (ROM 30639–30643) on 25 July 1938 at Favourable Lake, Kenora District. One of these young (ROM 30641) looked considerably smaller and younger than the others and was obviously flightless.

The only nest record so far was on 19 June 1973 and was an undocumented observation of a female entering a presumed nest hole 7 m (23 ft) high in a dead poplar sp. 12 m (40 ft) in height, located in a poplar-black spruce forest in Tidewater Provincial Park, Charles Island, near Moosonee, Cochrane District.

Breeding Distribution

The bufflehead (Fig. 162B) is a western tree-nesting species of scattered breeding occurrence east of Alberta. The very few breeding records of the bufflehead available for Ontario scarcely indicate the real distribution of this species (Erskine, 1971). Although apparently very thinly distributed, it may well breed throughout the forested portions of northern Ontario. It was reported formerly as breeding in the extreme south of Ontario, but such records were never verified and were considered erroneous (Baillie and Harrington, 1936).

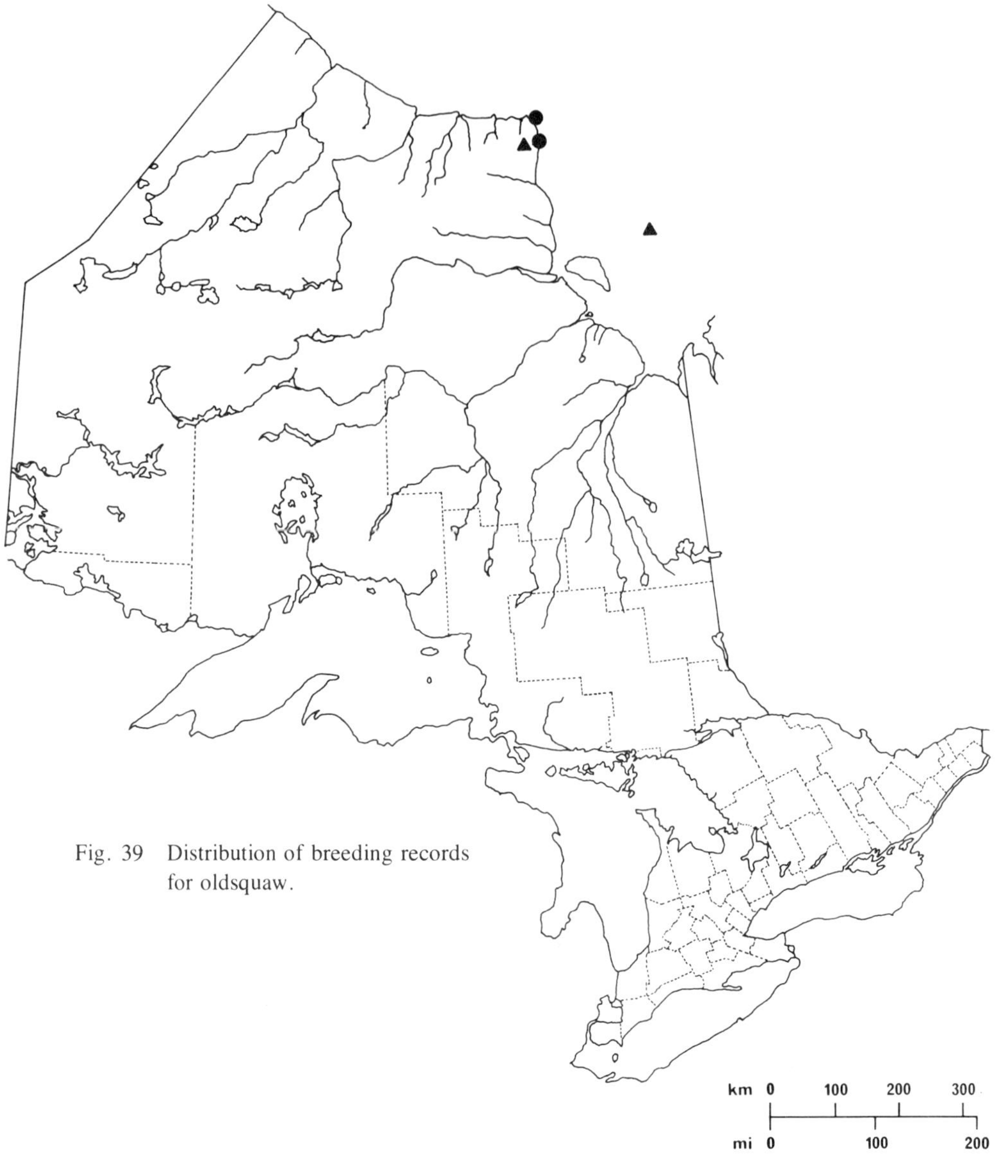

Fig. 39 Distribution of breeding records for oldsquaw.

Oldsquaw, *Clangula hyemalis* (Linnaeus)

Nidiology

RECORDS 4 nests representing 1 provincial region and an adjacent island in James Bay (NWT). All 4 recorded nests were in tundra regions; 1 nest was in a clump of arctic willow at the edge of a tundra slough, and another was on a windswept hilltop in a patch of bog birch, on a large island (NWT).

One nest was composed of mosses, leaves, sedges, and down.

Two nests contained an unspecified number of eggs that were due to hatch about 26 July.

EGGS 2 nests with 3 and 5 eggs.

INCUBATION PERIOD No information.

EGG DATES 2 nests, 10 July, 15 July, and ca 25 July (3 dates).

Breeding Distribution

A bird of the arctic tundra, the oldsquaw (Fig. 170B) apparently breeds in Ontario only in the area near Cape Henrietta Maria. The first nests in Ontario were found in 1947 (Manning, 1952), although this species was probably breeding in the Cape area for many years before that date. The inaccessibility of the area and the difficulty in locating nests keep us, even today, largely ignorant of its nesting habits in the province. It may breed sparingly along the entire coast of Hudson Bay where tundra conditions prevail (Fig. 169).

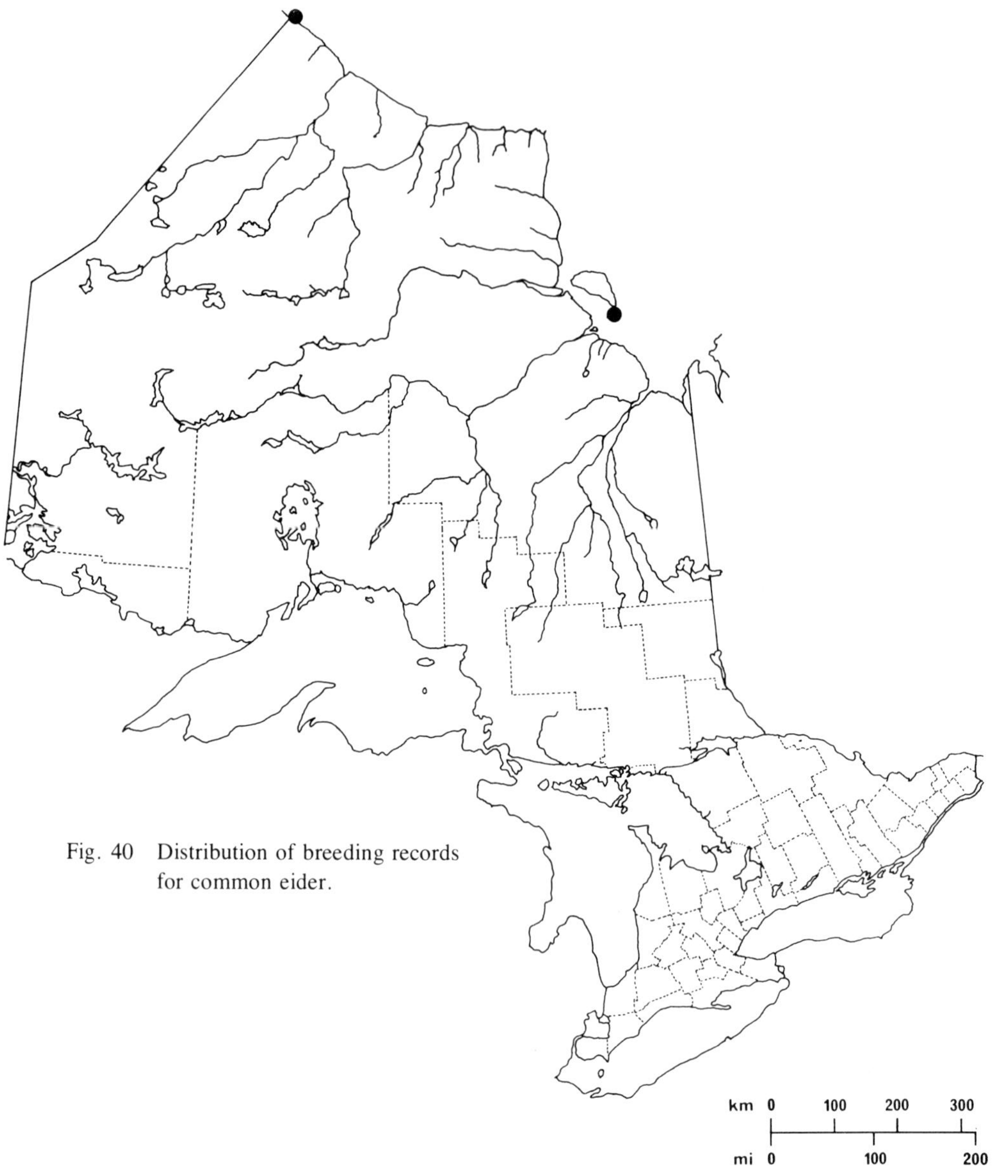

Fig. 40 Distribution of breeding records for common eider.

Common Eider, *Somateria mollissima* (Linnaeus)

Nidiology

RECORDS 3 nests representing 1 provincial region and an adjacent island in James Bay (NWT). This northern breeding duck nests on islands off the subarctic tundra coast of Hudson and James bays.

Two nests containing eggs and 2 broods of newly hatched ducklings with adults were found on a small coastal island. The nests were in sandy, old beach-line areas sparsely overgrown with grasses. Both nests were placed among grasses and 1 was under a willow bush. Both nests were lined with grass stems and down.

EGGS 3 nests; 1 with 5 eggs and 2 with 6 eggs.

INCUBATION PERIOD No information.

EGG DATES 3 nests, 15 July, 15 July, and 16 July.

Breeding Distribution

The first report of the common eider breeding in Ontario came from East Pen Island, which is still the only location in Ontario where the species is known to breed. In 1960 a female with young was observed and in 1963 specimens were collected (Baillie, 1963). There are also reported observations of birds near Cape Henrietta Maria. This is not outside the species' breeding range because it is known to breed on islands in James Bay. Possibly the common eider will be found nesting on several other islands near the Hudson Bay and James Bay coasts in the future.

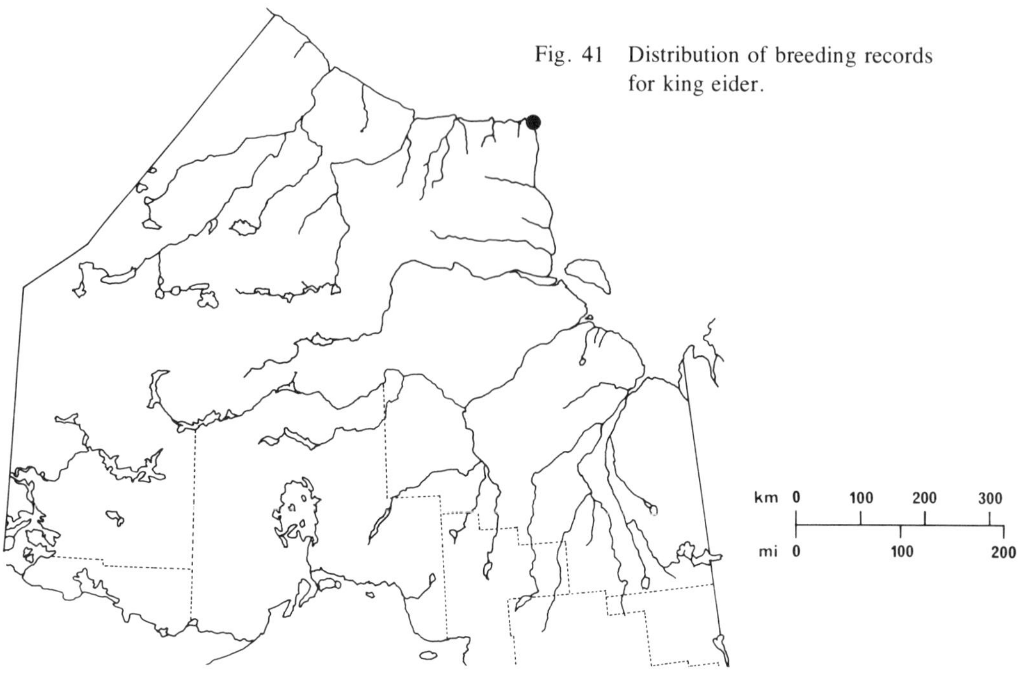

Fig. 41 Distribution of breeding records for king eider.

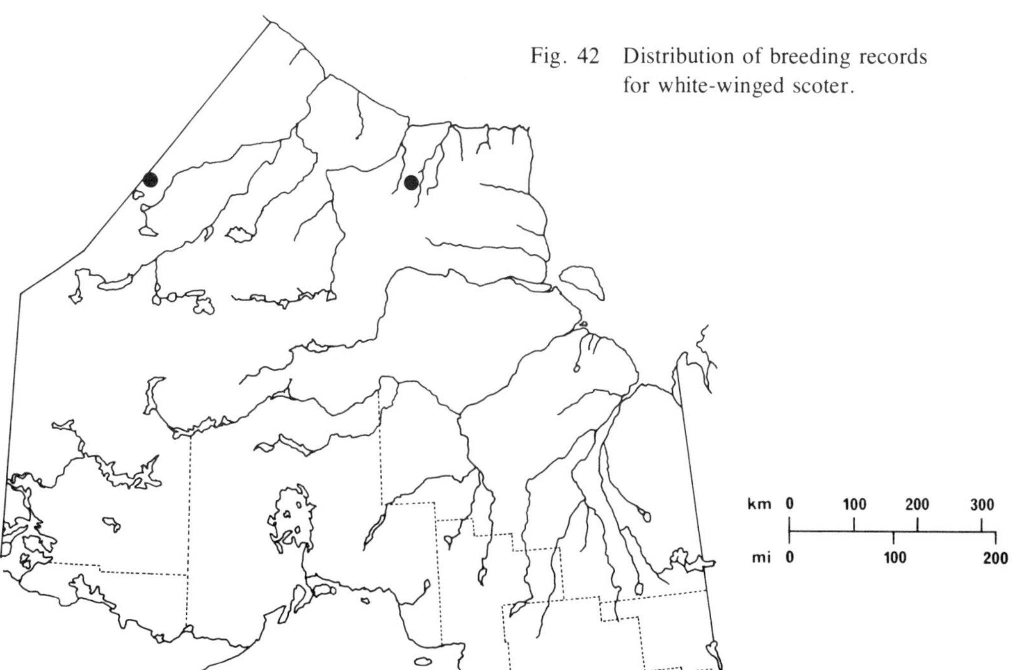

Fig. 42 Distribution of breeding records for white-winged scoter.

King Eider, *Somateria spectabilis* (Linnaeus)

Nidiology

No nests of this eider have been found in Ontario to date, although proof of its breeding has been definitely established. On 23 August 1944 at Cape Henrietta Maria, Kenora District, a female attending 6 young was collected (ROM 71442). Also at Cape Henrietta Maria, on 23 and 24 July 1947, a female with a brood of week-old young and another female with a brood of newly hatched young were seen; some of the young were collected (NMC) (Manning, 1952).

Breeding Distribution

The breeding status of the king eider is based on the above-mentioned collection of birds in the Cape Henrietta Maria region in 1944 and 1947 (Baillie, 1961). It is not known to nest at any other place in the province. Summering birds have been observed as far west as the Sutton River, and possibly the species breeds along the stretch of the Hudson Bay coast that lies between Cape Henrietta Maria and the Sutton River.

White-winged Scoter, *Melanitta deglandi* (Bonaparte)

Nidiology

RECORDS 1 nest representing 1 provincial region.
This species was first found breeding in the province on 8 August 1936, when a female with downy young was observed on Ney Lake, Kenora District, near the Manitoba border (Baillie, 1939). Photographs of the downy young were taken (ROM PR 104).

On 31 July 1965 a nest with 7 eggs was discovered on the muskeg at Cool Lake, west of Hawley Lake, Kenora District. It was in reindeer lichen at the base of a black spruce, at some distance from water. One egg in an advanced stage of incubation was collected by T. G. Harrison, along with some down, and both were presented to the ROM (ROM 11577).

Breeding Distribution

Although Manning (1952) reported the white-winged scoter to be summering along the Hudson Bay coast, this species is apparently now seen there only rarely (H. G. Lumsden, pers. comm.) and its breeding range in Ontario remains almost a complete mystery. The first record in 1936 from Ney Lake is close to its known range in western Canada. The above nest record from Cool Lake suggests that this species may nest all across northern Ontario, but because of the inaccessibility of much of the area it may be many years before we have an adequate idea of its true breeding distribution.

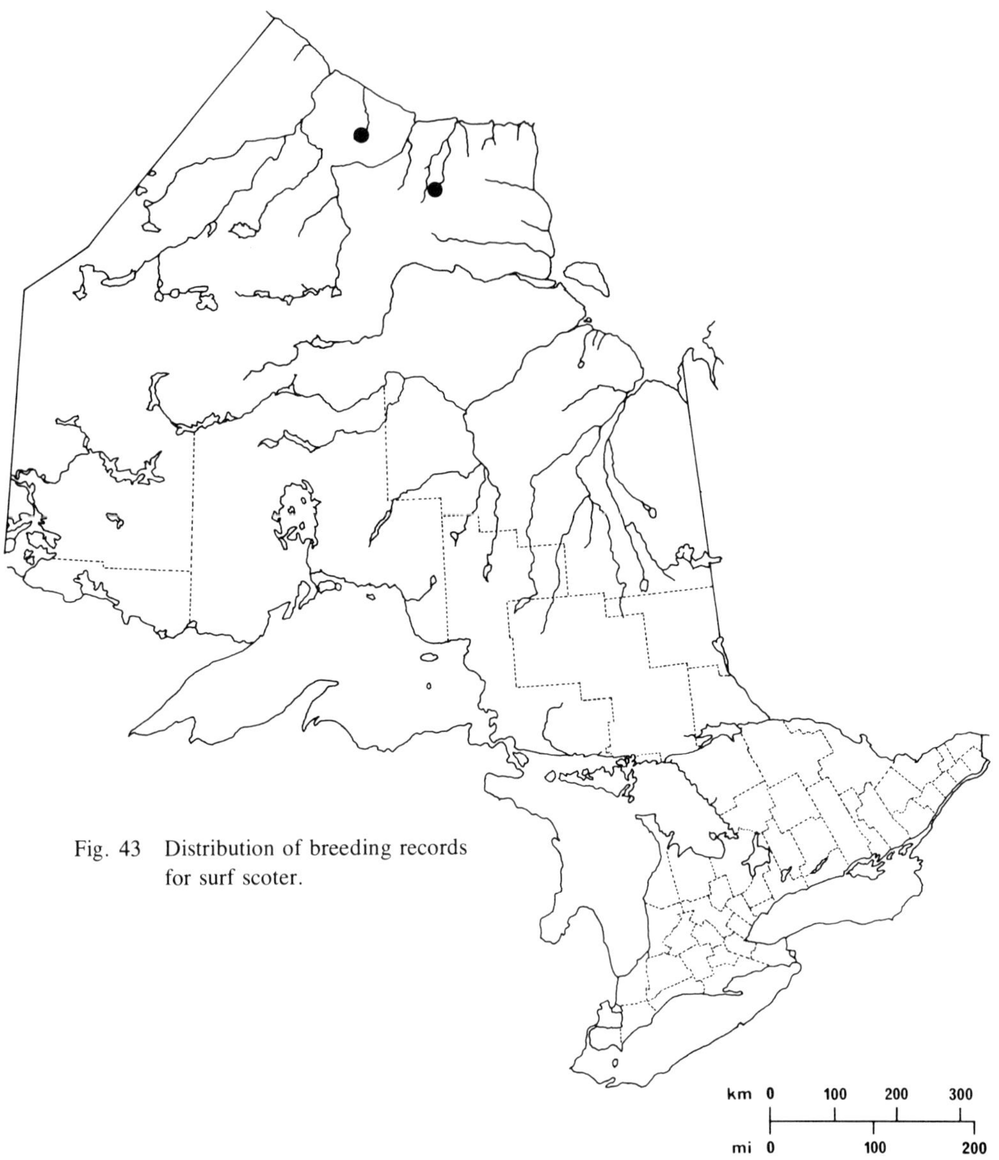

Fig. 43 Distribution of breeding records for surf scoter.

Surf Scoter, *Melanitta perspicillata* (Linnaeus)

Nidiology

This northern breeding duck was placed on the Ontario breeding-bird list with the collection of an adult female and 1 of 6 flightless young (ROM 90960) on 3 August 1960 just north of Shagamu Lake, between Fort Severn and Winisk, Kenora District (Simkin, 1963). The second provincial breeding record was established in July 1980, when 3 downy young were collected near Aquatuk Lake, Kenora District (ROM 137131–137133).

Breeding Distribution

No nests of the surf scoter have ever been found, nor are there any other reports of breeding in the province. Summer records indicate that the species may nest in small numbers inland near the coasts of Hudson and James bays.

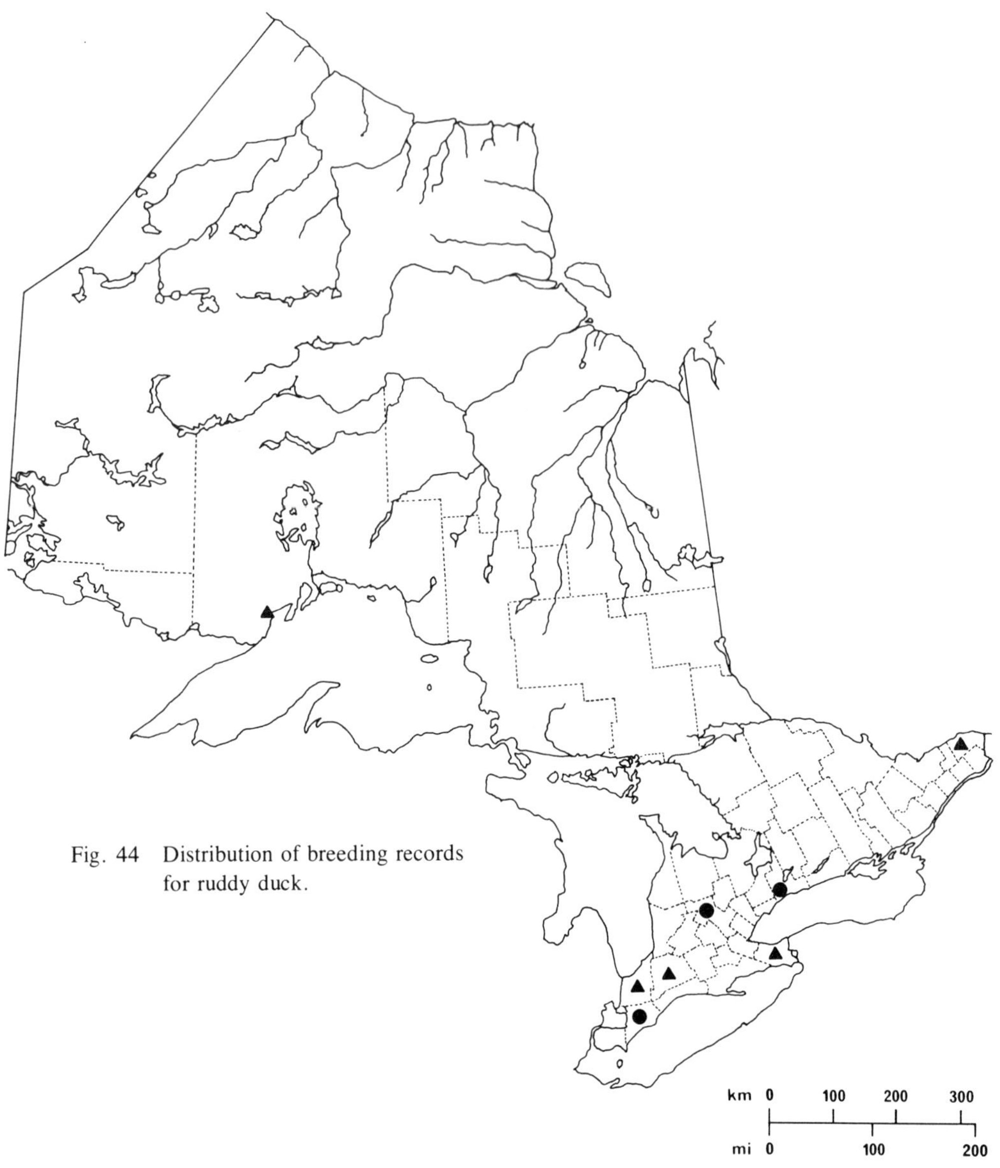

Fig. 44 Distribution of breeding records for ruddy duck.

Ruddy Duck, *Oxyura jamaicensis* (Gmelin)

Nidiology

RECORDS 19 (27 nests) representing 5 provincial regions.
Breeds in freshwater marshes or the marshy borders of lakes, sewage lagoons, and islands. Nests were invariably over water, from 3 to 27 m (10 to 90 ft) from open water.

Nests were situated in cattail, bulrush, and sedge borders. They were woven platforms anchored to the surrounding vegetation (cattails, bulrushes, sedges, and bur-reeds). Nests were 15 to 20.5 cm (6 to 8 inches) above water, which ranged in depth from 20.5 to 76 cm (8 to 30 inches). Some nests had sloping ramps, and 1 nest was built on top of an old nest of American coot.

Nests were basket-shaped platforms or mounds of woven dead and living cattails, bulrushes, sedges, and pondweeds, and were variously lined with shredded cattail leaves, dried sedges, down, and feathers. An outside diameter of 24 by 25.5 cm (9.5 by 10 inches), an outside depth of 12.5 cm (5 inches), an inside diameter of 15 cm (6 inches), and an inside depth of 5 cm (2 inches) were reported for different nests.

EGGS 18 nests with 1 to 16 eggs; 1E (1N), 3E (1N), 4E (3N), **5E** (1N), **6E** (4N), **8E** (1N), **9E** (3N), **10E** (1N), 11E (1N), 13E (1N), 16E (1N).
Average clutch range 5 to 9 eggs (9 nests).
A nest of American bittern with 5 eggs also contained 1 egg of ruddy duck; a ring-necked duck nest with 6 eggs also held 1 ruddy duck egg.

INCUBATION PERIOD No information.

EGG DATES 27 nests, 24 May to 20 August (32 dates), 14 nests, 4 June to 28 June.

Breeding Distribution

Despite the fact that reports of breeding by the ruddy duck and descriptions of nests in the Lake St. Clair area were published in the 1880s (Baillie, 1962), no specimen evidence of breeding was obtained from that area until 1949 (Lumsden, 1951). Since then a few nests or breeding reports have come from widely scattered areas south of the Canadian Shield and from Thunder Bay (Richards, 1977). However, nesting is very irregular and in small numbers.

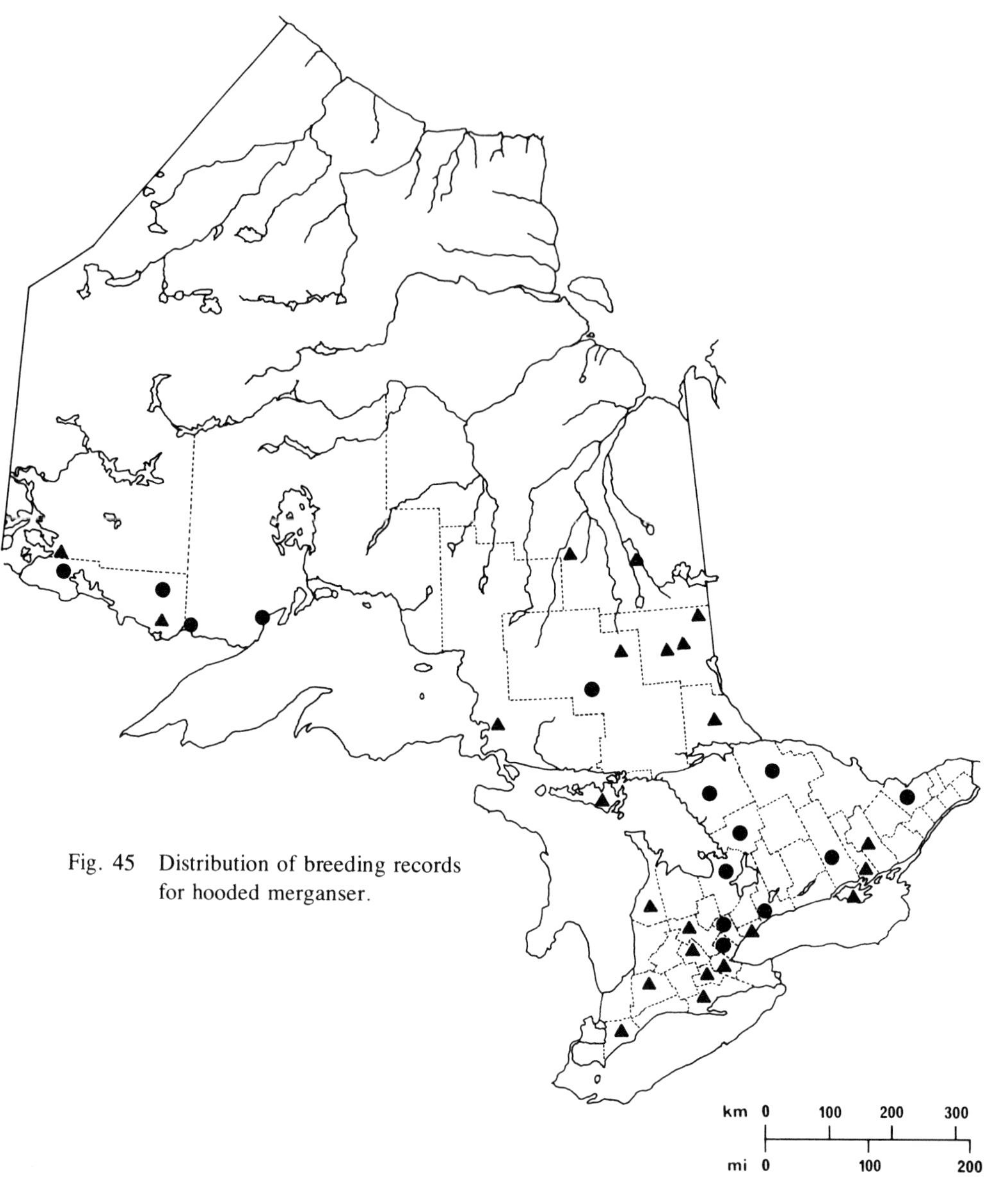

Fig. 45 Distribution of breeding records for hooded merganser.

Hooded Merganser, *Lophodytes cucullatus* (Linnaeus)

Nidiology

RECORDS 50 nests representing 14 provincial regions.
Breeds in freshwater beaver ponds, in wooded swamps, on wooded shores of lakes and rivers, on wooded islands, and in marsh areas (1 nest was recorded on an open, grassy river flood plain near an urban development). Nest sites were usually at the edge of or over water, but 1 nest was at a distance of 15 m (50 ft) from it.

Nests were in natural cavities (7 nests) or woodpecker cavities (4 nests) in either living (5 nests) or dead (4 nests) coniferous and deciduous trees. Nest boxes (36 nests) in appropriate habitats were often used. Tree diameter at 1 nest cavity was 18 to 20.5 cm (7 to 8 inches); another nest-tree diameter was 45.5 cm (18 inches). Heights of 8 nests in natural sites ranged from 4 to 12 m (12 to 40 ft), with 4 averaging 6 to 10.5 m (20 to 35 ft).

Nest cavities were lined only with down and feathers, which increased in amount as clutches were completed.

EGGS 43 nests with 1 to 20 eggs; 1E (6N), 2E (2N), 4E (1N), 5E (3N), **6E** (4N), **7E** (4N), **8E** (4N), **9E** (6N), **10E** (4N), **11E** (1N), **12E** (1N), **13E** (2N), **14E** (1N), 15E (1N), 19E (2N), 20E (1N).
Average clutch range 7 to 13 eggs (22 nests).
Two nests containing 5 and 9 eggs of hooded merganser held 5 and 1 eggs of common goldeneye respectively; 1 nest with 10 to 15 eggs of hooded merganser also contained 4 eggs of common merganser; 1 nest with 16 eggs of wood duck contained 8 eggs of hooded merganser. Three large clutches (19 and 20 eggs) were known to be "dump" nests and probably involved laying by more than 1 female.

INCUBATION PERIOD 4 nests: 2 of at least 26 days, 1 of 31 days, 1 of 32 days.

EGG DATES 41 nests, 8 April to 28 June (53 dates); 21 nests, 12 May to 25 May.

Breeding Distribution

Despite previous reports of the hooded merganser's breeding along the James Bay coast (Macoun and Macoun, 1909), no evidence exists that the species ever nested at that latitude. Records indicate that it is largely a breeding bird of southern Ontario, but extends into the north at widely scattered locations, at least as far as 50°N. However, summer records available as far north as the Albany River, and beyond, possibly indicate a somewhat more extensive range in the north.

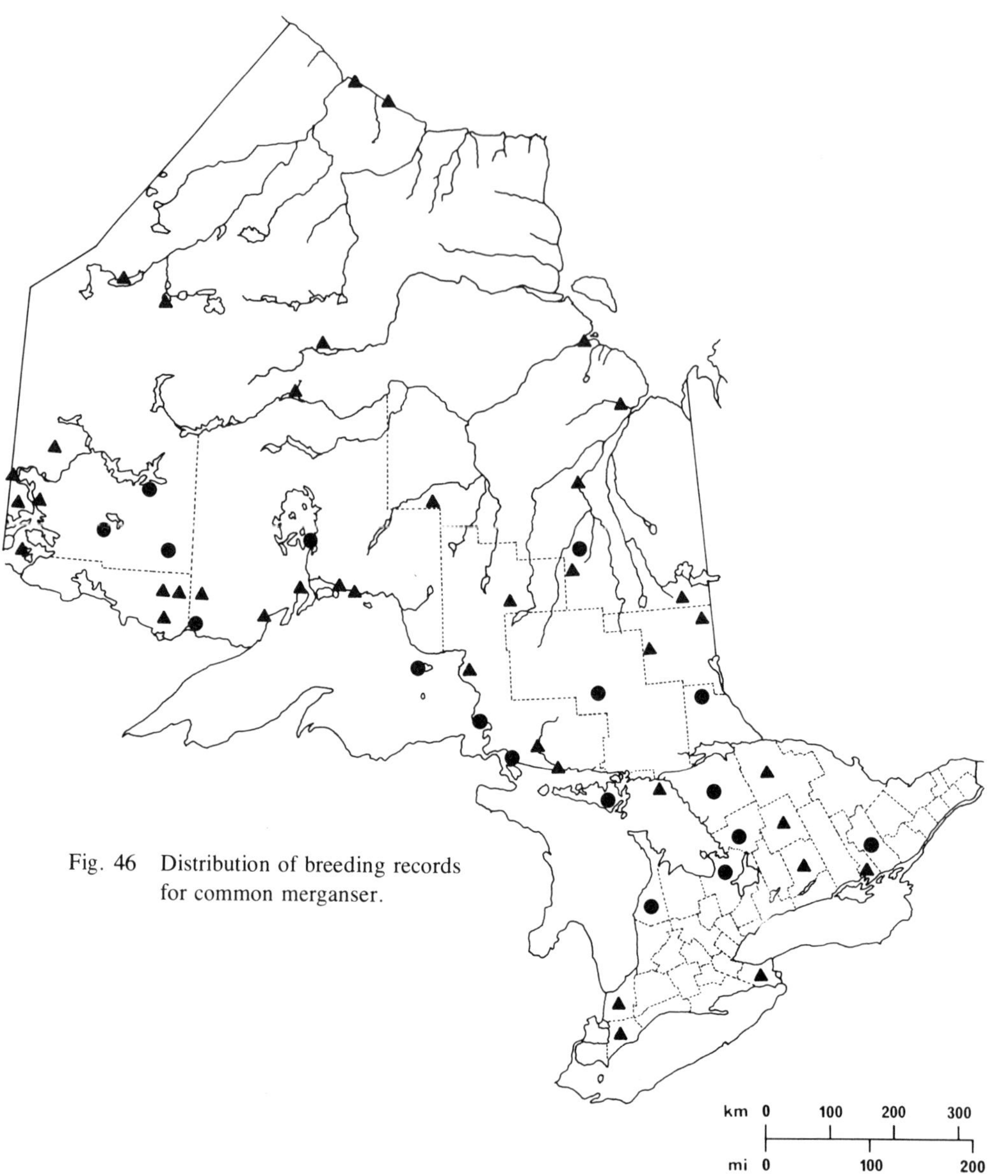

Fig. 46 Distribution of breeding records for common merganser.

Common Merganser, *Mergus merganser* Linnaeus

Nidiology

RECORDS 93 (97 nests) representing 16 provincial regions.
Breeds along the wooded and rocky shores of lakes and rivers and on rocky islands (Fig. 156B) that usually have some trees and ground cover. Abandoned or seasonally used buildings such as campground buildings, fishing shacks, and lighthouses often served as nest sites. Nests were sometimes located in crevices and ledges of rocky cliff faces and were often at water's edge but occasionally at distances of more than 180 m (600 ft) from water. Some island nest sites also harboured gull and tern colonies.

Nests were variously situated on the ground under bushes, roots, branches, stumps, and fallen trees, among or under boulders, and in reed and grass clumps (34 nests); in chimneys and walls, on floors, or under buildings, tent platforms, and lighthouses (14 nests); in natural cavities and woodpecker cavities of dead or living coniferous and deciduous trees—with 2 nests in horizontal, hollow logs (24 nests); and on cliffs in rocky crevices—1 nest was on a ledge in an old raven nest (7 nests). For several years a nest was recorded in a cavity in a tree that also held an active nest of bald eagle. Heights of 26 tree nests ranged from 0.6 to 17 m (2 to 55 ft), with 13 averaging 2 to 9 m (6 to 30 ft).

Ground nests were hollows or depressions lined with grasses, weeds, pine needles, twigs and bark, dead leaves, lichens and dried mosses, and down and feathers; tree-cavity nests usually had no lining except for down and feathers.

EGGS 75 nests with 2 to 19 eggs; 2E (3N), 3E (5N), 4E (3N), 6E (1N), **7E** (4N), **8E** (11N), **9E** (11N), **10E** (6N), **11E** (10N), **12E** (7N), **13E** (3N), 14E (3N), 15E (3N), 16E (3N), 18E (1N), 19E (1N).
Average clutch range 8 to 12 eggs (45 nests).
The clutches of 18 and 19 eggs were known to be from 2 or more females and were probably "dump" nests. One nest with 10 to 15 eggs of hooded merganser also contained 4 eggs of common merganser.

INCUBATION PERIOD 3 nests: 2 of 32 days, 1 of 34 days.

EGG DATES 77 nests, 18 April to 27 July (96 dates); 39 nests, 29 May to 20 June.

Breeding Distribution

The common merganser can be found breeding throughout Ontario, but it is very infrequently reported south of the Canadian Shield or in the extreme north of the province.

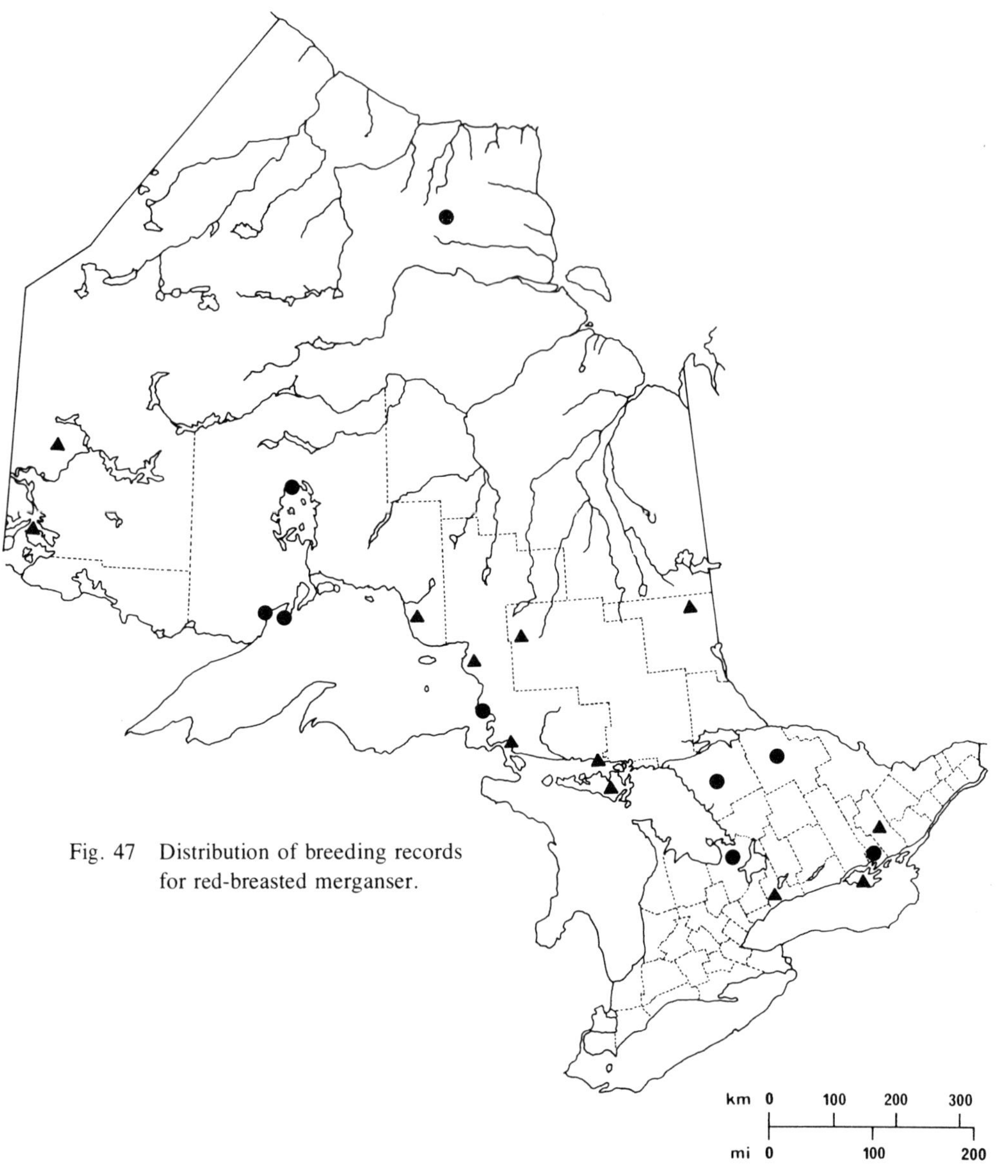

Fig. 47 Distribution of breeding records for red-breasted merganser.

Red-breasted Merganser, *Mergus serrator* Linnaeus

Nidiology

RECORDS 24 nests representing 12 provincial regions.
Breeds on small and large islands and occasionally on the shores of lakes and rivers. Islands chosen were either well vegetated or relatively rocky and barren, but river and lakeshore sites were treed. Two island sites were also occupied by herring gull colonies; 1 of these sites also had nesting double-crested cormorants. Other island sites were also occupied by colonies of ring-billed gull and common and Caspian terns. Nest sites were at water's edge or as far as 70 m (228 ft) distant from it.

Nests were variously positioned under coniferous trees (cedar, spruce), under bushes (blueberry, nettles), under overturned roots, in grasses and weeds, among rocks or in rocky crevices, and once under a board. One nest was placed within 1.5 m (5 ft) of a herring gull nest.

Nests were usually formed of nearby grasses, leaves, and various vegetable materials. They were lined with down and feathers, which increased in amount as the clutch neared completion.

EGGS 21 nests with 1 to 10 eggs; 1E (1N), 2E (2N), 4E (1N), 6E (1N), **7E** (2N), **8E** (6N), **9E** (3N), **10E** (5N).
Average clutch range 7 to 9 eggs (11 nests).

INCUBATION PERIOD 1 nest, 35 days.

EGG DATES 20 nests, 26 May to 6 August (25 dates); 10 nests, 17 June to 29 June.

Breeding Distribution

The red-breasted merganser is decidedly less common than the common merganser, and only in recent years have we obtained a clearer idea of the extent of its range in Ontario. The available records indicate that it breeds throughout the province except in the extreme south.

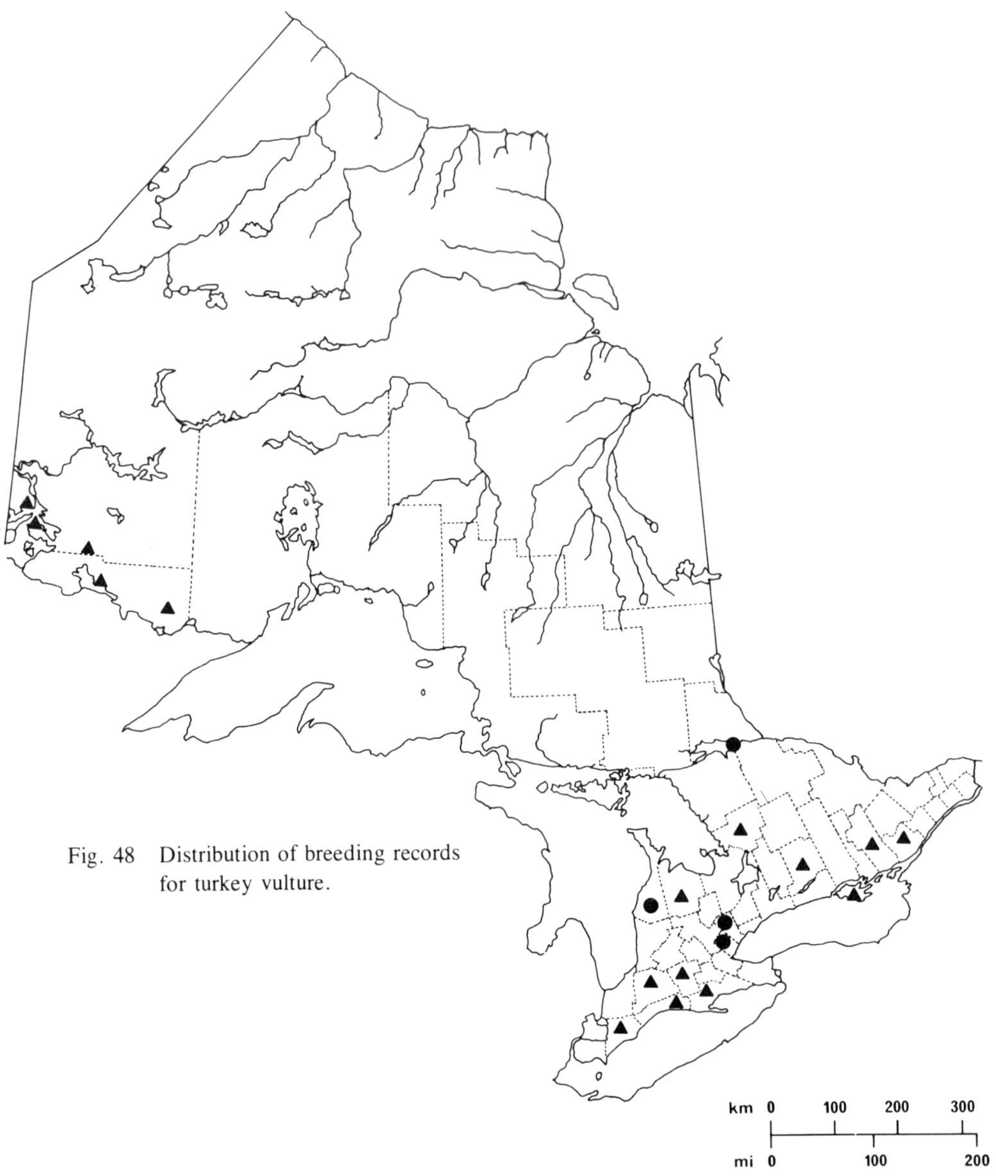

Fig. 48 Distribution of breeding records for turkey vulture.

Turkey Vulture, *Cathartes aura* (Linnaeus)

Nidiology

RECORDS 29 nests representing 14 provincial regions.
Breeds on cliffs and rocky outcroppings of shield and escarpment areas, as well as in deciduous and mixed woods. In the former situation nestings recorded were in caves and crevices; in the latter they were in hollow logs and stumps and 1 nest was on the ground beside a fallen log.

The cave and crevice sites (15 nests) slightly outnumbered the woodland nests in logs and stumps (11 nests). Nests were located both near water and at a considerable distance from it.

Nests were in various locations from near the top of cliffs to near the bottom of talus slopes, and 1 nest was noted in a hollow log 3.5 m (12 ft) from the opening and another in a standing stump, 2 m (7 ft) below the cavity entrance, at ground level. The eggs were placed usually on bare ground or rock and occasionally on wood chips or among dead grasses, weeds, and stones.

EGGS 27 nests with 1 to 2 eggs; **1E** (4N), **2E** (23N).
Average clutch range 2 eggs (23 nests).

INCUBATION PERIOD No information. The young remained in nest more than 55 days.

EGG DATES 19 nests, 2 May to 15 July (23 dates); 10 nests, 17 May to 1 June.

Breeding Distribution

No Ontario nesting records are available for the turkey vulture (Fig. 154A) before 1900. Probably land clearing and settlement in eastern North America was favourable to a northward expansion of the species into Ontario. Recent reports of breeding at North Bay and near Ottawa indicate that it may now be found throughout the southern part of the province.
It probably also breeds along the north shore of Lake Huron where summer reports are numerous from a few localities. In northern Ontario, nesting is largely confined to the more settled areas between Thunder Bay and Kenora.

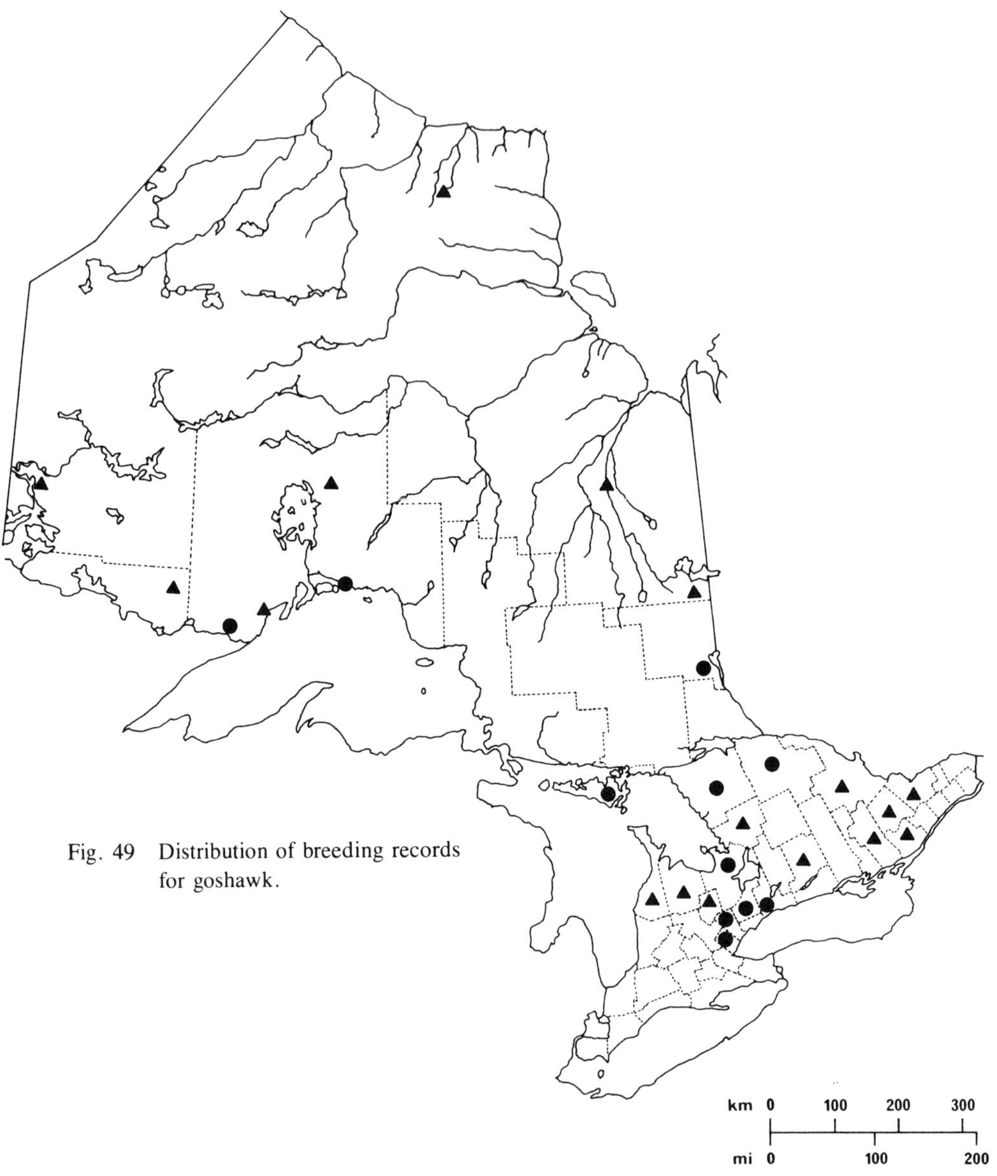

Fig. 49 Distribution of breeding records for goshawk.

Goshawk, *Accipiter gentilis* (Linnaeus)

Nidiology

RECORDS 45 nests representing 22 provincial regions.
Breeds in deciduous, mixed, and coniferous woods of various sizes, although large, dense stands were favoured. Older coniferous reforestation plots were occasionally used.

Nests were situated more commonly in deciduous trees (7 spp., 24 nests), and less often in coniferous trees (4 spp., 10 nests). Birch spp. (10 nests), pine spp. (10 nests), poplar spp. (7 nests), beech (3 nests), and maple spp. (3 nests) were the preferred trees. Nests were positioned in forks of branches at the trunk or in main crotches. Heights of 29 nests ranged from 7.5 to 23 m (24 to 75 ft), with 15 averaging 9 to 12 m (30 to 40 ft).

Nests were bulky structures of twigs and branches, some up to 90 cm (3 ft) in length, with diameters of 2.5 cm (1 inch). Nests had shallow cups and were variously lined with fresh sprigs of hemlock, pine and cedar, dried and fresh leaves, grasses, mosses, feathers, clay, and bark chips (the last of these reported in 8 nests). Outside diameters of 6 nests ranged from 43 to 106.5 cm (17 to 42 inches); inside diameters of 2 nests ranged from 23 to 53.5 cm (9 to 21 inches). The largest inside-diameter figure may have been the result of nest flattening by growing young.

EGGS 26 nests with 2 to 4 eggs; **2E** (9N), **3E** (15N), **4E** (2N).
Average clutch range 3 eggs (15 nests).

INCUBATION PERIOD No information. Incubation commenced before the clutch was complete, and young of various sizes have been reported.

EGG DATES 16 nests, 16 April to 2 June (20 dates); 8 nests, 30 April to 6 May.

Breeding Distribution

The secretive nature of the goshawk (Fig. 156A) makes it very difficult to obtain sufficient information to properly define its breeding range. However, the species probably breeds sparingly throughout the forested portions of Ontario. It apparently no longer breeds in the extreme south of the province probably because forest habitat is no longer available over large enough areas.

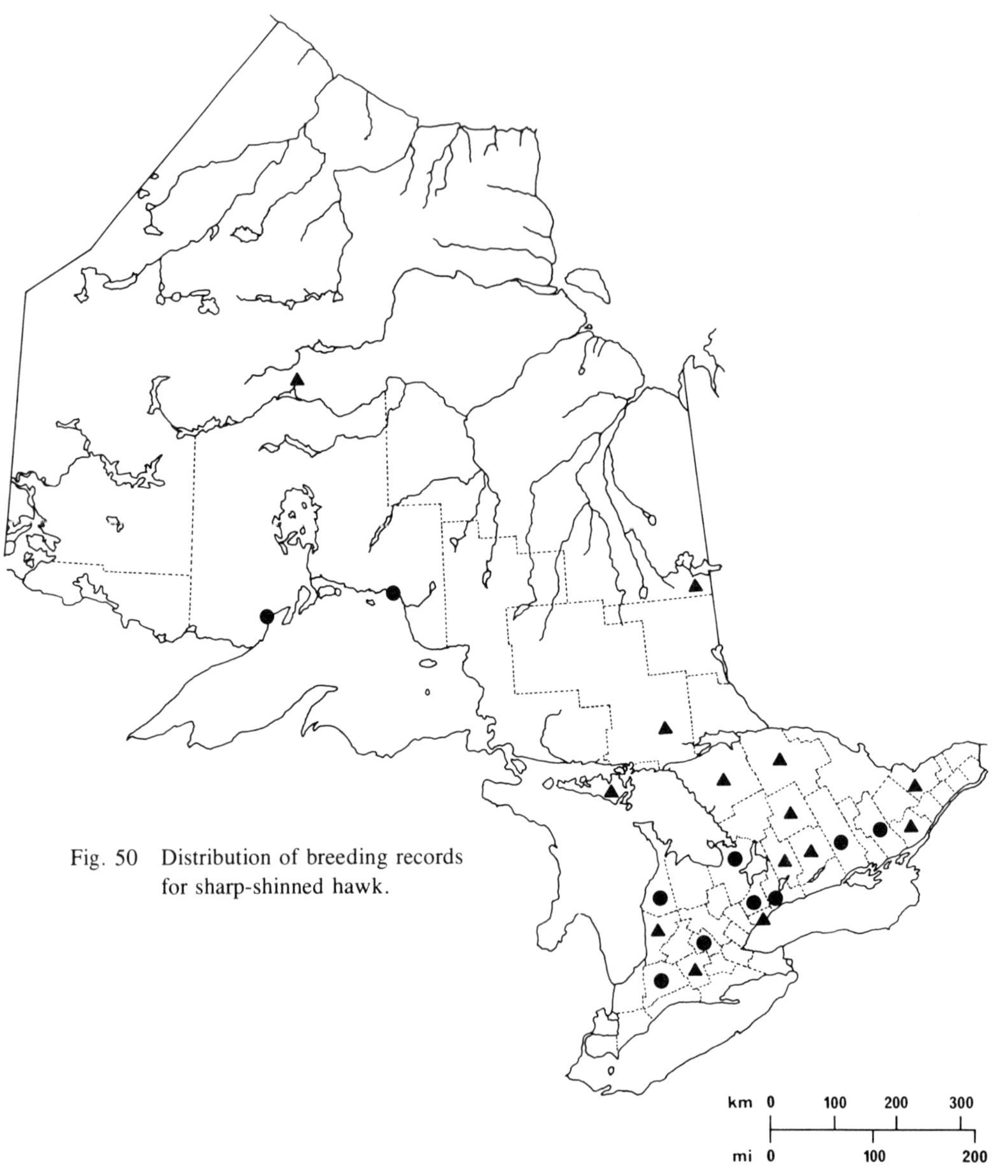

Fig. 50 Distribution of breeding records for sharp-shinned hawk.

Sharp-shinned Hawk, *Accipiter striatus* Vieillot

Nidiology

RECORDS 47 nests representing 18 provincial regions.
Breeds usually in dense, wet, coniferous woods and bogs and occasionally in mixed woods. One recorded nest site was on a wooded island in a lake.

Nests were invariably situated in coniferous trees (7 spp.), with spruce spp. (14 nests), hemlock (7 nests), cedar (6 nests), and pine spp. (5 nests) used most often.

Nests were positioned usually against the trunk, or sometimes near the trunk, and were supported by horizontal branches. One nest was in a natural cavity in a pine stub. Heights of 33 nests ranged from 3.5 to 15 m (12 to 50 ft), with 17 averaging 6 to 10.5 m (20 to 35 ft). Nests used by the species in previous breeding seasons were frequently noted near its active nests.

Nests were large, untidy accumulations of twigs and small branches up to 8 mm (0.3 inches) in diameter, from both deciduous and coniferous tree species. Nest bowls were usually shallow, and were lined with fine twigs, dead and fresh, from both deciduous and coniferous trees, with bark chips of hemlock and cedar, and with rootlets. Once a crow nest was used. Outside diameters of 5 nests ranged from 45.5 to 63.5 cm (18 to 25 inches); outside depths of 2 nests ranged from 15 to 30.5 cm (6 to 12 inches); inside diameters of 4 nests ranged from 15 to 23 cm (6 to 9 inches); inside depths of 3 nests ranged from 3.5 to 7.5 cm (1.3 to 3 inches).

EGGS 33 nests with 3 to 6 eggs; **3E** (4N), **4E** (12N), **5E** (16N), **6E** (1N).
Average clutch range 4 to 5 eggs (28 nests).

INCUBATION PERIOD No information.

EGG DATES 32 nests, 30 April to 30 June (34 dates); 16 nests, 30 May to 12 June.

Breeding Distribution

The sharp-shinned hawk is the commonest of the three accipiters in Ontario, yet very few nesting records are available from northern Ontario. These few reports plus summer sightings indicate that the species reaches the Albany River, but probably does not range much farther north. Its range extends into southern Ontario but there are few records from recent years.

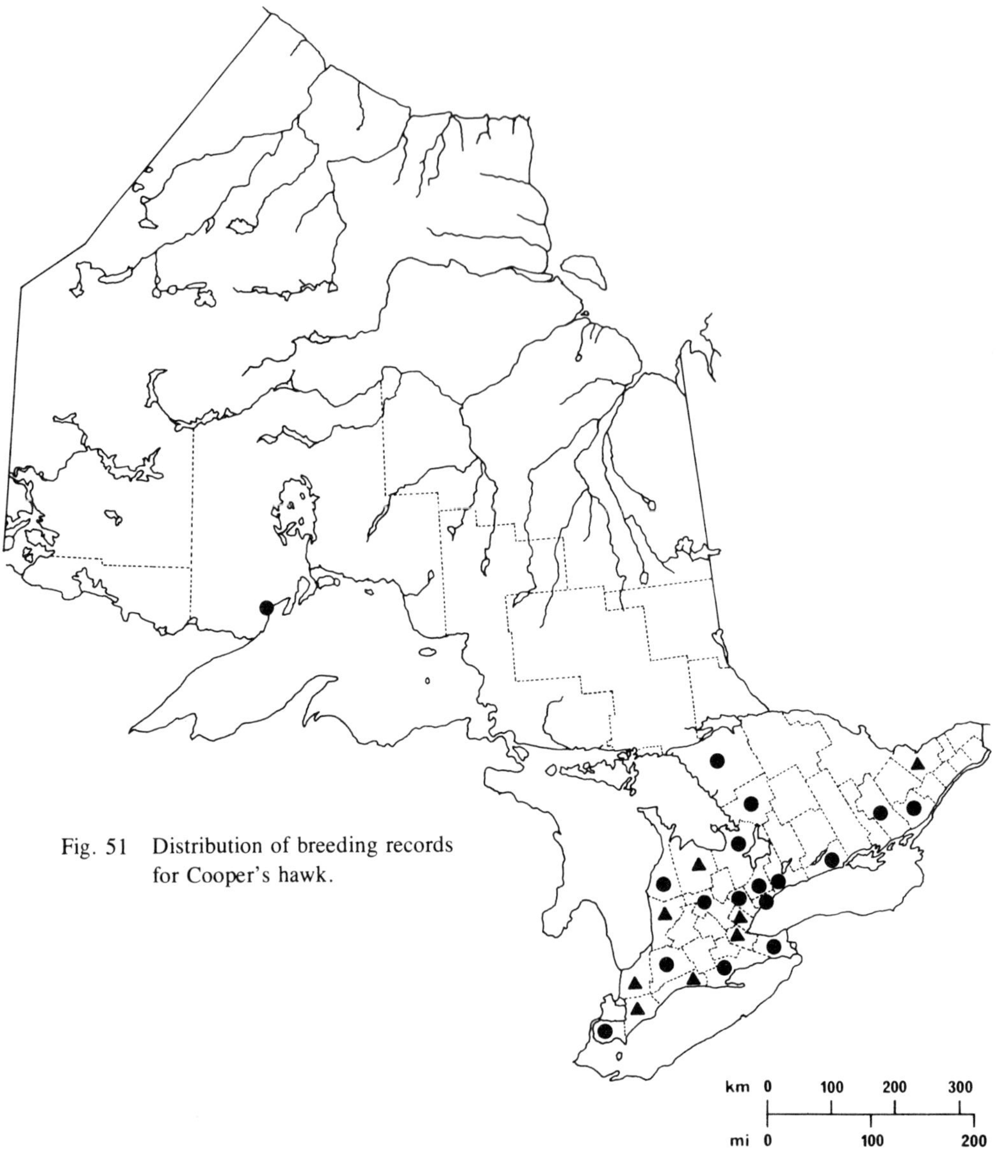

Fig. 51 Distribution of breeding records for Cooper's hawk.

Cooper's Hawk, *Accipiter cooperii* (Bonaparte)

Nidiology

RECORDS 65 nests representing 21 provincial regions.
Breeds usually in small or large deciduous woods in dry situations, and sometimes in mixed stands. One recorded nest was in a group of small willows in an open field pasture.

Nests were situated mostly in deciduous trees (10 spp., 30 nests) and less often in coniferous trees (3 spp., 14 nests). The favoured trees were maple spp. (7 nests), beech (6 nests), and hemlock (6 nests). Nests were positioned against the main trunk or in main crotches, and heights of 46 nests ranged from 6 to 26.5 m (20 to 87 ft), with 23 averaging 9 to 13.5 m (30 to 45 ft).

This species sometimes rebuilt and relined old nests of common crow (5 nests) and red-shouldered hawk (1 nest). Most nests were substantial structures of sticks and twigs with shallow bowls. Nests were variously lined with small deciduous twigs, sprigs of hemlock and pine, bark strips and chips (a characteristic material), grasses, weeds, and green leaves. Outside diameter of 1 nest measured 63.5 by 76 cm (25 by 30 inches) and its inside diameter was 33 cm (13 inches). Inside diameter of another nest was 40.5 cm (16 inches).

EGGS 52 nests with 1 to 7 eggs; 1E (1N), 2E (3N), **3E** (22N), **4E** (16N), **5E** (9N), 7E (1N).
Average clutch range 3 to 4 eggs (38 nests).

INCUBATION PERIOD No information.

EGG DATES 49 nests, 27 April to 8 July (55 dates); 25 nests, 13 May to 2 June.

Breeding Distribution

The most southerly distributed of the accipiters, the Cooper's hawk probably ranges only as far north as the latitude of northern Lake Superior. In the extreme south, where forests have largely been cut, its population is now very sparse.

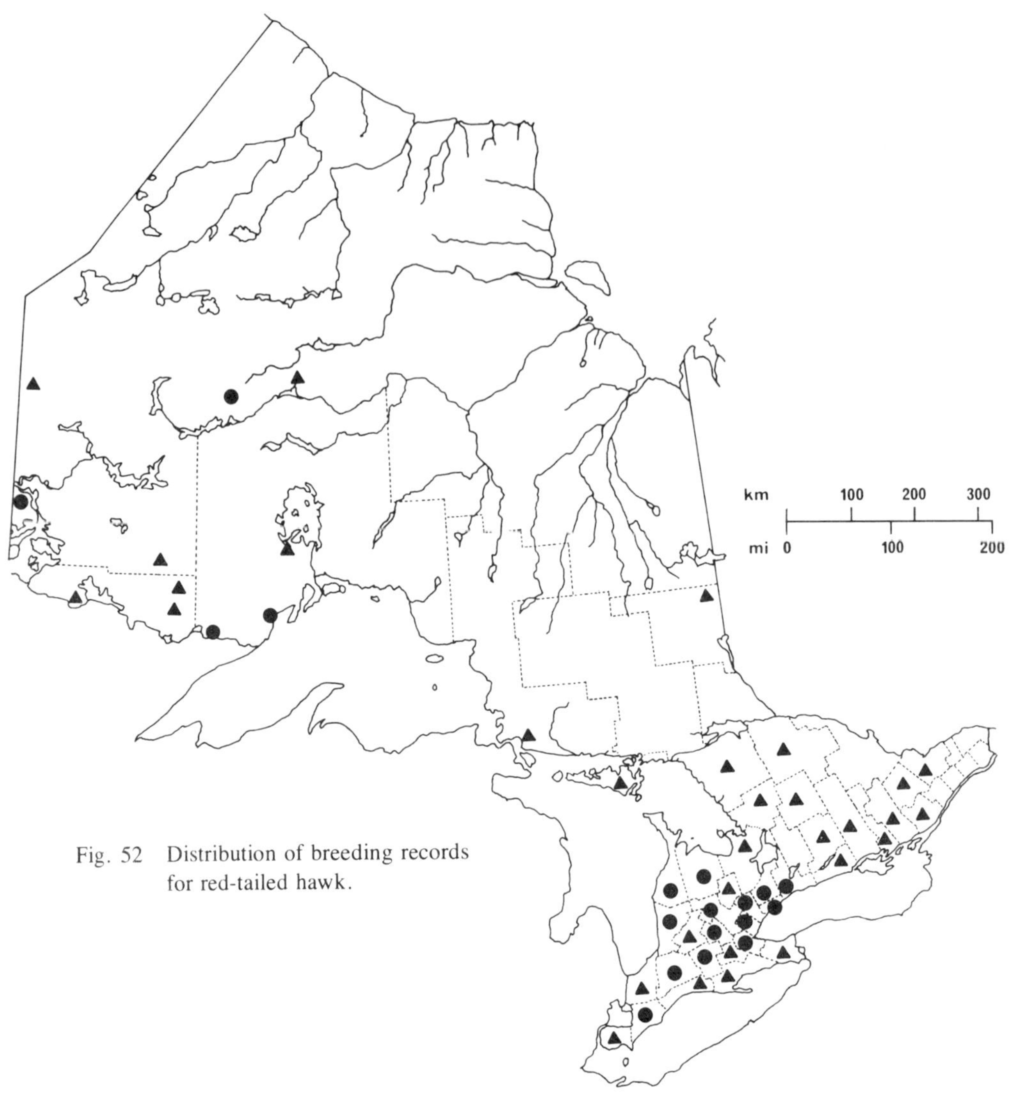

Fig. 52 Distribution of breeding records for red-tailed hawk.

Red-tailed Hawk, *Buteo jamaicensis* (Gmelin)

Nidiology

RECORDS 723 nests representing 37 provincial regions.
Breeds almost invariably in or near open areas in fields, pastures, and fence rows on agricultural land, on the edges or in clearings of wood lots, beside railroads and highways, in industrial areas, near or overlooking water, on islands, and on hills and cliffs. Occasionally

nests are located well inside larger woodlands, but nearby open country for hunting is always a prerequisite.

Nests were usually in trees (699 nests) that were most often living and sometimes dead. Occasionally nests were situated on cliff ledges (14 nests) and hydro or radar towers (10 nests). Deciduous trees (16 spp., 437 nests) were preferred to coniferous trees (5 spp., 107 nests), and those used most often were elm spp. (153 nests), pine spp. (100 nests), maple spp. (70 nests), beech (66 nests), oak spp. (65 nests), and ash spp. (26 nests).

Nests were situated usually near or in the tree crown, often in main crotches. They were commonly placed against the trunk, but in larger trees they were at times saddled on, or supported by, horizontal branches at some distance from the trunk. Heights of 363 tree nests ranged from 4.6 to 27 m (15 to 90 ft), with 181 averaging 12 to 18 m (40 to 60 ft). Cliff nests have been recorded as high as 80 m (270 ft).

Nests were small to bulky platforms made up of dead and fresh branches and twigs, which ranged in diameter from 6 to 40 mm (0.25 to 1.5 inches). Nests were often reused and material added over a period of years. Great horned owl and red-shouldered hawk occupied some nests of this species in some breeding seasons. Five records show active red-tailed hawk nests situated from 45 to 270 m (150 to 900 ft) away from active nests of great horned owl. Former nests of squirrels (5 nests), crows (4 nests), and ravens (2 nests) were occasionally used. House sparrow (5 records) and white-footed mouse (*Peromyscus leucopus*) (1 record) occasionally built their nests into the structure of active red-tailed hawk nests.

Nests were lined, in order of frequency, with bark strips, fresh or old conifer sprigs, deciduous twigs, leaves, corn leaves and cobs, grasses and straw, pine needles, feathers, paper, rootlets, horsehair and fur, mud, plastic bags, mosses, binder twine, and mouse and oriole nests. Outside diameters of 72 nests ranged from 30 to 150 cm (12 to 60 inches), with 36 averaging 60 to 90 cm (24 to 36 inches); outside depths of 31 nests ranged from 15 to 120 cm (6 to 47 inches), with 16 averaging 43 to 60 cm (17 to 24 inches); inside diameters of 26 nests ranged from 15 to 48 cm (6 to 19 inches), with 13 averaging 25 to 30 cm (10 to 12 inches); 27 nests ranged from being flat to having inside depths of 25 cm (10 inches), with 14 averaging 5 to 10 cm (2 to 4 inches).

EGGS 229 nests with 1 to 4 eggs; **1E** (31N), **2E** (86N), **3E** (105N), **4E** (7N).
Average clutch range 2 to 3 eggs (191 nests).

INCUBATION PERIOD 3 nests, ca 29 to 33 days.

EGG DATES 165 nests, 3 March to 15 July (197 dates); 82 nests, 5 April to 23 April.

Breeding Distribution

Although records are largely lacking in the north, the red-tailed hawk (Fig. 152B) probably breeds throughout the province as far north as forest trees are found.

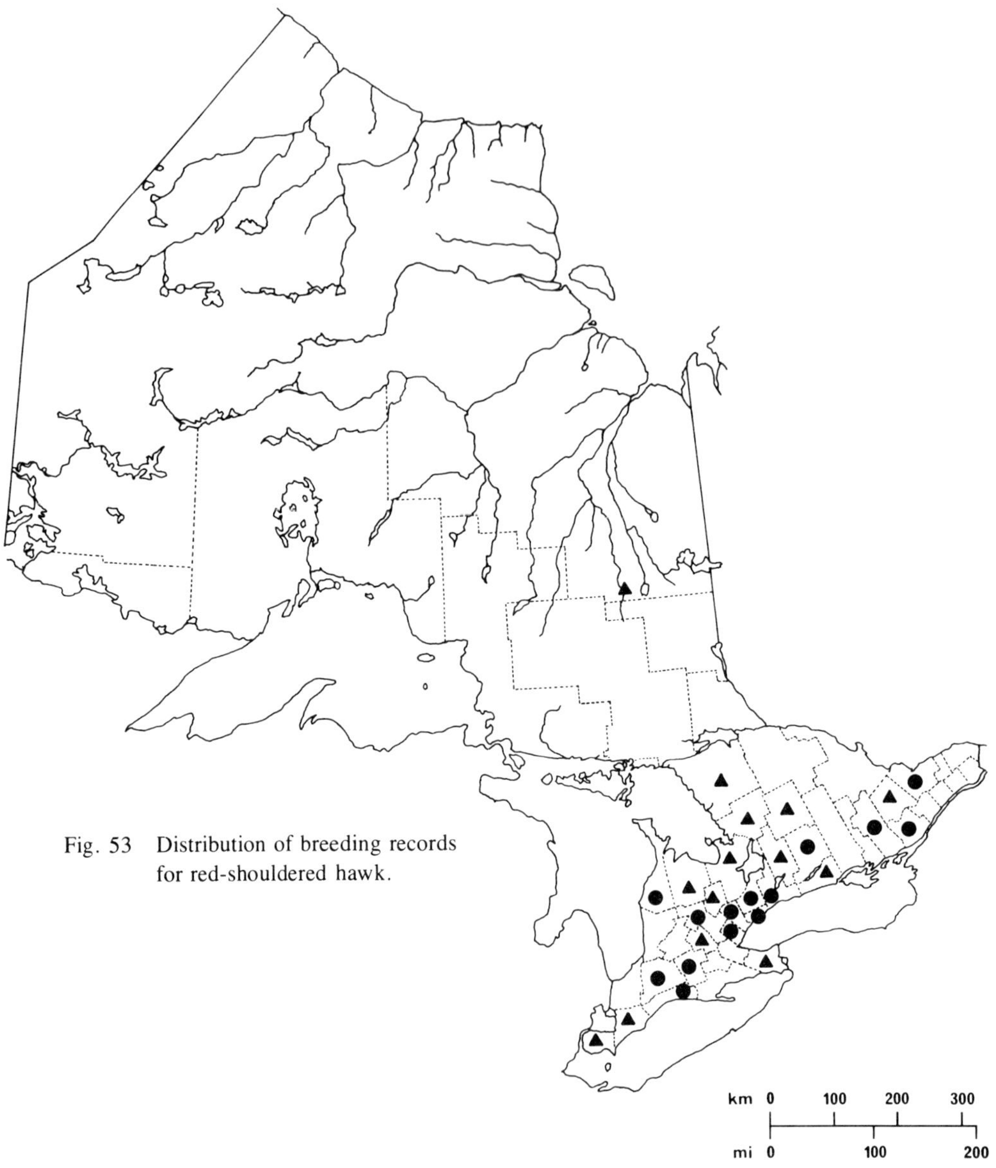

Fig. 53 Distribution of breeding records for red-shouldered hawk.

Red-shouldered Hawk, *Buteo lineatus* (Gmelin)

Nidiology

RECORDS 229 nests representing 25 provincial regions.
Breeds almost invariably in mixed and deciduous woods (1 nest recorded in a pine plantation). Nest sites were usually in fairly dense woodlands, although open woods and swampy or wet woods were sometimes used.

Nests were invariably in trees, which were usually living, although 1 tree was dead. Deciduous trees (9 spp., 190 nests) were greatly preferred to coniferous trees (3 spp., 10 nests). Beech (88 nests), maple spp. (60 nests), birch spp. (12 nests), ash spp. (10 nests), and pine spp. (8 nests) were selected most commonly.

Nests were placed in forks, at or near the trunk, or in main crotches, with only a few nests located near the crown of the tree. Heights of 206 nests ranged from 5 to 24.5 m (16 to 80 ft), with 103 averaging 11.5 to 15.5 m (38 to 51 ft).

Nests ranged from poorly built to bulky platforms of sticks and twigs and were sometimes added to and reused in successive years. Occasionally crow nests (3 records) and squirrel dreys (2 records) were used. Nests were rarely unlined; they were lined most commonly with green hemlock, pine, cedar, balsam fir and spruce sprigs, and other unspecified green vegetation; and with leaves, bark, grass tufts, straw and weed stalks, down and feathers, pine needles, caterpillar webs, corn stalks and cobs, binder twine, and old nests of birds (passerines). Outside diameters of 7 nests ranged from 45.5 to 91.5 cm (18 to 36 inches); outside depths of 2 nests were 45.5 and 48.5 cm (18 and 19 inches); inside diameters of 5 nests ranged from 19.5 to 35.5 cm (7.75 to 14 inches); inside depths of 3 nests ranged from 5 to 9 cm (2 to 3.5 inches). One nest of this species was taken over during the breeding season by a Cooper's hawk.

EGGS 152 nests with 1 to 6 eggs; 1E (2N), **2E** (28N), **3E** (79N), **4E** (38N), 5E (4N), 6E (1N).
Average clutch range 3 eggs (79 nests).

INCUBATION PERIOD No information.

EGG DATES 134 nests, 1 April to 9 July (150 dates); 67 nests, 20 April to 28 April.

Breeding Distribution

Once considered more numerous than the red-tailed hawk in southern Ontario, this species has become very scarce in the province as land has been cleared. While reports of summering red-shouldered hawks extend north as far as Thunder Bay and Lake Abitibi, such observations are very uncommon and we have only one breeding record in northern Ontario.

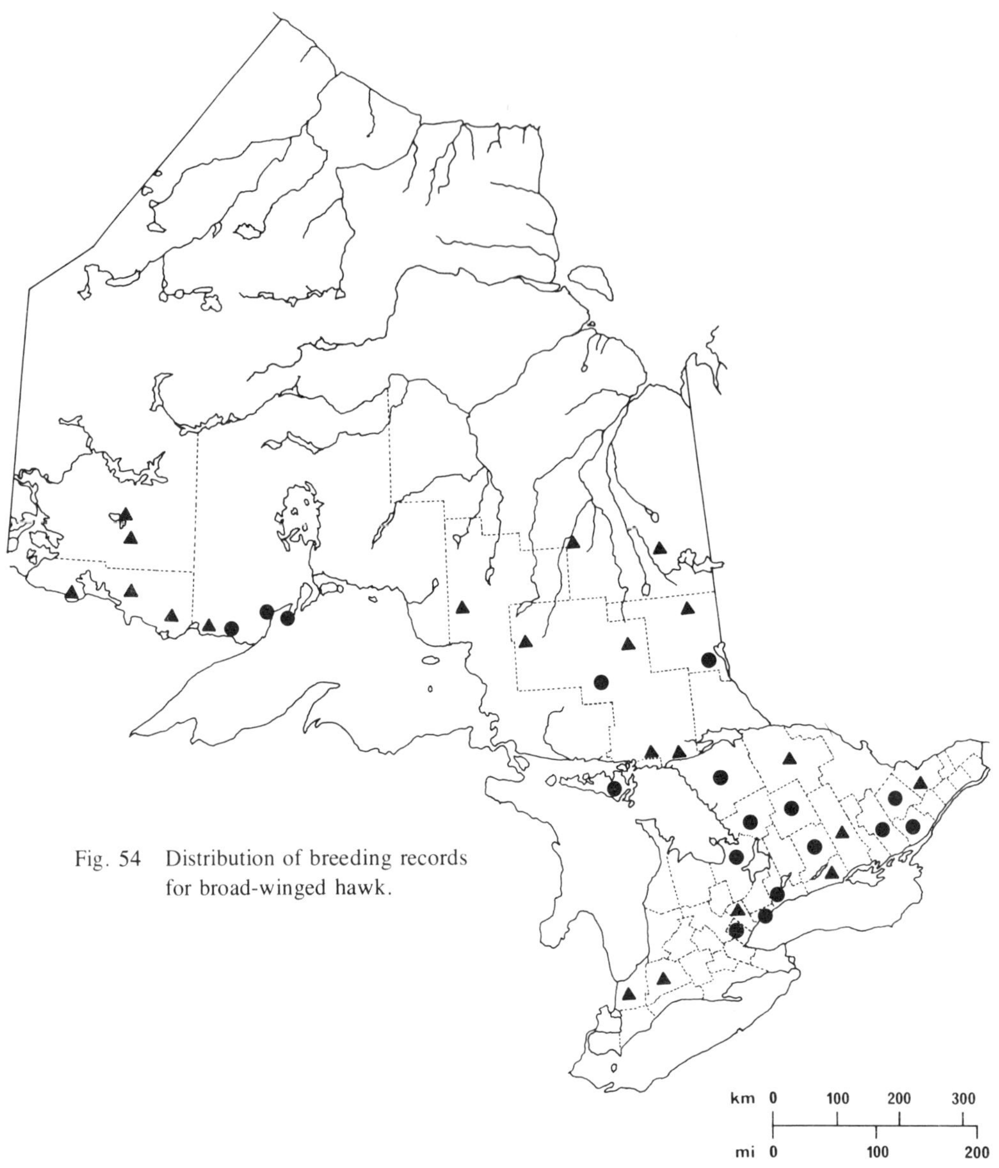

Fig. 54 Distribution of breeding records for broad-winged hawk.

Broad-winged Hawk, *Buteo platypterus* (Vieillot)

Nidiology

RECORDS 113 nests representing 28 provincial regions.
Usually breeds in dense woods: mixed (31 nests), deciduous (27 nests), and pure coniferous stands (1 nest); occasionally breeds in clearings and open areas near forest edges. Nesting woods were sometimes wet (12 nests) and some nests were near lakes and rivers (11 nests).

Nests were usually in living trees but 4 nests were in dead or partially dead trees. Deciduous trees (10 spp., 83 nests) were preferred to coniferous trees (4 spp., 11 nests). Birch spp. (38 nests), poplar spp. (15 nests), maple spp. (12 nests), oak spp. (6 nests), and pine spp. (6 nests) were used most often. The diameter (DBH) of 1 nest tree was 36 cm (14 inches).

Nests were positioned in crotches and main forks, usually near the trunk, and only a few were near the crown of the tree. Heights of 86 nests ranged from 4.5 to 18.5 m (15 to 60 ft), with 43 averaging 7.5 to 12 m (25 to 40 ft).

Nests ranged from small to bulky platforms of sticks and twigs, commonly lined with bark chips and occasionally with down and feathers. Many nests were decorated on the rim and bowl with green poplar leaves, deciduous branches with attached leaves, or coniferous sprigs (hemlock, cedar, and spruce). Fresh vegetation such as this was frequently added during the period of nest occupation. Outside diameters of 3 nests ranged from 30.5 to 60 cm (12 to 24 inches); outside depths of 2 nests were 30.5 and 45.5 cm (12 and 18 inches); inside diameters of 2 nests were 15 and 21.5 cm (6 and 8.5 inches); inside depths of 2 nests were 5 and 5.5 cm (2 and 2.25 inches). Nests were sometimes reused in successive breeding seasons (2 records).

EGGS 64 nests with 1 to 4 eggs; 1E (6N), **2E** (30N), **3E** (26N), **4E** (2N).
Average clutch range 2 to 3 eggs (56 nests).

INCUBATION PERIOD 2 nests, 24 to 25 days and 28 to 31 days.

EGG DATES 38 nests, 22 April to 2 July (43 dates); 19 nests, 26 May to 10 June.

Breeding Distribution

The broad-winged hawk breeds throughout southern Ontario, but very rarely in the extreme southwest where suitable habitat is limited. In northern Ontario summer sight records indicate that it probably breeds at least as far north as the Albany River area.

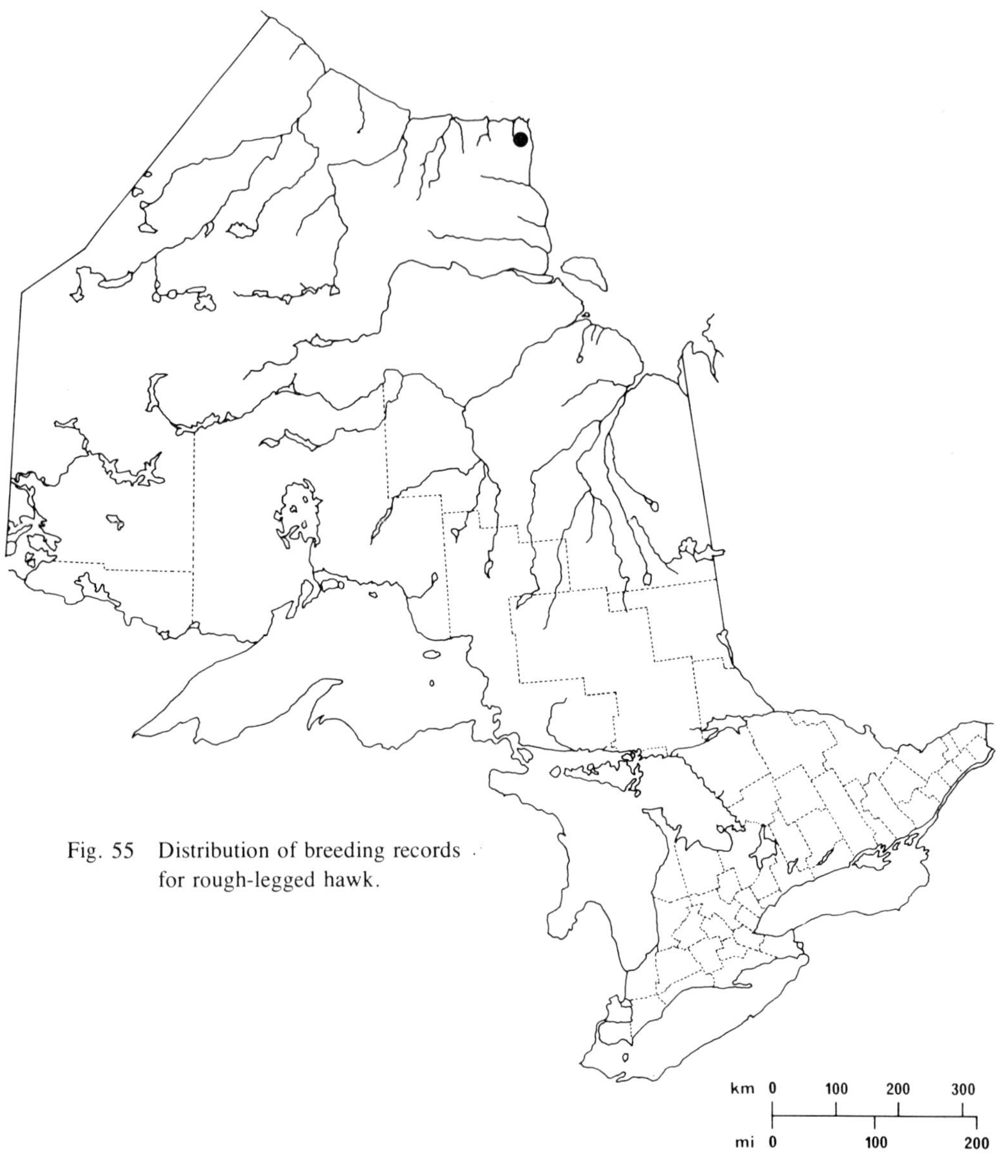

Fig. 55 Distribution of breeding records for rough-legged hawk.

Rough-legged Hawk, *Buteo lagopus* (Pontoppidan)

Nidiology

RECORDS 5 nests representing 1 provincial region.

This species breeds in the Subarctic and Arctic and was recently found nesting in Ontario on the narrow strip of tundra a few miles inland from both James and Hudson Bays. The tundra in this area is virtually treeless and of a heath-lichen complex (Fig. 167).

In this flat terrain 4 nests were positioned on part of an abandoned radar tower or on radar screens and 1 nest on the roof of an abandoned metal building. These nests were probably all built and previously occupied by ravens. Heights of the 5 nests were 4.5, 6, 9, 12, and ca 15 m (14, 20, 30, 40, and 50 ft).

The nests were constructed of branches and twigs of dwarf willow and birch and were lined—1 nest with lichen and another with plant stems.

All 5 nests contained young when found: 3 nests with 2 young, 2 nests with 3 young.

The dates on which the nests were found ranged from 29 June to 26 July.

Breeding Distribution

The rough-legged hawk was first reported breeding in the area of Cape Henrietta Maria in 1958 (Baillie, 1963), although the report was never verified. The first documentation of nesting in Ontario was only in 1976, when photographs were taken of a nest and young (Peck, 1976). The 1976 nest and several subsequent nests have all been placed on structures at abandoned radar sites (sites 415 and 416) near Cape Henrietta Maria. These structures provide the only elevated places in the area for this species to nest.

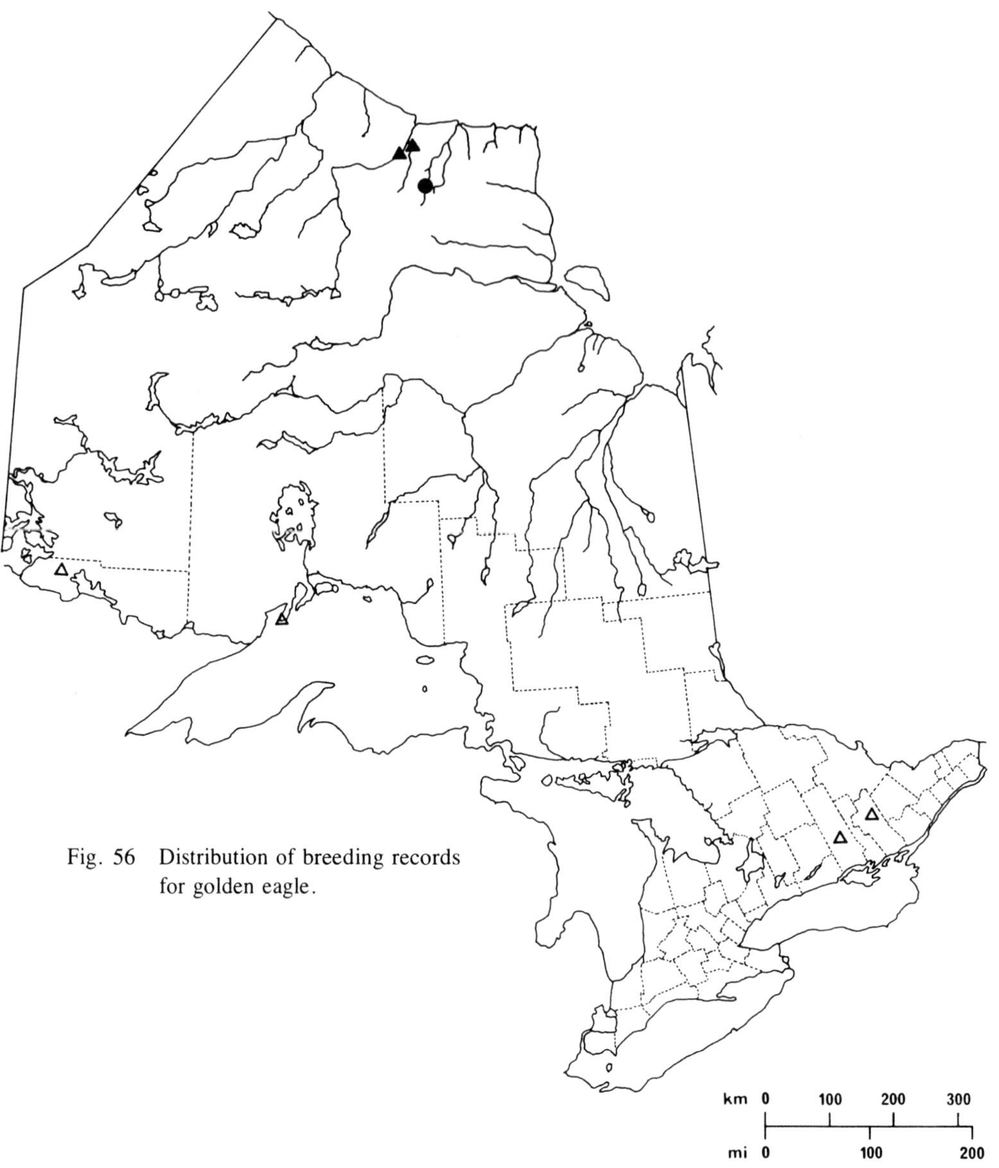

Fig. 56 Distribution of breeding records for golden eagle.

Golden Eagle, *Aquila chrysaetos* (Linnaeus)

Nidiology

RECORDS 7 nests representing 1 provincial region.
This species has been found nesting in Ontario at 3 locations in Kenora District, all within 75 km (46 mi) of each other. Other more southerly nesting sites all remain unsubstantiated and hypothetical. The 3 Kenora sites were in black spruce boreal forest areas within 95 km (60 mi) of Hudson Bay. Two of the sites were on the low rocky banks of the Shamattawa and Winisk rivers and the third was on the sheer west side of the Sutton Gorge between Sutton and Hawley lakes.

The nests were on rocky ledges at heights of 6 to ca 30 m (20 to 100 ft). The Sutton Gorge nest was on a ledge beneath an overhanging rock, 4.5 m (15 ft) below the cliff top. The nests were large structures of spruce sticks and were typically reoccupied in successive or subsequent years.

The nests were occupied by young when found between 15 June and 15 July; 3 nests contained 1 young and 3 nests held 2 young.

Breeding Distribution

Unverified nesting reports from Frontenac and Hastings counties and from Thunder Bay and western Rainy River districts, together with the far northern records, suggest that the golden eagle, although never common, was once more widespread in the province. Today the only area where nesting has been confirmed is in the far north, where only the three nest sites in Kenora District are known, and not until 1959 was one of these nests with young photographed (Lumsden, 1964).

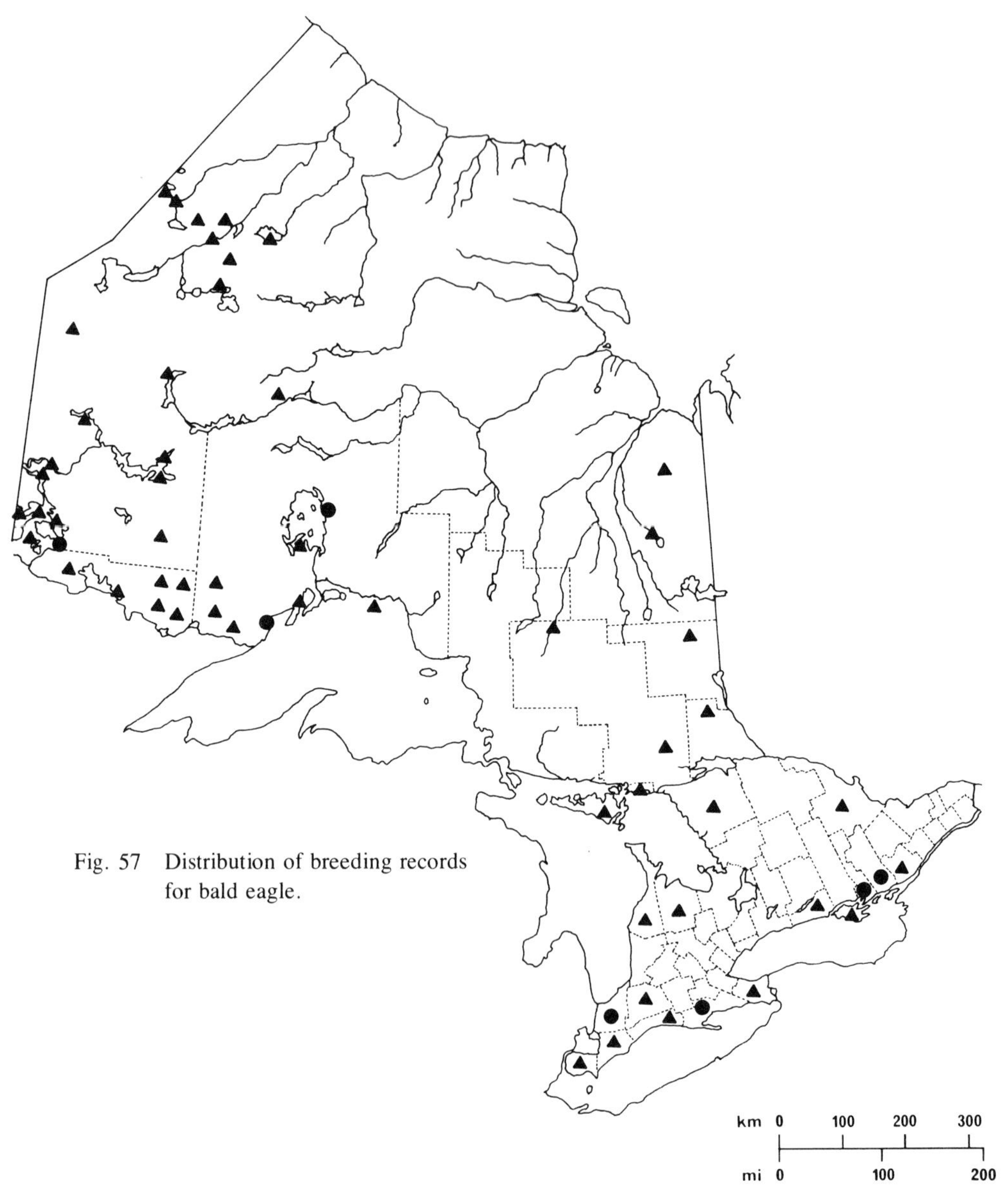

Fig. 57 Distribution of breeding records for bald eagle.

Bald Eagle, *Haliaeetus leucocephalus* (Linnaeus)

Nidiology

RECORDS 195 nests representing 16 provincial regions.
Breeds in coniferous, deciduous, and mixed woodlands, almost invariably near water, (usually near lakes and less often near rivers). The majority of nests were on wooded islands in lakes and large rivers; 3 nests were situated at distances of 0.4 to 1.6 km (0.25 to 1 mi) from water. A few nests were in clearings or in open agricultural land.

All nests were in trees and were more often in living than in dead trees. Conifers (3 spp., 68 nests) were slightly preferred to deciduous (10 spp., 48 nests) trees.

Nests were usually placed in tree crotches at or near the crown. Three nests were on lateral branches against the trunk, some distance below the top. Heights of 113 nests ranged from 9 to 27.5 m (30 to 90 ft), with 57 averaging 15 to 21.5 m (50 to 70 ft).

Nests were huge, cup-shaped platforms with flat tops and were made of dead branches and sticks, variously lined with grasses, mosses, shredded bark, twigs, leaves, and feathers. One active eagle nest contained at least a dozen occupied house sparrow nests and a starling nest. Outside diameters of 8 nests ranged from 1.2 to 3 m (4 to 10 ft), with 4 averaging 1.5 to 2.5 m (5 to 7.5 ft); outside depths of 6 nests were from 1.2 to 3 m (4 to 10 ft) depending on the extent of additions over years of use inside diameters of 3 nests ranged from 0.7 to 1.1 m (2.3 to 3.5 ft).

EGGS 43 nests with 1 to 3 eggs; **1E** (6N), **2E** (29N), **3E** (8N).
Average clutch range 2 eggs (29 nests).

INCUBATION PERIOD No information.

EGG DATES 19 nests, 3 April to 28 June; 10 nests, 10 April to 7 May.

Breeding Distribution

The bald eagle was formerly found near the larger lakes and rivers throughout Ontario, except in the Hudson Bay Lowland. It is now thinly distributed in the north, and in this century the species has all but disappeared from its former haunts in the southern portions of the province. A remaining high concentration exists at Lake of the Woods.

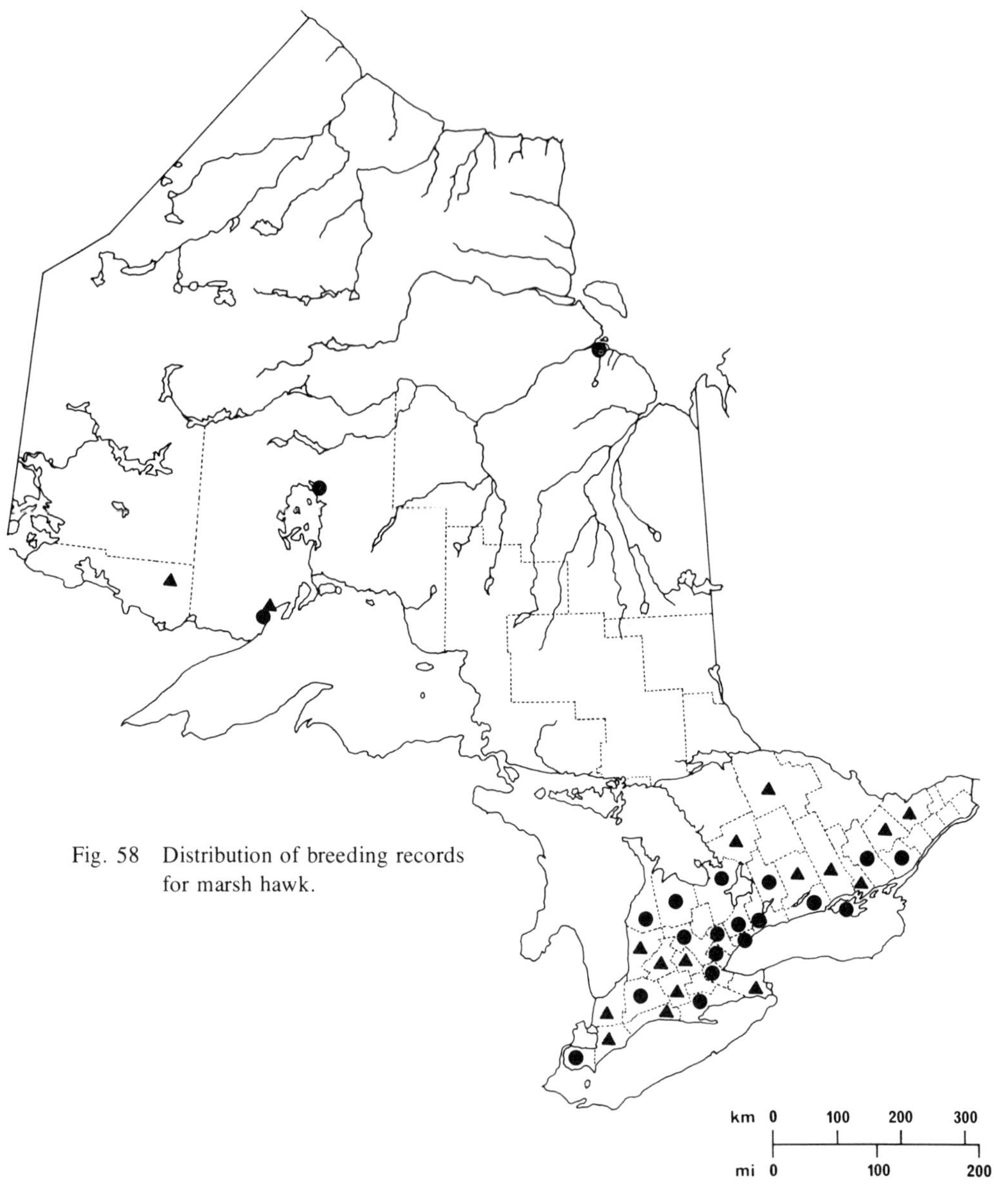

Fig. 58 Distribution of breeding records for marsh hawk.

Marsh Hawk, *Circus cyaneus* (Linnaeus)

Nidiology

RECORDS 192 nests representing 32 provincial regions.
Breeds in marshes, bogs, swales, open swamps, beaver meadows, and marshy edges of lakes and rivers; on agricultural land in hay and grain fields, pastures, wet meadows, coniferous plantations, and abandoned gravel pits; and occasionally in open woods, hydro slashings, and berry patches. Open areas are preferred, and even in wet locations the actual nest sites are more often dry than wet.

Nests were positioned on or near the ground in grasses and weeds; in cattails, sedges, common reed grasses, bulrushes, and mosses; among shoots of willow, dogwood, tamarack, swamp birch, cedar, blueberry, and spirea; occasionally under dead limbs and in brush piles; and once on the end of a log. A few nests were raised to 30.5 cm (12 inches) above ground in surrounding vegetation. Some nests in wet situations were built in marsh vegetation 15 to 30.5 cm (6 to 12 inches) above water, which ranged in depth from 30.5 to 45.5 cm (12 to 18 inches). A number of marsh nests were on hummocks, and field nests were sometimes on knolls. Two nests were situated 15 to ca 60 m (50 to 200 ft) from nests of American bittern.

Nests were bulky structures of sticks and branches from 1 cm (0.4 inch) to more than 2.5 cm (1 inch) in diameter, with the largest sticks in the nest base and smaller ones at the depressions and some had slightly hollowed bowls. Outside diameters of 8 nests ranged from 30.5 to 51 cm (12 to 20 inches); outside depths of 3 nests ranged from 5 to 20.5 cm (2 to 8 inches); inside diameters of 2 nests were 15 and 19 cm (6 and 7.5 inches) and their inside depths were 2.5 and 4.5 cm (1 and 1.75 inches).

In order of frequency, nesting materials were grasses, marsh vegetation, sticks and twigs, weed stalks, rootlets, leaves, feathers, bark, mosses, and pine needles. Nests were often homogeneous structures of grasses or other vegetation, but sometimes had sticks and twigs on the exterior. They were lined with fine grasses, rootlets, and weed stalks.

EGGS 171 nests with 1 to 7 eggs; 1E (5N), 2E (5N), 3E (17N), **4E** (37N), **5E** (74N), **6E** (31N), 7E (2N).
Average clutch range 4 to 5 eggs (111 nests).
One nest contained 5 eggs of marsh hawk and an unidentified duck egg.

INCUBATION PERIOD 9 nests, 27 to 32 days.

EGG DATES 160 nests, 26 April to 7 July (201 dates); 80 nests, 21 May to 7 June.

Breeding Distribution

The marsh hawk nests most commonly in southern Ontario. Despite the almost complete absence of breeding records from northern Ontario for this species, it is seen along the James Bay and Hudson Bay coasts in considerable numbers. It probably breeds along these coasts, and perhaps throughout the rest of northern Ontario in suitable habitat.

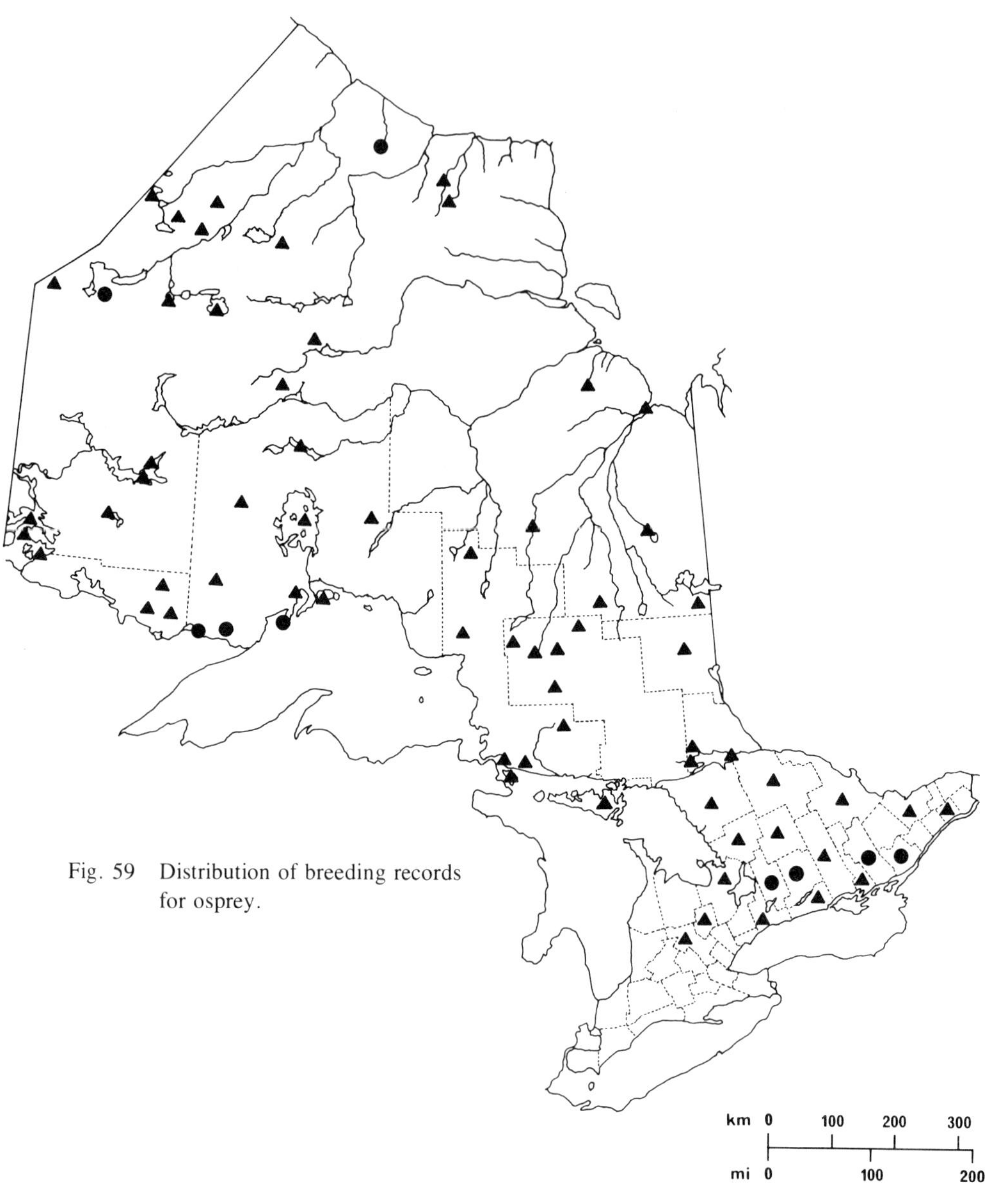

Fig. 59 Distribution of breeding records for osprey.

Osprey, *Pandion haliaetus* (Linnaeus)

Nidiology

RECORDS 260 nests representing 25 provincial regions.
Breeds in marshes, swamps, flooded areas, and bogs, and, somewhat less often, on the shores of lakes and rivers or on islands. Nesting habitats are most often mixed woodland areas and less often deciduous and coniferous stands. Nesting was sometimes recorded in active and deserted heronries.

Nests were usually in trees, either coniferous (85 nests) or deciduous (14 nests) and either dead (94 nests) or living (15 nests). Most nest trees were standing in water, but some were on dry land near water, and a very small number were at some distance from the nearest water. Hydro poles and towers and manmade platforms (17 nests) were also used to support nests.

Nests were commonly built in the tops of trees, but sometimes were placed in main crotches. A number (20 nests) were on broken stubs or stumps, not far above water or ground. Heights of 137 nests ranged from 0.6 to 30 m (2 to 100 ft), with 69 averaging 9 to 18 m (30 to 60 ft).

Nests were bulky structures of sticks and branches from 1 cm (0.4 inch) to more than 2.5 cm (1 inch) in diameter, with the largest sticks in the nest base and smaller ones at the top. Because material was added during a nesting season and in succeeding nestings, nests sometimes became quite large. Occasionally old nests of herons were used. A common grackle nest was reported in an active nest of this species. Nests were variously lined with grasses, straw, bark (usually cedar), mosses, corn stalks, cow manure, brush and weeds, fine sticks, twine, mud, rootlets, and feathers. Outside diameters of 15 nests ranged from 0.8 to 2.5 m (2.5 to 8 ft), with 8 averaging 0.9 to 1.5 m (3 to 5 ft); outside depths of 6 nests ranged from 0.5 to 2 m (1.5 to 6 ft); inside depths of 9 to 12.5 cm (3.5 to 5 inches) were reported.

EGGS 46 nests with 1 to 5 eggs; 1E (4N), **2E** (11N), **3E** (29N), 5E (2N).
Average clutch range 2 to 3 eggs (40 nests).

INCUBATION PERIOD No information.

EGG DATES 37 nests, 9 May to 21 June (41 dates); 18 nests, 22 May to 7 June.

Breeding Distribution

The osprey breeds in Ontario as far north as the limit of trees and throughout the Canadian Shield. Breeding occasionally occurs in a few regions south of the Shield.

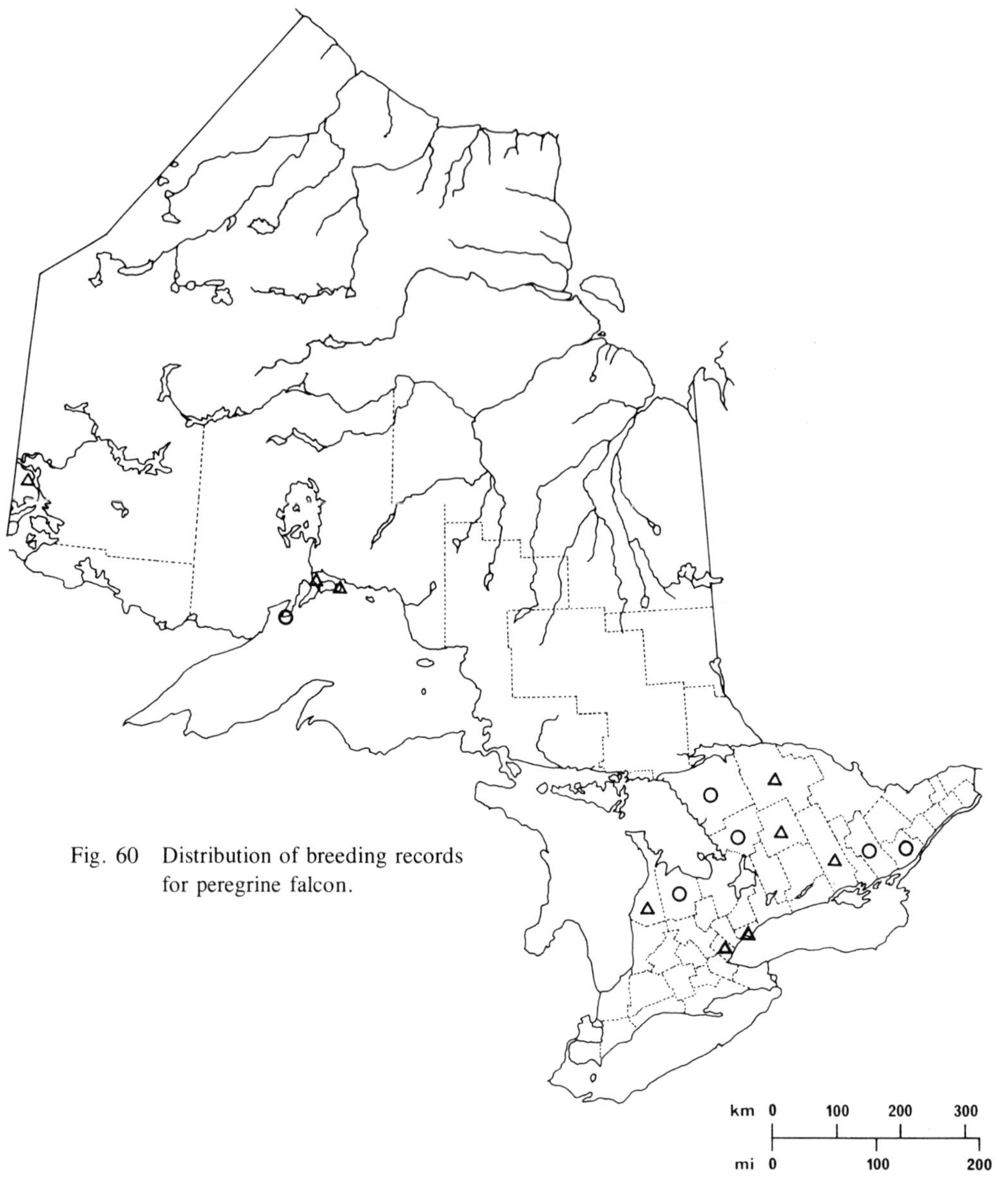

Fig. 60 Distribution of breeding records for peregrine falcon.

Peregrine Falcon, *Falco peregrinus* Tunstall

Nidiology

RECORDS 48 nests representing 11 provincial regions.
Bred formerly (no breeding since 1963) on cliffs overlooking lakes or on island cliffs in lakes and rivers. Two records of nests on cliffs on the Niagara Escarpment were the only recorded sites not adjacent to water. Usually the nesting cliffs were in wooded areas.

Nests were invariably on cliff ledges (1 nest was described as being in a cleft). Heights of 37 nests ranged from 9 to 52 m (30 to 170 ft), with 19 averaging 11 to 23 m (35 to 75 ft).

Nests were mere depressions on rocky, grassy, mossy, or clay-covered cliff ledges.

EGGS 33 nests with 2 to 5 eggs; 2E (7N), **3E** (8N), **4E** (16N), 5E (2N).
Average clutch range 3 to 4 eggs (24 nests).

INCUBATION PERIOD No information.

EGG DATES 24 nests, 23 April to 9 June (27 dates); 12 nests, 1 May to 20 May.

Breeding Distribution

The peregrine falcon (Fig. 154B) was apparently never very common in Ontario in this century (Baillie and Harrington, 1936) and has now disappeared as a breeding bird in the province. Shooting, egg collecting, and, most significantly, chemical pesticides in the environment effected this result. The falcon formerly nested throughout most of southern Ontario at locations where cliffs afforded nest sites and in northern Ontario predominantly along the north shore of Lake Superior. Undoubtedly, however, we lack records of all the nesting locations it used.

In 1977 attempts were begun in Algonquin Provincial Park to reintroduce the species from captive breeding stock (McKeating, 1978). The success of these efforts remains to be seen.

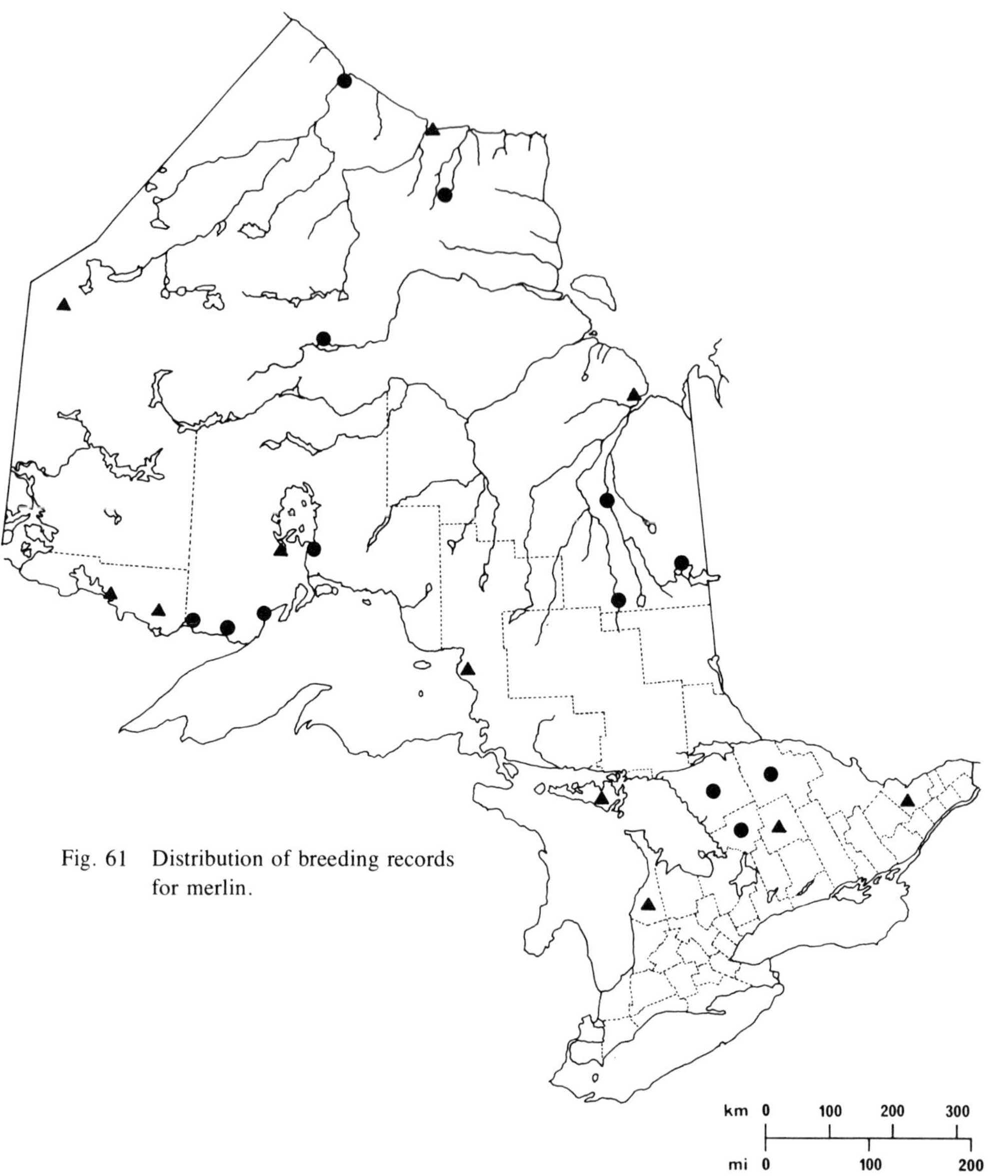

Fig. 61 Distribution of breeding records for merlin.

Merlin, *Falco columbarius* Linnaeus

Nidiology

RECORDS 31 nests representing 10 provincial regions.
Breeds in mature and second-growth coniferous forests, in pine plantations, and in mixed woods. Tree and ground cover is sometimes dense, more often semi-open. Nests were found on islands, shores of lakes, and hillsides (1 nest was in a small park on the outskirts of a city).

Most nests were in coniferous trees such as spruce spp. (18 nests), pine spp. (5 nests), and cedar (3 nests); 1 nest was on a cliff ledge.

Nests were usually placed near or against the trunk, not far below the tree-top. One nest was on a large branch 1.8 m (6 ft) from the trunk. The cliff-ledge nest was under shrubs. Heights of 23 tree nests ranged from 3 to 15 m (10 to 50 ft), with 11 averaging 6 to 9 m (20 to 30 ft). The cliff nest was at a height of 49 m (160 ft).

Tree nests were platforms, mainly composed of twigs and sticks, and sometimes of leaves and bark. Old nests of common crow were sometimes used (3 nests), and once an old squirrel drey was used. Linings of bark, mosses, fibre, feathers, bits of paper, and scraps of fish nets were reported, but some nests contained little or no added material. The cliff nest was a scrape in existing debris containing bits of twigs. Nests were as small as 30.5 cm (12 inches) in width, and a large nest measured 89 cm (35 inches) in width and 12.5 cm (5 inches) in depth.

EGGS 21 nests with 3 to 5 eggs; 3E (4N), **4E** (9N), **5E** (8N).
Average clutch range 4 to 5 eggs (17 nests).

INCUBATION PERIOD No information.

EGG DATES 18 nests, 16 May to 5 July (20 dates); 9 nests, 30 May to 13 June.

Breeding Distribution

Despite the scarcity of records, summer sightings indicate that the merlin breeds throughout the forested parts of northern Ontario. In the south only a few widely scattered records are available.

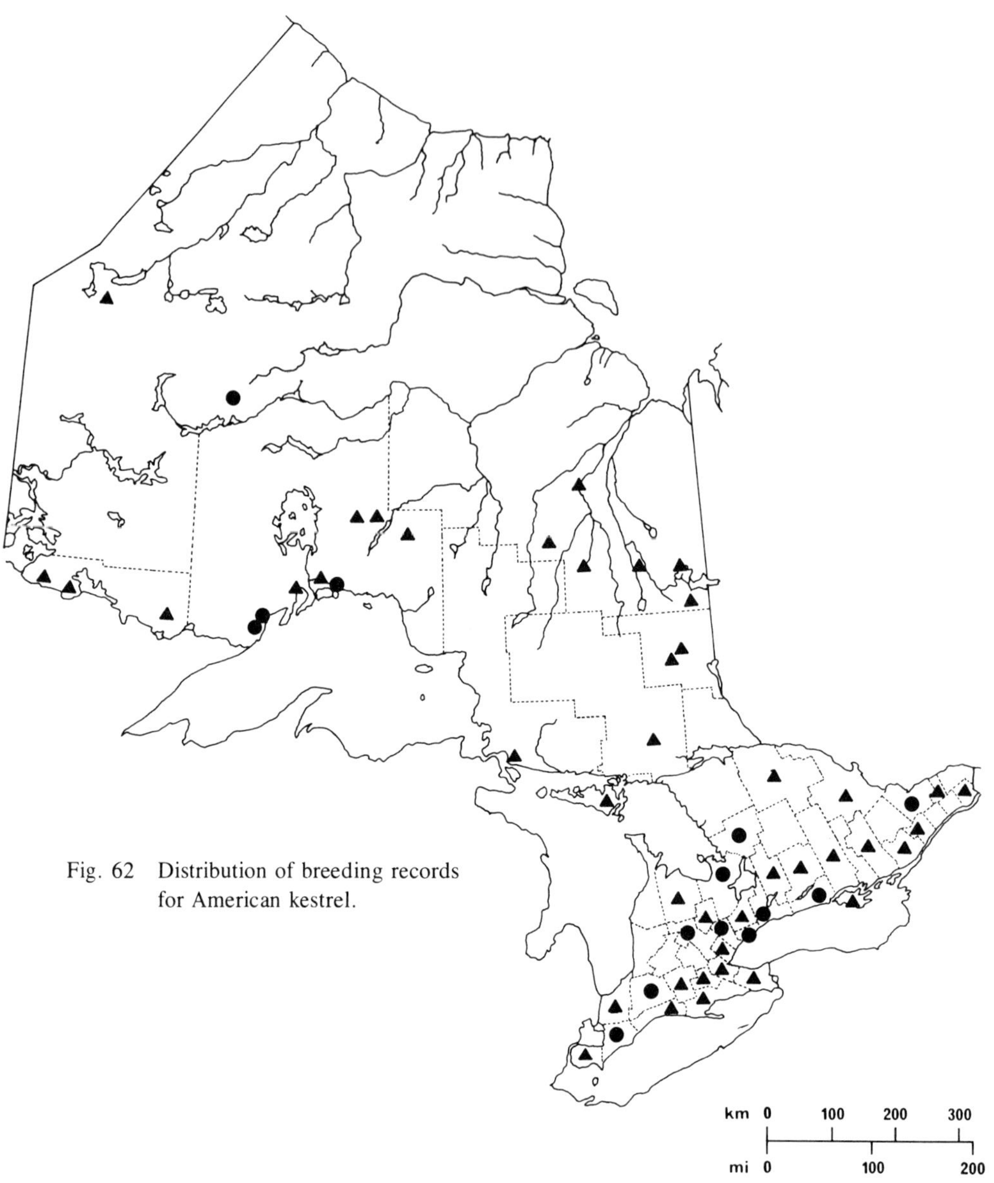

Fig. 62 Distribution of breeding records for American kestrel.

American Kestrel, *Falco sparverius* Linnaeus

Nidiology

RECORDS 233 nests representing 38 provincial regions.
Breeds usually in open areas in fields and pastures, at roadsides, and near buildings; frequently in slashings and woodlands (usually at their edges); and occasionally in bogs, marshes, and swamps.

Nests traditionally were situated in deciduous (63 nests) or coniferous (8 nests) trees, which were usually dead (104 nests) and occasionally living (3 nests). They were in old woodpecker cavities (26 nests) or in natural cavities (12 nests). Where available, nest boxes (95 nests) were often used; occasionally nests were in buildings (18 nests) and in hydro poles (4 nests); 1 nest was in an earthen burrow, and another in a fence post. Heights of 121 nests (excluding those in nest boxes) ranged from 1.8 to 23 m (6 to 75 ft), with 61 averaging 5 to 12 m (16 to 40 ft).

Nesting cavities usually either contained pre-existing materials or were unlined. Materials recorded were wood chips, bark, straw, fur, grasses, weeds, wire, and cardboard. Diameters of 5 cavity openings ranged from 5 to 20.5 cm (2 to 8 inches); depths of 13 cavities ranged from 10 to 61 cm (4 to 24 inches), with 7 averaging 20.5 to 30.5 cm (8 to 12 inches).

EGGS 151 nests with 1 to 6 eggs; 1E (4N), 2E (7N), **3E** (16N), **4E** (39N), **5E** (80N), **6E** (5N).
Average clutch range 3 to 5 eggs (135 nests).
One record indicated the strong possibility of 2 females having laid eggs in 1 nest.

INCUBATION PERIOD No information.

EGG DATES 125 nests, 11 April to 16 July (151 dates); 62 nests, 18 May to 6 June.

Breeding Distribution

The American kestrel (Fig. 158B) breeds throughout the province at least as far north as Sandy Lake and Moosonee.

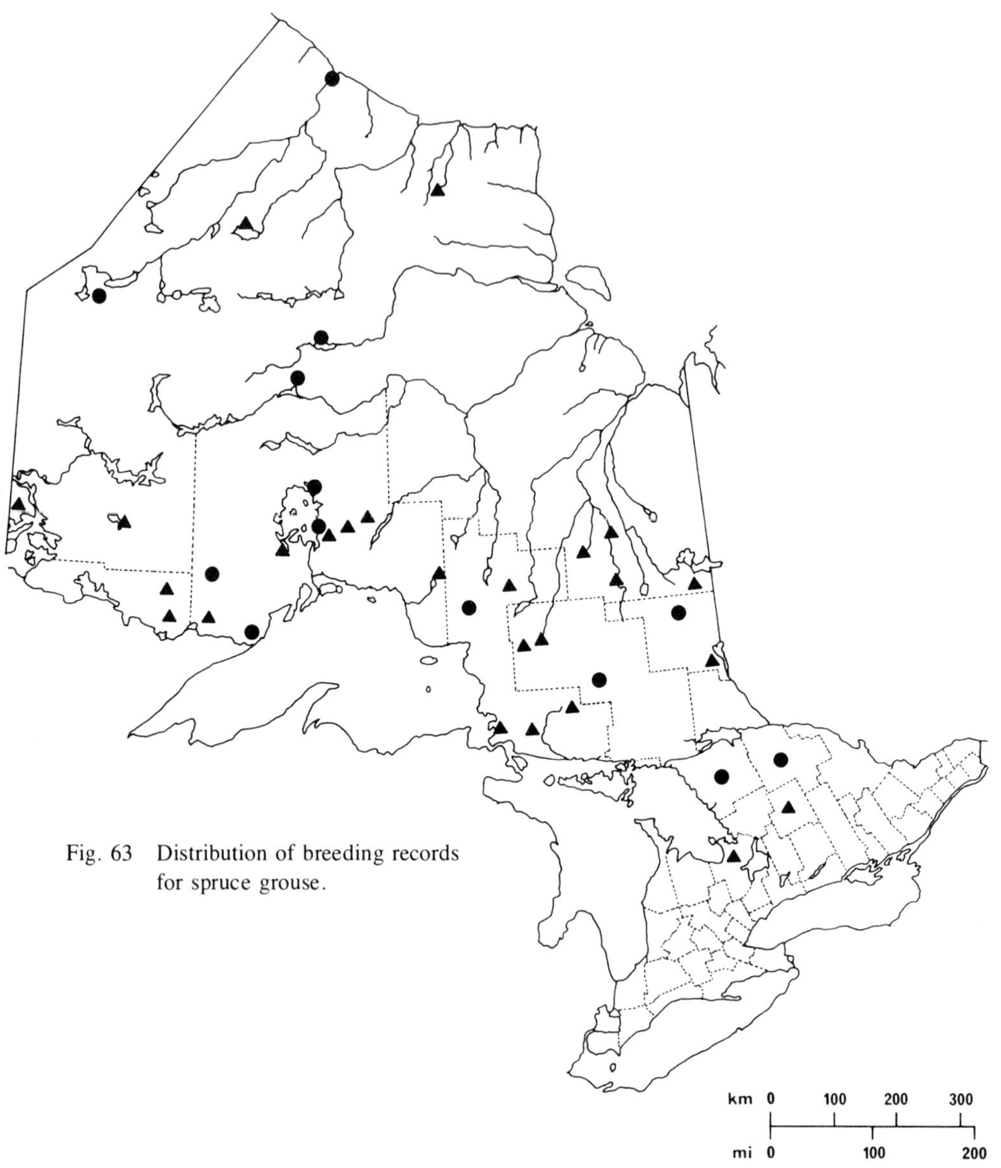

Fig. 63 Distribution of breeding records for spruce grouse.

Spruce Grouse, *Canachites canadensis* (Linnaeus)

Nidiology

RECORDS 29 nests representing 7 provincial regions.
This species breeds in wet areas such as bogs with black spruce and tamarack, in dry areas such as sandy jack pine stands, and in other coniferous and mixed woodlands. The presence of conifers, especially spruce and jack pine, appears to be a prerequisite in its nesting habitat.

Nests were on the ground and were usually placed against or near the bases of trees and stumps and under branches and logs. One nest was between the roots of a balsam fir. In addition to the presence of coniferous trees, overhead cover of shrubs and small trees such as serviceberry spp., cherry spp., and willow spp., was reported, as was ground cover of blueberry, grasses, bunchberry, bush honeysuckle, and various other plants. Almost no ground cover was reported in the immediate vicinity of 1 nest. Some nests were situated on small mounds or hummocks (4 nests). Nests were usually shallow depressions in sphagnum, grasses, and conifer needles. One nest was raised ca 8 cm (3 inches) above a moss-covered surface. One nest had an inside depth of 3.8 cm (1.5 inches).

Nests were variously lined with conifer needles, grasses, feathers, mosses, and dead vegetation.

EGGS 27 nests with 1 to 8 eggs; 1E (1N), 2E (1N), 3E (1N), **4E** (5N), **5E** (8N), **6E** (6N), **7E** (4N), 8E (1N).
Average clutch range 4 to 6 eggs (19 nests).

INCUBATION PERIOD 2 nests, ca 25 days.

EGG DATES 26 nests, 9 May to 25 June (32 dates); 13 nests, 29 May to 13 June.

Breeding Distribution

The spruce grouse (Fig. 160A) breeds throughout the forested portions of northern Ontario. Below Lake Nipissing it is scarce, although there are records from northern Simcoe County and Haliburton District. The species probably also breeds in northern Bruce County and the Ottawa area.

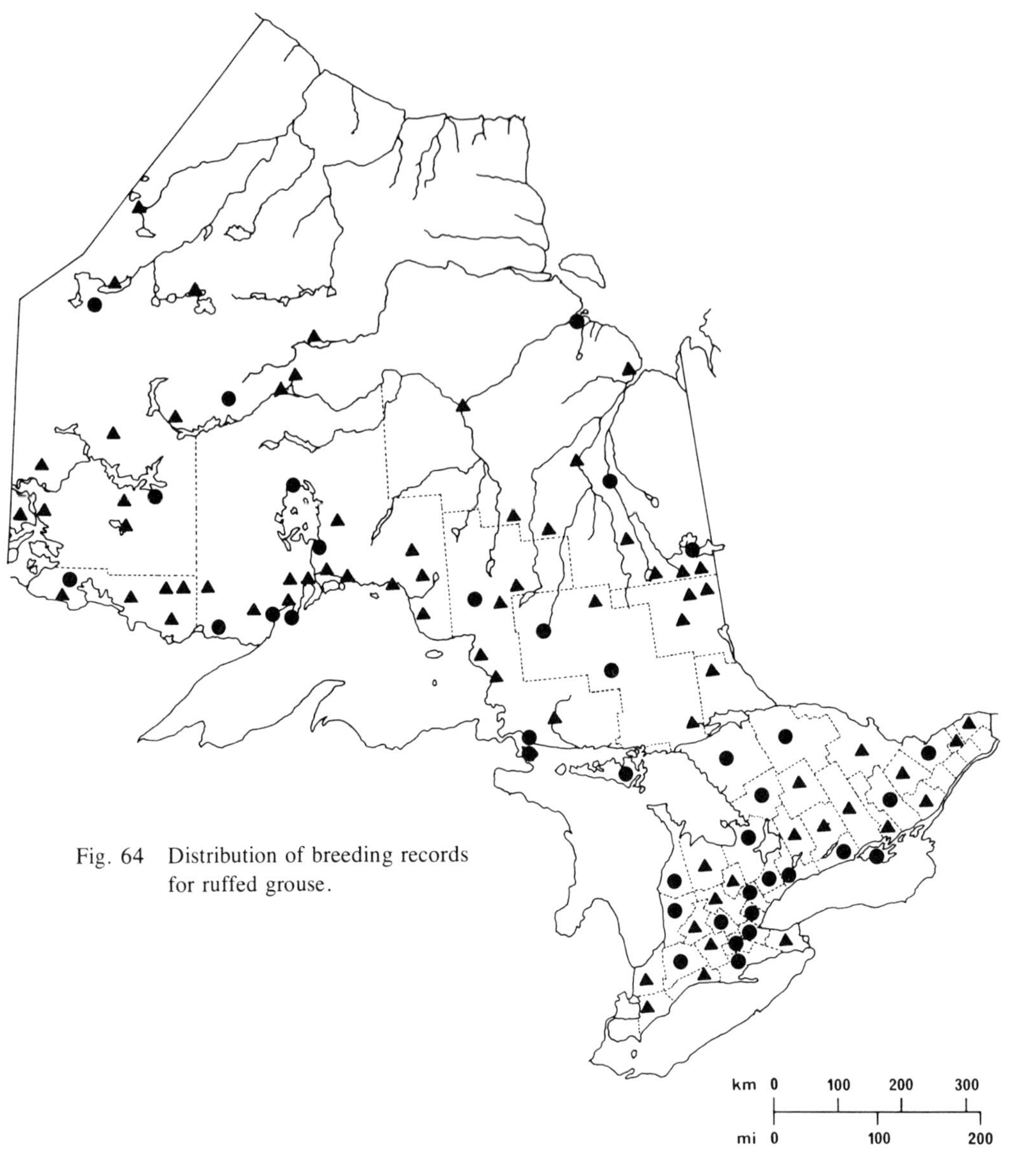

Fig. 64 Distribution of breeding records for ruffed grouse.

Ruffed Grouse, *Bonasa umbellus* (Linnaeus)

Nidiology

RECORDS 245 nests representing 44 provincial regions.
Usually breeds in wood lots and forests. Recorded tree stands were mixed (67 nests), deciduous (30 nests), coniferous (19 nests), and unspecified (54 nests). Such stands were dense or open, and often were wet (40 nests). Nests were sometimes situated deep in the woods, but were more often at forest edges or in or near clearings. Some nests were situated on wooded slopes or hills. Occasionally nests were on small islands (2 nests) and on railroad and hydro right-of-ways (2 nests); 1 nest was in a field.

Although woodland nests were sometimes in open situations, frequently they were positioned beneath or at the base of living and dead trees and shrubs (95 nests), under fallen trees and logs (26 nests), beside stumps and logs (24 nests), in or under dead branches (19 nests), and between tree roots (5 nests). Some were also well hidden by nearby undergrowth. Two active nests were 15 m (50 ft) apart, and another nest was noted within 45 m (150 ft) of a nest used the previous year.

Nests were depressions in the ground or ground cover and usually they were shallow structures. The majority were placed in deciduous leaves, but a few were placed in grasses and conifer needles. One nest was noted on rocks and another in a mossy hollow on a decaying log. Nests were variously lined, in order of preference, with leaves (a characteristic material), feathers and down, grasses, conifer needles, twigs, plant fibres, and bark. Diameters of 3 nests were 15, 20.5, and 25.5 cm (6, 8, and 10 inches). In 3 nests eggs were covered with leaves in the female's absence.

EGGS 239 nests with 2 to 20 eggs; 2E (3N), 3E (5N), 4E (3N), 5E (5N), 6E (10N), **7E** (15N), **8E** (16N), **9E** (20N), **10E** (28N), **11E** (38N), **12E** (42N), **13E** (27N), **14E** (17N), 15E (7N), 16E (1N), 17E (1N), 20E (1N).
Average clutch range 9 to 12 eggs (128 nests).
One nest with 10 ruffed grouse eggs contained 1 egg of ring-necked pheasant, and another nest with 9 ruffed grouse eggs contained 3 eggs of ring-necked pheasant.

INCUBATION PERIOD 6 nests, 23 to 26 days. Incubation began with the laying of the last egg and hatching was simultaneous.

EGG DATES 238 nests, 15 April to 17 July (306 dates); 119 nests, 14 May to 28 May.

Breeding Distribution

The ruffed grouse breeds throughout Ontario at least as far north as Big Trout Lake and Fort Albany, and possibly somewhat farther. It is scarce now in the extreme south where wooded areas have been decimated.

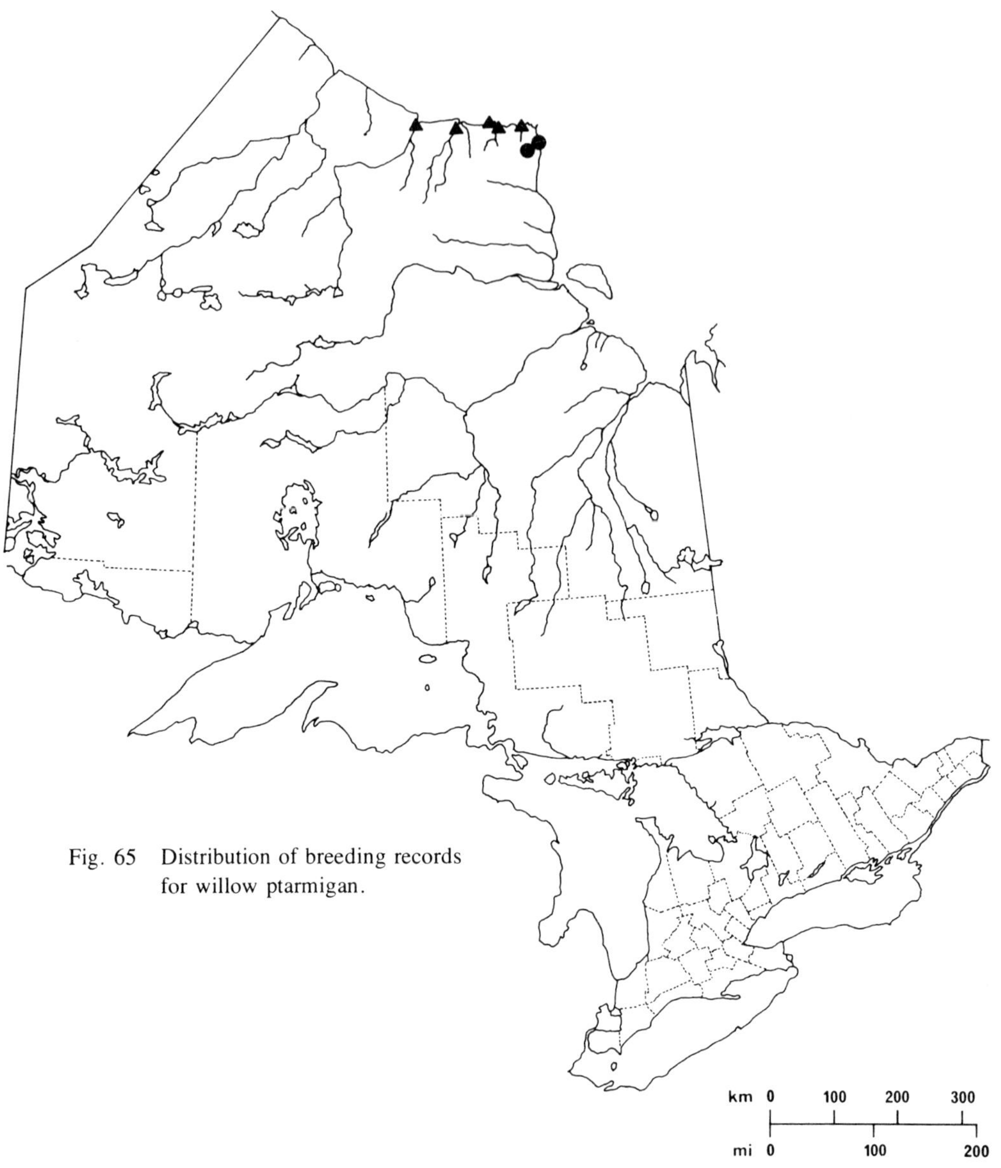

Fig. 65 Distribution of breeding records for willow ptarmigan.

Willow Ptarmigan, *Lagopus lagopus* (Linnaeus)

Nidiology

RECORDS 20 nests representing 1 provincial region.
Breeds on the subarctic tundra that borders parts of the coasts of James and Hudson bays. Both wet (sedge-grass, tussock tundra) and dry (heath-lichen tundra) areas are used as nesting sites. In wet areas nests were situated on the numerous characteristic hummocks, and in dry areas they were often on raised former beach lines. In both areas scattered arctic willows were customarily present, along with some dwarf birch, bog laurel, rhododendron, leatherleaf, and occasionally stunted tamarack and black spruce.

Nests were most often positioned in sedge and grass clumps and frequently were beside or under small willows. Nest depressions were nearly always lined with grass and sedge stems, and often with leaves of birch and willow, twigs, and a few feathers as well. One incomplete clutch of eggs was covered in the absence of the female.

EGGS 20 nests with 1 to 12 eggs; 1E (1N), 2E (1N), 4E (2N), **5E** (2N), **6E** (1N), **7E** (4N), **8E** (3N), **9E** (4N), **11E** (1N), 12E (1N).
Average clutch range 5 to 9 eggs (14 nests).

INCUBATION PERIOD No information. Eggs in 1 nest were laid at ca 24 hour intervals.

EGG DATES 20 nests, 23 June to 17 July (22 dates); 10 nests, 28 June to 15 July.

Breeding Distribution

Available nest records indicate that the willow ptarmigan (Fig. 168B) breeds in the areas south and west of Cape Henrietta Maria. However, summer sight records suggest that it probably breeds all along the Hudson Bay coast wherever tundra conditions prevail (Fig. 167).

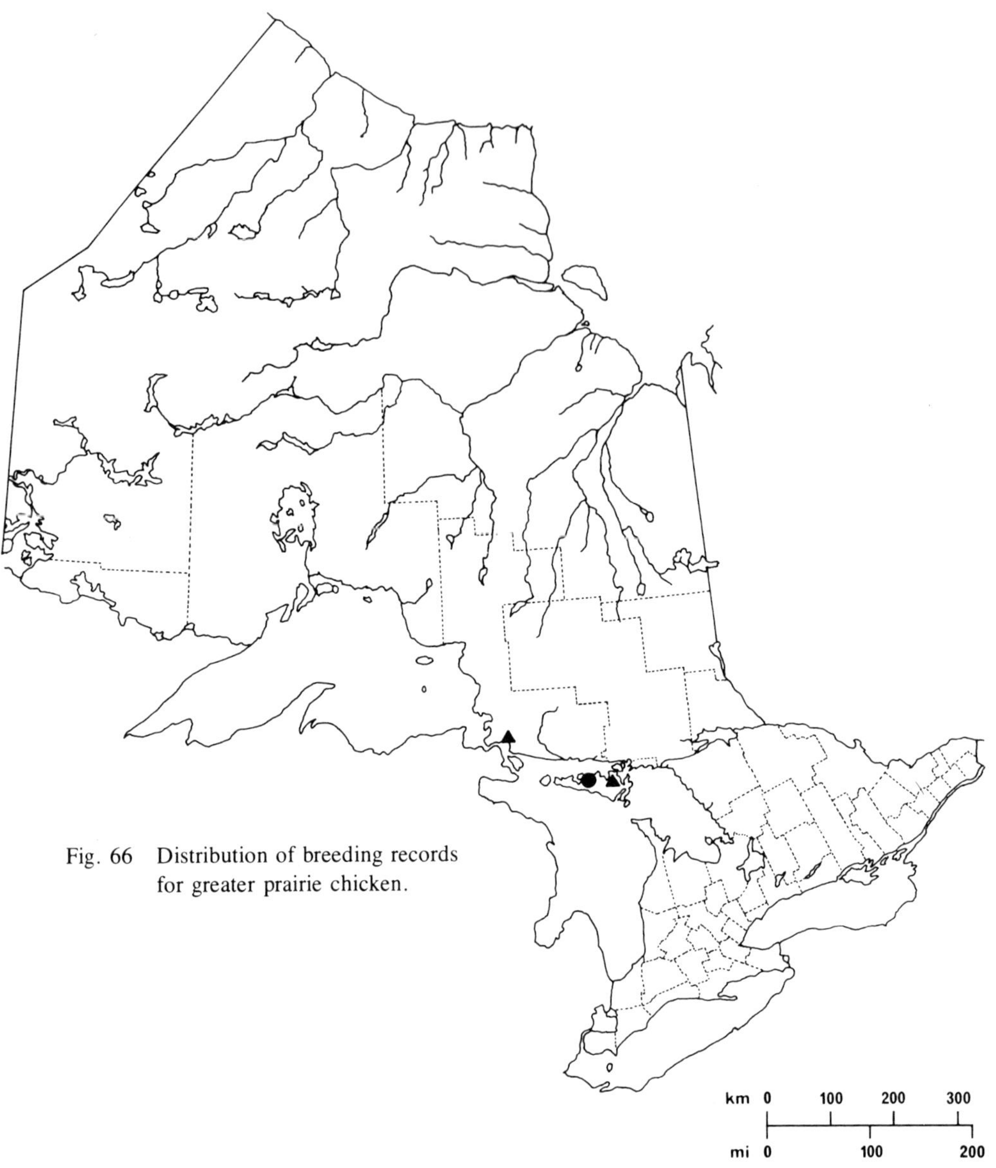

Fig. 66 Distribution of breeding records for greater prairie chicken.

Greater Prairie Chicken, *Tympanuchus cupido* (Linnaeus)

Nidiology

RECORDS 2 nests representing 1 provincial region.
A review of the published evidence of the Ontario breeding of this species (Baillie, 1961) indicates that it was first included on the breeding-bird list on the basis of hearsay evidence of nesting in 1946 on Manitoulin Island. This undocumented record involved a nest with 13 eggs found in June, and all of these eggs were reported later to have hatched. The species had immigrated to the island from Michigan about 7 years earlier. In July 1948 it was properly documented with photographs of one-third grown immatures taken in Billings Township, Manitoulin Island (ROM PR 174 and 175). A set of 3 eggs (ROM 5148) was collected on 20 June 1953 at an airport 11.5 km (7 mi) west of Gore Bay, Manitoulin Island (Devitt and Miller, 1954).

The 3 eggs were from a nest in a dry, stony flatland area of limestone, thinly covered with an alkaline soil. The nest depression in the ground was sparsely lined with dried grasses; a tuft of grass 38 cm (15 inches) in length arched over the nest.

Breeding Distribution

The greater prairie chicken apparently occurred naturally in the prairie-like areas of Essex, Kent, and Lambton Counties prior to 1900 (Lumsden, 1966), but has long since disappeared from the southern region of the province. It expanded farther north in Ontario along the north shore of Lake Huron in about 1925 and onto Manitoulin Island in 1938 or 1939 (Baillie, 1947). However, the species soon hybridized extensively with sharp-tailed grouse and, in the absence of a continued influx of pure prairie chicken stock, the greater prairie chicken disappeared from Ontario by the early 1960s. Records also exist of the occurrence of this species in the Thunder Bay area (Denis, 1961), although no indication of breeding was ever found.

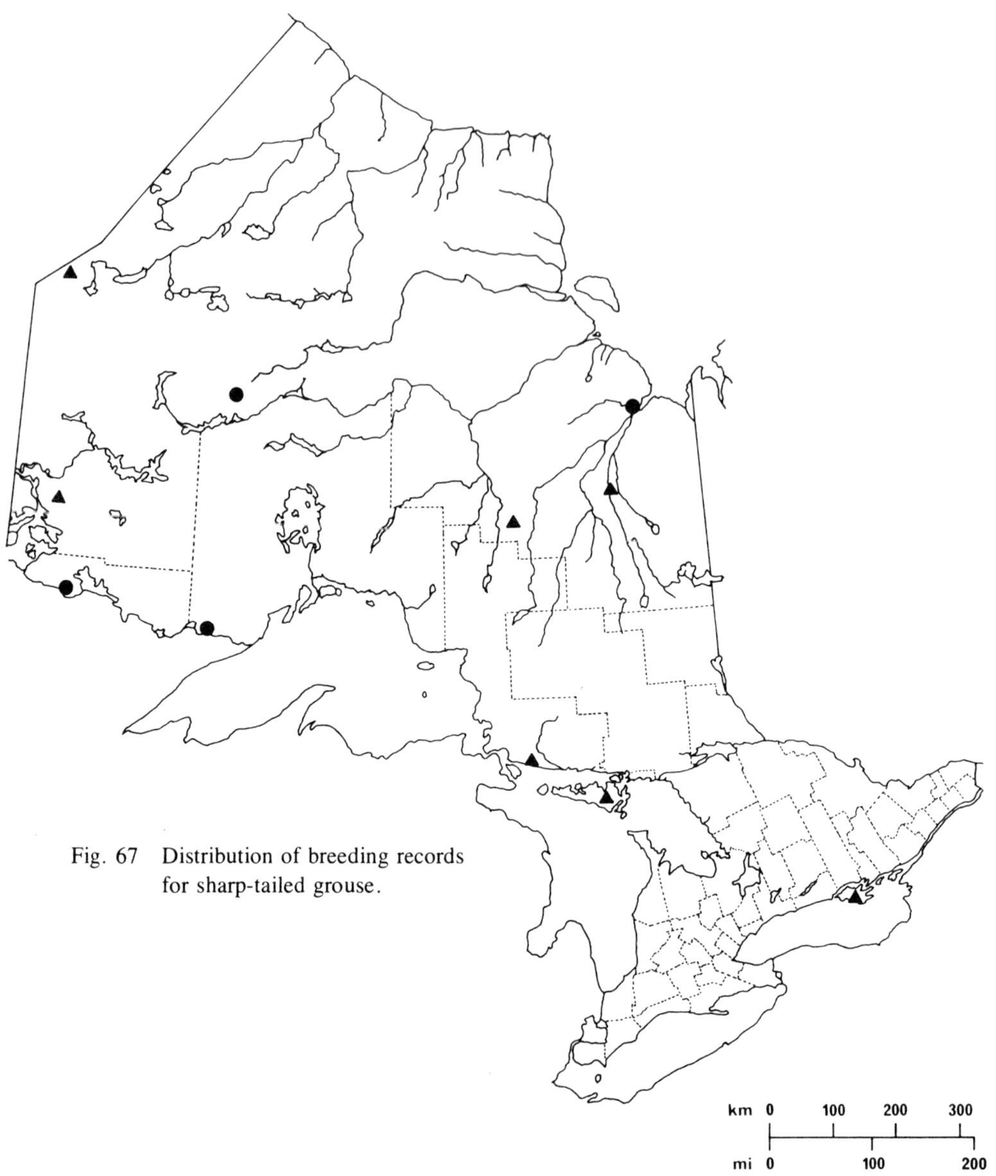

Fig. 67 Distribution of breeding records for sharp-tailed grouse.

Sharp-tailed Grouse, *Pedioecetes phasianellus* (Linnaeus)

Nidiology

RECORDS 2 nests representing 2 provincial regions.
This species is a relatively rare permanent resident of northern regions of the province, and in the south there are local occurrences and releases reported as far east as Prince Edward County. To date only 2 nests have been reported in Ontario. This grouse hybridizes with the greater prairie chicken, and on Manitoulin Island it was instrumental in effecting the demise of the latter species.

The first reported nest in Ontario was found on 9 June 1929 and contained 7 eggs, 2 of which were destroyed by crows. The remaining 5 eggs were collected (ROM 1860). The nest was in a partial clearing near a road. The second nest was found on 24 July 1930 on the ground under a hazel bush. This nest was composed of leaves and grasses and contained 9 addled eggs (ROM 10993).

Breeding Distribution

Although breeding records are few, the sharp-tailed grouse probably nests throughout northern Ontario, particularly in grassy fens (Fig. 163). However, it is not found on the northern coasts nor in a strip extending from eastern Lake Superior to Quebec that includes central Algoma and Sudbury districts, and southern Timiskaming and northern Nipissing districts. The natural breeding area of this grouse extends into the south only on Manitoulin Island. The species has been released in several counties in southern Ontario, but successful breeding has been reported only from Prince Edward County, in 1974.

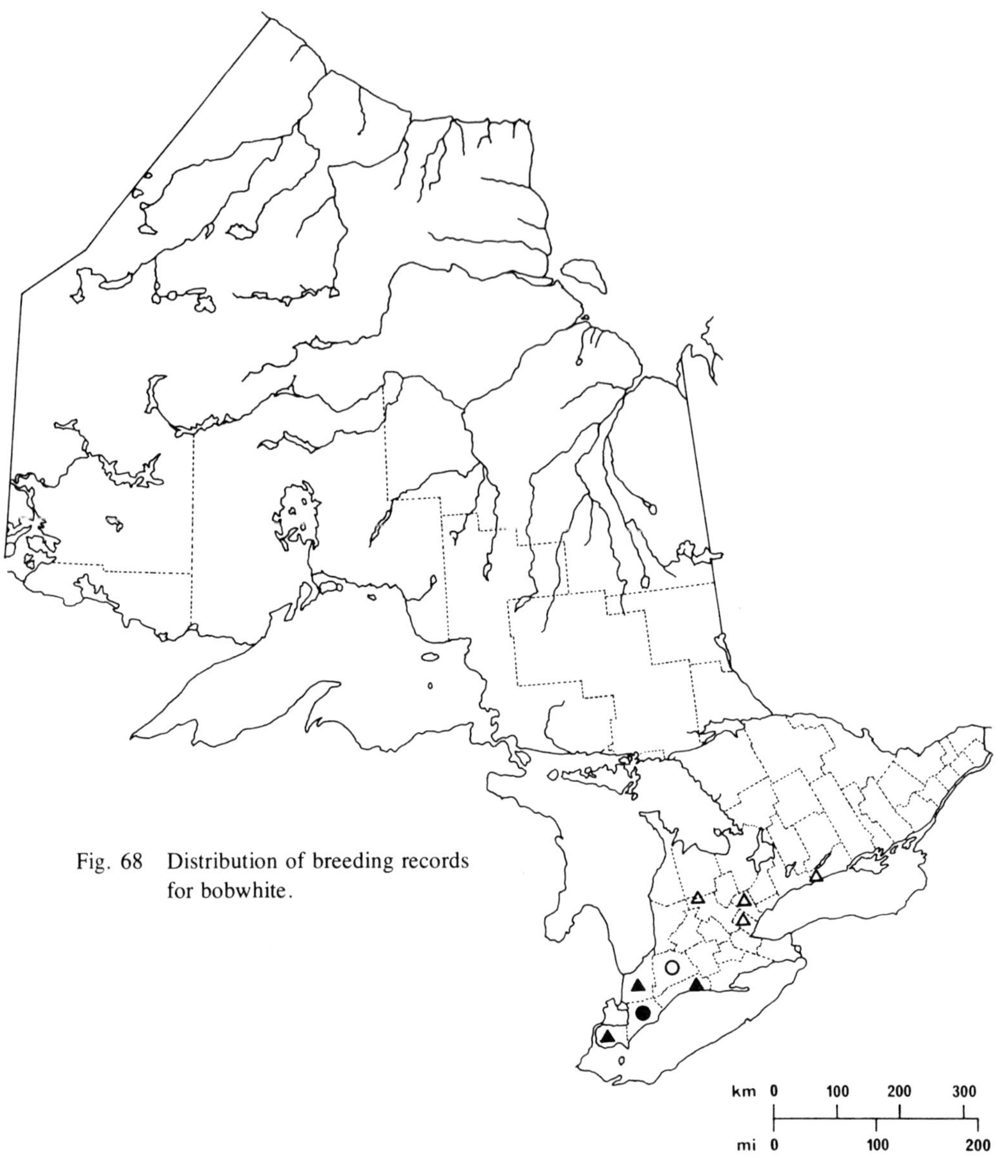

Fig. 68 Distribution of breeding records for bobwhite.

Bobwhite, *Colinus virginianus* (Linnaeus)

Nidiology

RECORDS 11 nests representing 4 provincial regions.
Breeds in agricultural areas in hay fields, fence rows, and roadsides (1 record in a strawberry patch in an orchard); in city parks; and at the edges of golf courses. Small open wood lots and heavy thickets were adjacent to several nests.

Characteristically, nests were shallow, grass-lined hollows placed in long grasses, and 3 nests were found, respectively, beside a fence, a telephone pole, and a small tree. Nearby grasses sometimes arched over the nests.

EGGS 11 nests with 7 to 18 eggs; 7E (2N), **11E** (2N), **13E** (2N), **14E** (1N), **15E** (1N), **16E** (1N), 17E (1N), 18E (1N).
Average clutch range 11 to 15 eggs (6 nests).

INCUBATION PERIOD 1 nest, 24 days.

EGG DATES 9 nests, 21 May to 19 September (12 dates); 5 nests, 4 July to 31 July.

Breeding Distribution

With the establishment of settlers in southern Ontario and the initial clearing of forests (Fig. 151), the bobwhite became a relatively common breeding resident at least as far north as southern Georgian Bay, and as far east as Northumberland County, and possibly Kingston in Frontenac County (Clarke, 1954). But with the advent of modern clean farming its cover and food supply were removed and it has retreated back into three or four counties in the extreme south of Ontario. Although the bobwhite has been released in many places in the south, it does not seem to survive long outside its limited range.

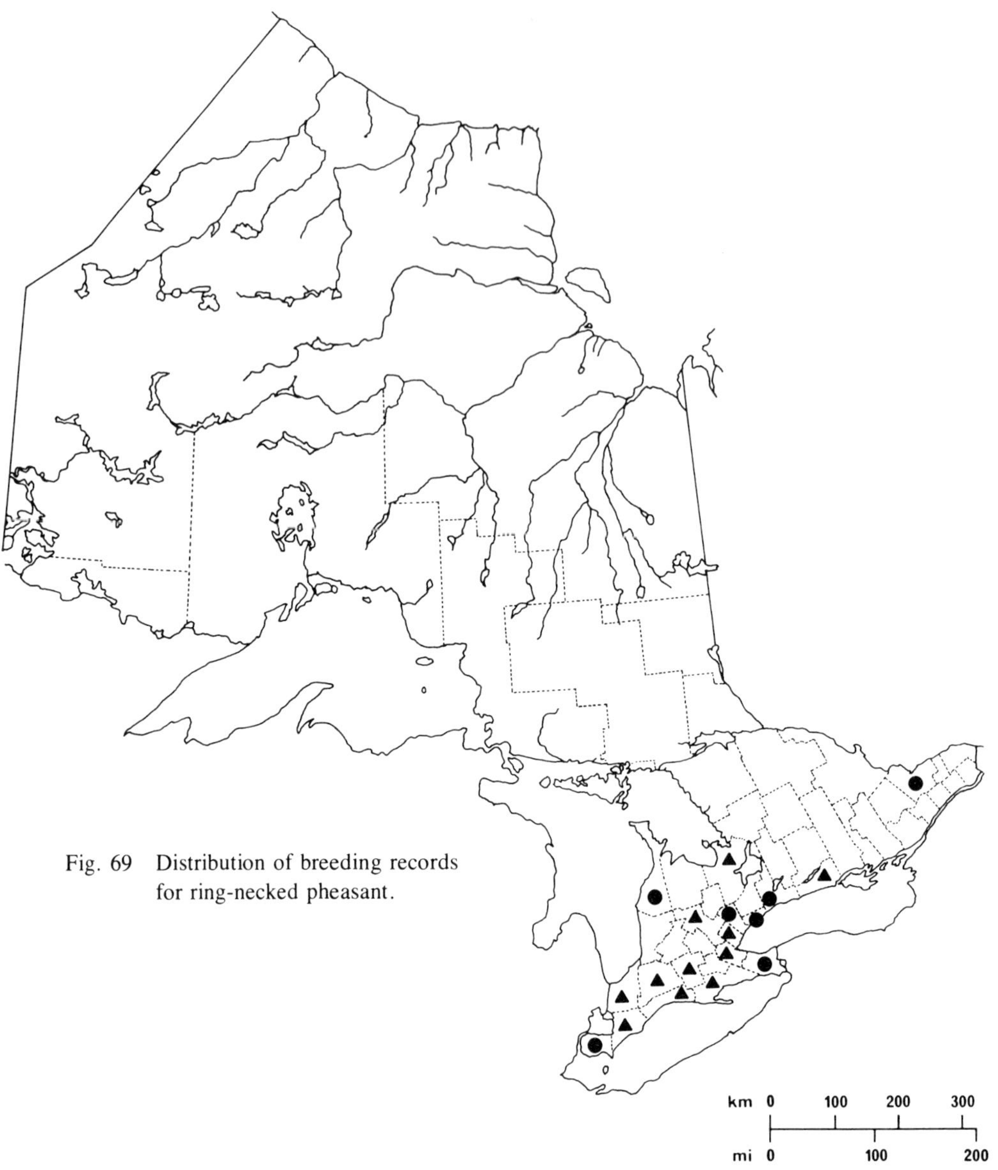

Fig. 69 Distribution of breeding records for ring-necked pheasant.

Ring-necked Pheasant, *Phasianus colchicus* Linnaeus

Nidiology

RECORDS 127 (176 nests) representing 15 provincial regions.
This Asiatic species was introduced and is now widely established in southern Ontario. A number of local populations are maintained for sport hunting.

It breeds in agricultural areas (Fig. 151) in grass and grain fields, and orchards and vineyards; in marshes; in mixed and deciduous woods; in wooded swamps and willow swales; in gardens and parks in residential areas; in rail and road cuts and in drainage ditches; in abandoned quarries; and in sand-beach areas.

Nests were located in tall grasses, grains, weeds, rushes, and ferns; in vines and shrubs; and under trees, saplings, bushes, brush piles, and logs. One nest was on a muskrat house and another under a basket in a greenhouse. Nests were frequently near water and 2 nests were 4 and 10 cm (1.5 and 4 inches) above water.

Nests were hollows or depressions on the ground and formed of grasses, rushes, grain and weed stalks, sticks, and twigs. Most nests were variously lined with grasses, weeds, rushes, feathers and down, leaves, and pine needles. Some nests were canopied by grasses, weeds, or grapevines. Outside diameter of 1 nest was 15 by 20.5 cm (6 by 8 inches); inside depth of another was 7.5 cm (3 inches).

EGGS 116 nests with 1 to 36 eggs; 1E (4N), 2E (1N), 3E (1N), 4E (1N), 5E (6N), 6E (4N), **7E** (8N), **8E** (5N), **9E** (7N), **10E** (9N), **11E** (15N), **12E** (9N), **13E** (6N), **14E** (11N), **15E** (6N), **16E** (8N), 17E (3N), 18E (1N), 19E (2N), 21E (2N), 23E (1N), 24E (1N), 25E (1N), 26E (2N), 28E (1N), 36E (1N).
Average clutch range 8 to 14 eggs (62 nests).
Large sets (21 to 36 eggs) were probably "dump" nests. Eggs were occasionally laid in other species' nests: 2 mallard nests contained pheasant eggs; a blue-winged teal nest with 8 eggs had 2 pheasant eggs; 2 ruffed grouse nests with 10 and 9 eggs also contained 1 and 3 pheasant eggs respectively.

INCUBATION PERIOD 1 nest, ca 21 to 22 days.

EGG DATES 122 nests, 21 April to 28 August (135 dates); 61 nests, 13 May to 27 May.

Breeding Distribution

This hardy pheasant has been widely introduced into North America, and was first introduced into Ontario in the 1890s. Apparently natural populations sustained by wild breeding birds are found only in the Deciduous Forest region of southern Ontario. North of this area deep snows make survival difficult. Extensive releases of captive-bred birds maintain small populations as far north as Sault Ste Marie and Ottawa in the east and the Thunder Bay area in the west.

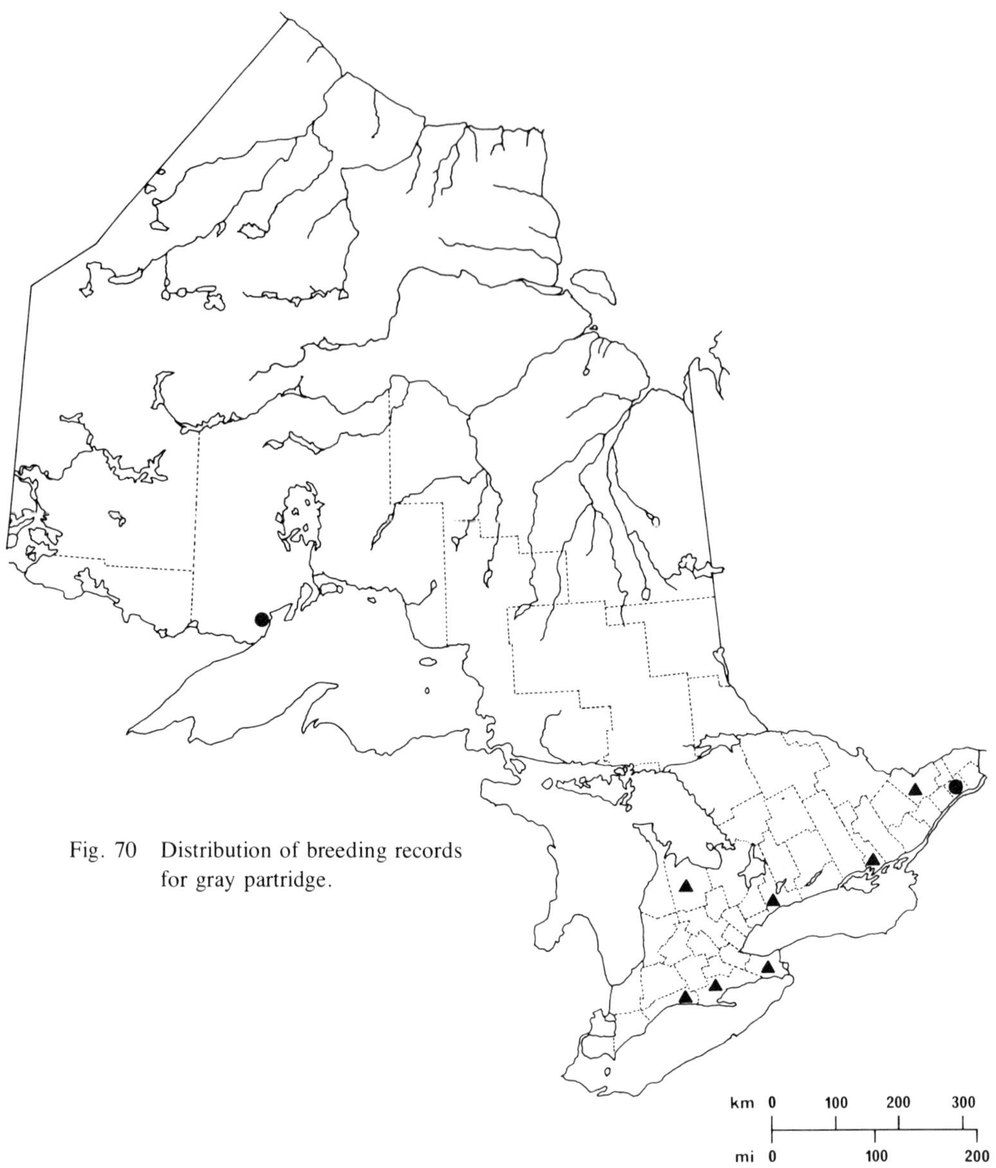

Fig. 70 Distribution of breeding records for gray partridge.

Gray Partridge, *Perdix perdix* (Linnaeus)

Nidiology

RECORDS 7 nests representing 7 provincial regions.
A Eurasian species that was introduced into Ontario, this partridge breeds in grassland areas. One nest was on a grassy hillside with a few pine and locust trees present and another was by a roadside fence.

Nests were depressions in the ground among heavy grass cover. They were woven structures of dried grasses and some contained leaves. At least 1 nest had a covering of dried grasses that formed a canopy over the nest. Outside diameter of 1 nest was 33 cm (13 inches); inside diameters of 2 nests were 15 cm (6 inches); inside depth of 1 nest was 7.5 cm (3 inches).

EGGS 7 nests with 12 to 25 eggs; **12E** (1N), **13E** (1N), **20E** (1N), **21E** (1N), **24E** (2N), 25E (1N).
Average clutch range 13 to 24 eggs (5 nests).
All the eggs in 1 set of 24 hatched successfully. In 1 nest an egg was laid each day.

INCUBATION PERIOD No information.

EGG DATES 5 nests, 7 May to 7 September (7 dates); 3 nests, 14 June to 19 June.

Breeding Distribution

The gray partridge is a native of Europe and Asia, where it is a widespread species. Between 1909 and 1938 this partridge was released in many counties of southern Ontario and several districts in central Ontario. Many of these early attempts to establish the species were failures, but others were locally successful. Today the species is found at widely scattered localities in southern Ontario. No recent evidence is available to indicate its continued presence in the northern part of the province where populations existed for some years at Sault Ste Marie, Haileybury, and Thunder Bay.

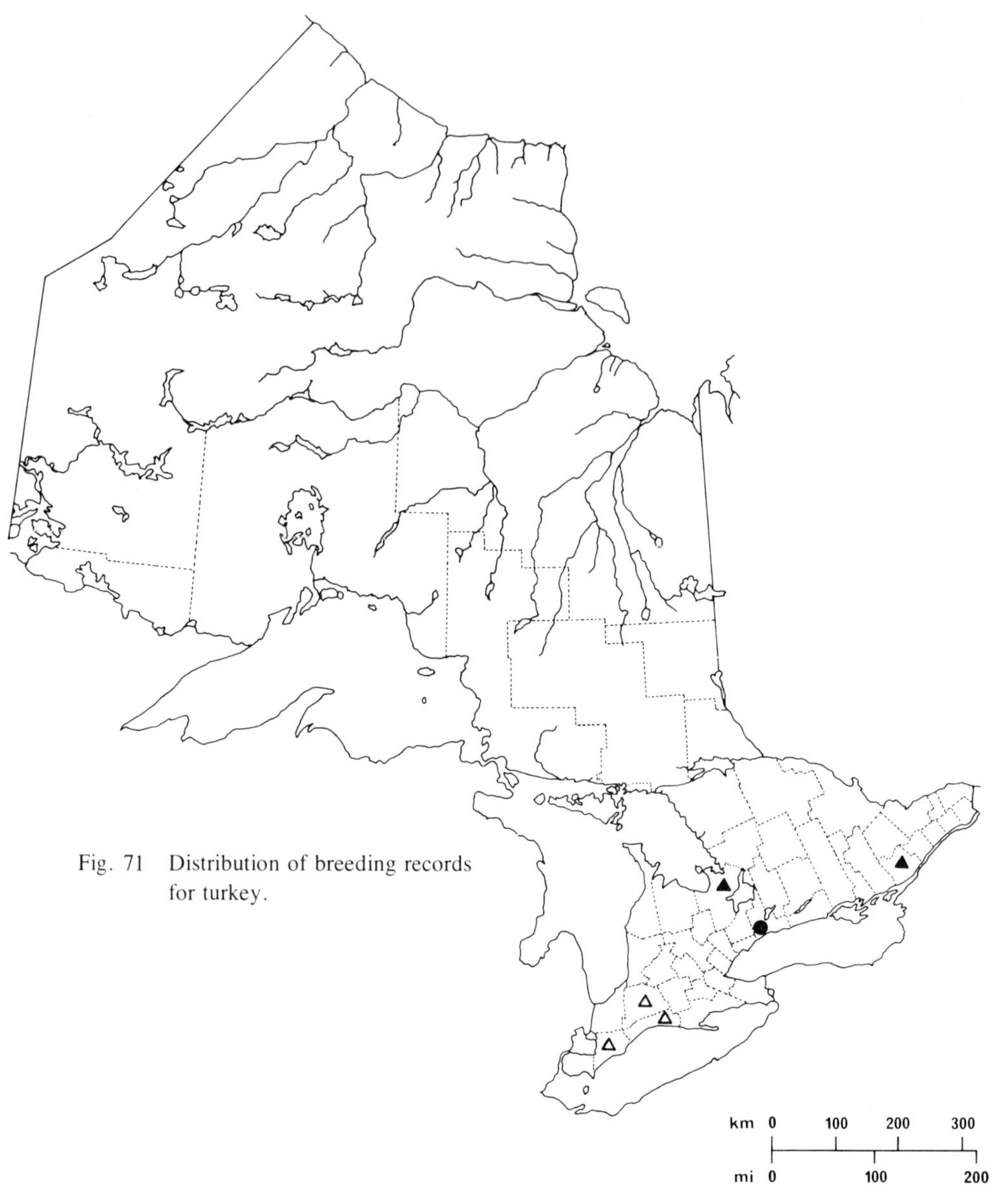

Fig. 71 Distribution of breeding records for turkey.

Turkey, *Meleagris gallopavo* Linnaeus

Nidiology

RECORDS 3 nests representing 3 provincial regions.
Before 1900, the turkey was reported to be a more or less common breeding species in Ontario and owed its place on the breeding list to just such hearsay evidence (Baillie and Harrington, 1936). Two early records are presumed to be authentic Ontario nests: a Kent County record of 1 egg (ROM 11960) collected in the late 1800s, and a set of 13 eggs from Delaware Township, Middlesex County, collected in 1878 (Saunders and Dale, 1933). However, by the turn of the century the turkey had been extirpated from the province. Later, a number of introductions took place (in Halton, Lambton, Leeds, Ontario, and other counties), and as a result of 1 of these introductions, a nest with 12 eggs was found on 25 July 1965, in Lot 11, Conc. 6, Pickering Township, Durham RM (Tozer and Richards, 1974). On 11 August 1965, an infertile egg was collected from this nest and it is now in the possession of J. M. Richards, R.R. 2, Orono, Ontario. This egg is the only existing and reliable material evidence of breeding by free-ranging turkeys in Ontario.

Breeding Distribution

Almost nothing is known of the nesting of the turkey in Ontario. A few birds may have survived the first decade of the 20th century in southern portions of their former range, which extended north as far as Georgian Bay and east as far as Durham RM (Alison, 1976). Introductions provided the first reliable material evidence of breeding in 1965 (Peck, 1976). Several other introductions, notably on Hill Island, Leeds County, have resulted in small local groups, but all are maintained through the winter by artificial feeding.

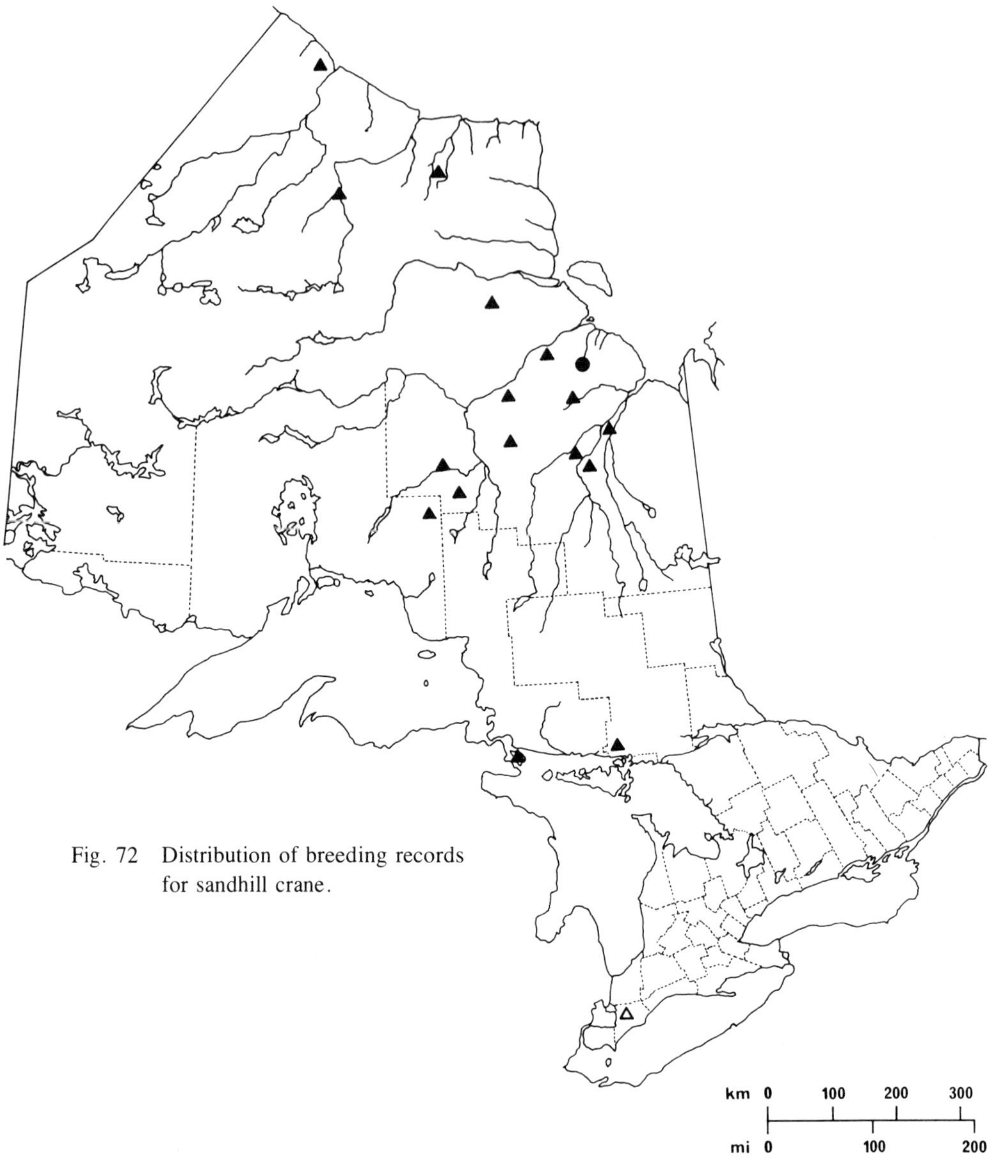

Fig. 72 Distribution of breeding records for sandhill crane.

Sandhill Crane, *Grus canadensis* (Linnaeus)

Nidiology

RECORDS 2 nests representing 1 provincial region.
Although summering adults and adults with young have been observed in several northern districts (Algoma, Cochrane, Kenora, Rainy River, Sudbury, and Thunder Bay) for more than 20 years, to date only 2 documented nests have been recorded, both at Kinoje Lake, Cochrane District.

These 2 nests were in wet areas of muskeg, which supported growths of sphagnum moss, heath (leatherleaf, Labrador tea), stunted black spruce, and tamarack.

One nest was located 9 m (30 ft) from the edge of a pond, and the other was on an islet in a string bog.

One nest, placed on a sphagnum moss substrate, consisted merely of a small collection of tamarack twigs, not numerous enough to cover the bottom of the nest. The other nest was on a pad of sphagnum moss in a stand of leatherleaf and Labrador tea.

EGGS 2 nests, each with 2 eggs.

INCUBATION PERIOD 1 nest, ca 28 days (27 days in an incubator).

EGG DATES 2 nests, 7 May to 9 May and 29 May to 4 June.

Breeding Distribution

Historical reports indicate that the sandhill crane (Fig. 164B) formerly bred in the Lake St. Clair marshes (McIlwraith, 1886). Early writings on the Hudson Bay area also indicate a regular occurrence of this crane in far northern Ontario. However, the overall decline of the species throughout North America in the 19th and early 20th centuries may have all but extirpated it from the province. In 1960 only four recent sightings in northern Ontario were known, and the first nest was not located until 1969 (Lumsden, 1971). The summary of observations reported by Lumsden (1971) indicates that this bird probably still breeds sparsely throughout the Hudson Bay Lowland. Additional undocumented breeding reports have recently come from the area between Sault Ste Marie and Lake Superior Provincial Park and from Manitoulin and southern Sudbury districts. Regular summering also occurs in the western Rainy River District. A continuing increase in populations of sandhill crane makes it possible that most of northern Ontario will some day harbour scattered groups of this species.

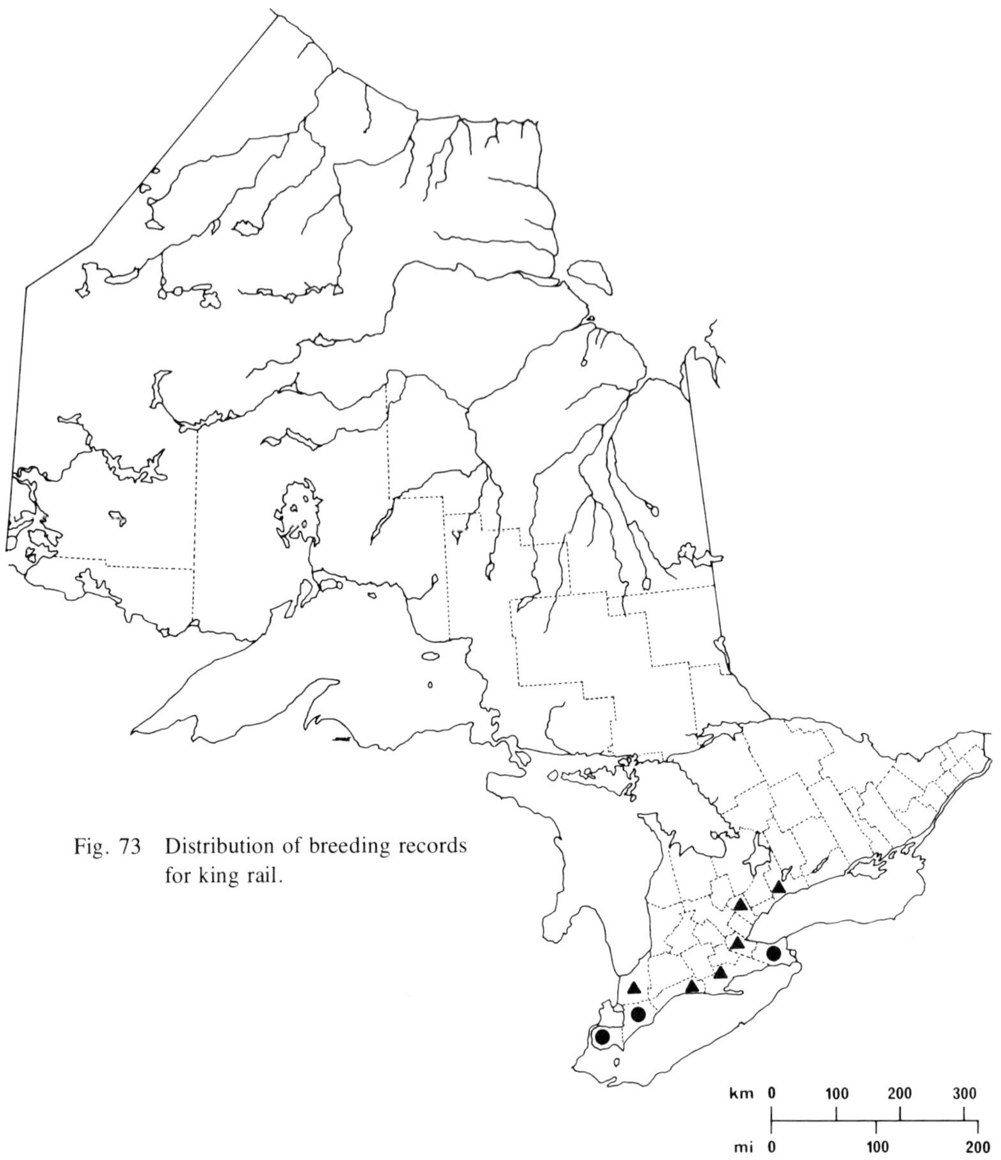

Fig. 73 Distribution of breeding records for king rail.

King Rail, *Rallus elegans* Audubon

Nidiology

RECORDS 12 nests representing 5 provincial regions.
Breeds in freshwater marshes of cattail, bulrush, and sedge, and in the marshy borders of lakes, ponds, rivers, and creeks.

Nests were located near water on ground that ranged from very wet to almost dry. They were positioned in cattails, bulrushes, and marsh ''grasses'' (the latter were green and 1.1 m [3.5 ft] in height). Heights above ground of 2 nests were 10 cm (4 inches) and 46 cm (18 inches).

Nests were made of dry, coarse rushes, and grasses. One nest was attached to live grass stalks. Nests were slightly cupped structures lined with finer rushes and grasses. One pair used a previous year's nest. Outside diameter of 1 nest was 28 cm (11 inches); inside diameters of 2 nests ranged from 16.5 to 25 cm (6.5 to 10 inches).

EGGS 12 nests with 2 to 13 eggs; 2E (1N), 6E (1N), **8E** (3N), **10E** (4N), **11E** (2N), 13E (1N).
Average clutch range 8 to 10 eggs (7 nests).

INCUBATION PERIOD No information. Incubation commenced before clutches were complete.

EGG DATES 11 nests, 18 May to 17 July (12 dates); 6 nests, 24 May to 26 June.

Breeding Distribution

The king rail is very thinly scattered in the Deciduous Forest region of southern Ontario. A few large marshes along Lake Erie and Lake St. Clair (Fig. 147) probably have regular breeding populations; elsewhere the species nests sporadically.

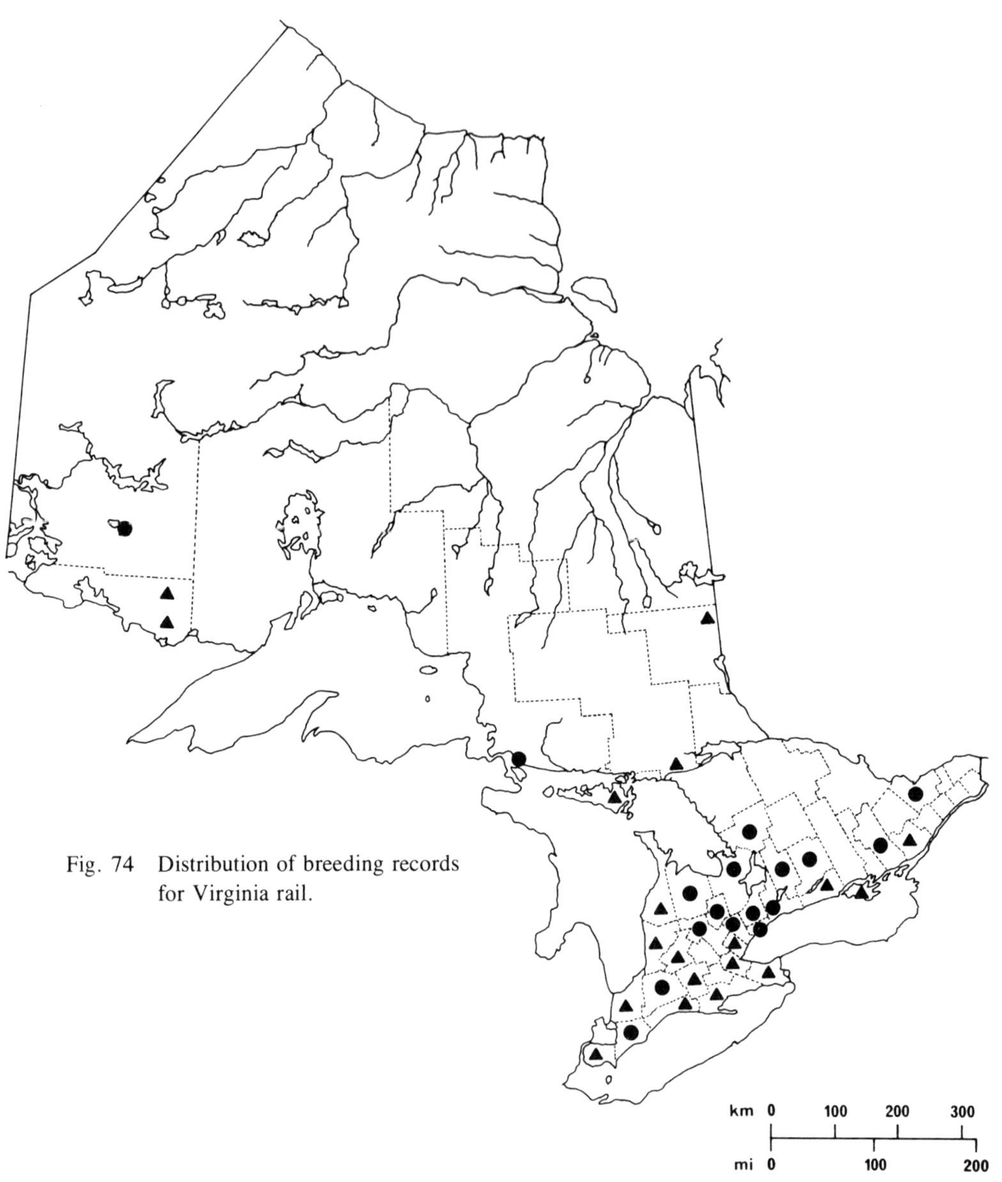

Fig. 74 Distribution of breeding records for Virginia rail.

Virginia Rail, *Rallus limicola* Vieillot

Nidiology

RECORDS 156 nests representing 27 provincial regions.
Breeds in freshwater cattail, bulrush, and sedge marshes; on marshy shores of lakes, ponds, and rivers; on floating islets of marsh vegetation and on grassy islands in lakes; in heath bogs and cedar and deciduous swamps; in drainage ditches in road and rail cuts; and in agricultural areas in willow swales, grassy fields, and wet meadows.

Most nests were in or near the borders of wet areas. They were usually situated in clumps of dead, and occasionally living, vegetation (grasses, sedges, cattails, bulrushes, bur-reeds, and ferns); often at the bases of clumps of willow, alder, and dogwood; occasionally in open areas of short grass; and on dead branches (2 records). Nests were often placed on hummocks of grass, sedge, fern, and moss. Positions of nests ranged from dry-land sites 9 m (30 ft) or more from water, to sites over water 1.5 to 4.5 m (5 to 15 ft) from open water. In wet situations, heights of 44 nests ranged from water level to 61 cm (24 inches) above water, with 22 averaging 10 to 23 cm (4 to 9 inches) above water. Water depths below nests ranged from 5 to 30.5 cm (2 to 12 inches). In dry situations (12 records) nests were usually on the ground, but some were elevated to a height of 30.5 cm (12 inches).

Nests were almost invariably woven cups or platforms, which were occasionally flat or only slightly cupped; a few nests were mere depressions in standing vegetation. Nest structures were woven of grasses, sedges, cattails, bulrushes, and weeds, and were sometimes unlined but usually lined with pieces of dead cattail, fine grasses, sedges, leaves, and other marsh vegetation. Some nests were well canopied with grasses, cattails, and sedges; others had scanty canopies or none at all. One record described a ramp of cattail stalks leading to a nest that was 45.5 cm (18 inches) above water. Outside diameters of 2 nests were 18 and 20.5 cm (7 and 8 inches); outside depth of 1 nest was 12.5 cm (5 inches).

EGGS 143 nests with 1 to 14 eggs; 1E (8N), 2E (4N), 3E (2N), 4E (10N), 5E (8N), 6E (10N), **7E** (10N), **8E** (20N), **9E** (39N), **10E** (20N), **11E** (9N), 12E (1N), 13E (1N), 14E (1N).
Average clutch range 7 to 10 eggs (89 nests).
One nest containing 8 eggs of Virginia rail also contained 1 egg of brown-headed cowbird.

INCUBATION PERIOD 4 nests, 16 to 18 days. Incubation commenced before clutches were complete.

EGG DATES 149 nests, 4 May to 19 July (187 dates); 75 nests, 27 May to 7 June.

Breeding Distribution

The Virginia rail breeds throughout southern Ontario but largely avoids the Canadian Shield, where suitable nesting habitat is scarce. In northern Ontario it is very thinly scattered as far north as Kenora in the west and Lake Abitibi in the east.

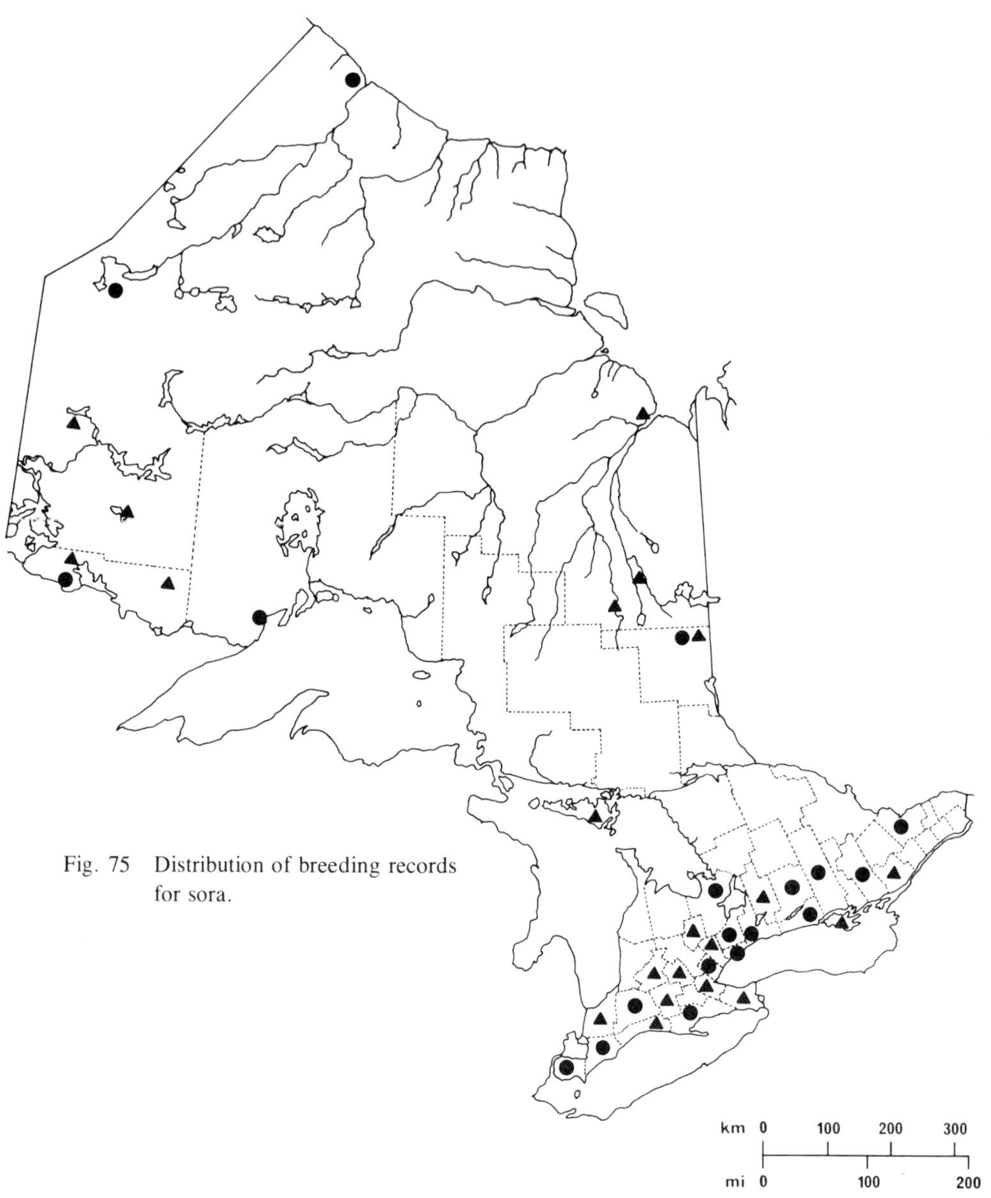

Fig. 75 Distribution of breeding records for sora.

Sora, *Porzana carolina* (Linnaeus)

Nidiology

RECORDS 168 nests representing 30 provincial regions.
Breeds in large and small freshwater marshes of cattail, bulrush, and sedge; in marshy borders of lakes, ponds, rivers, and creeks; in drainage ditches alongside road and rail cuts; in wet fields and pastures; and in swamps, beaver ponds and meadows, willow swales, and once in a cranberry bog. This species preferred a wetter habitat and nested more often in cattails than did Virginia rail.

Nests were usually in wet situations and were positioned in clumps of dead, and occasionally living, cattail, bulrush, sedge, and grass; at the bases of clumps of willow, alder, and other bushes; and sometimes on hummocks of sedge and grass. Nest situations ranged from on solid ground 12 m (40 ft) from water to over water 46 m (150 ft) from shore. One record described the use of an old red-winged blackbird nest. In wet situations heights of 62 nests ranged from water level to 45 cm (18 inches), with 31 averaging from water level to 15 cm (6 inches). Water depths below 26 nests ranged from 8 to 120 cm (3 to 48 inches), with 13 averaging 30 to 60 cm (12 to 24 inches). In dry situations (5 records) heights of nests ranged from ground level to 76 cm (30 inches), the latter being the height of a nest in a bush.

Nests were woven cups or platforms of cattail, grass, sedge, weeds, and bulrush, usually with shallow interiors. They were occasionally unlined, but more often lined with dead pieces of cattail, fine grasses, weed stalks, and leaves. Nests were often attached to and supported by standing vegetation and were commonly canopied with dead or living vegetation. Occasionally nests had ramps leading to them. Outside diameters of 4 nests ranged from 20 to 25 cm (8 to 10 inches); outside depths of 2 nests were 15 and 20 cm (6 and 8 inches); inside diameters of 3 nests ranged from 10 to 15 cm (4 to 6 inches); inside depth of 1 nest was 5 cm (2 inches).

EGGS 162 nests with 1 to 17 eggs; 1E (9N), 2E (6N), 3E (5N), 4E (6N), 5E (5N), 6E (9N), **7E** (12N), **8E** (18N), **9E** (18N), **10E** (22N), **11E** (17N), **12E** (16N), 13E (8N), 14E (7N), 15E (1N), 16E (2N), 17E (1N).
Average clutch range 7 to 11 eggs (87 nests).

INCUBATION PERIOD 5 nests: 1 of 18 to 19 days, 4 of ca 20 to 21 days. Because this species commenced incubation soon after the clutch was initiated, eggs in various stages of incubation were sometimes noted together with hatching young.

EGG DATES 155 nests, 4 May to 2 August (189 dates); 78 nests, 29 May to 7 June.

Breeding Distribution

The sora breeds throughout the agricultural south but becomes sparsely distributed in the southern Canadian Shield and throughout the northern part of the province, perhaps as far north as the Hudson Bay coast.

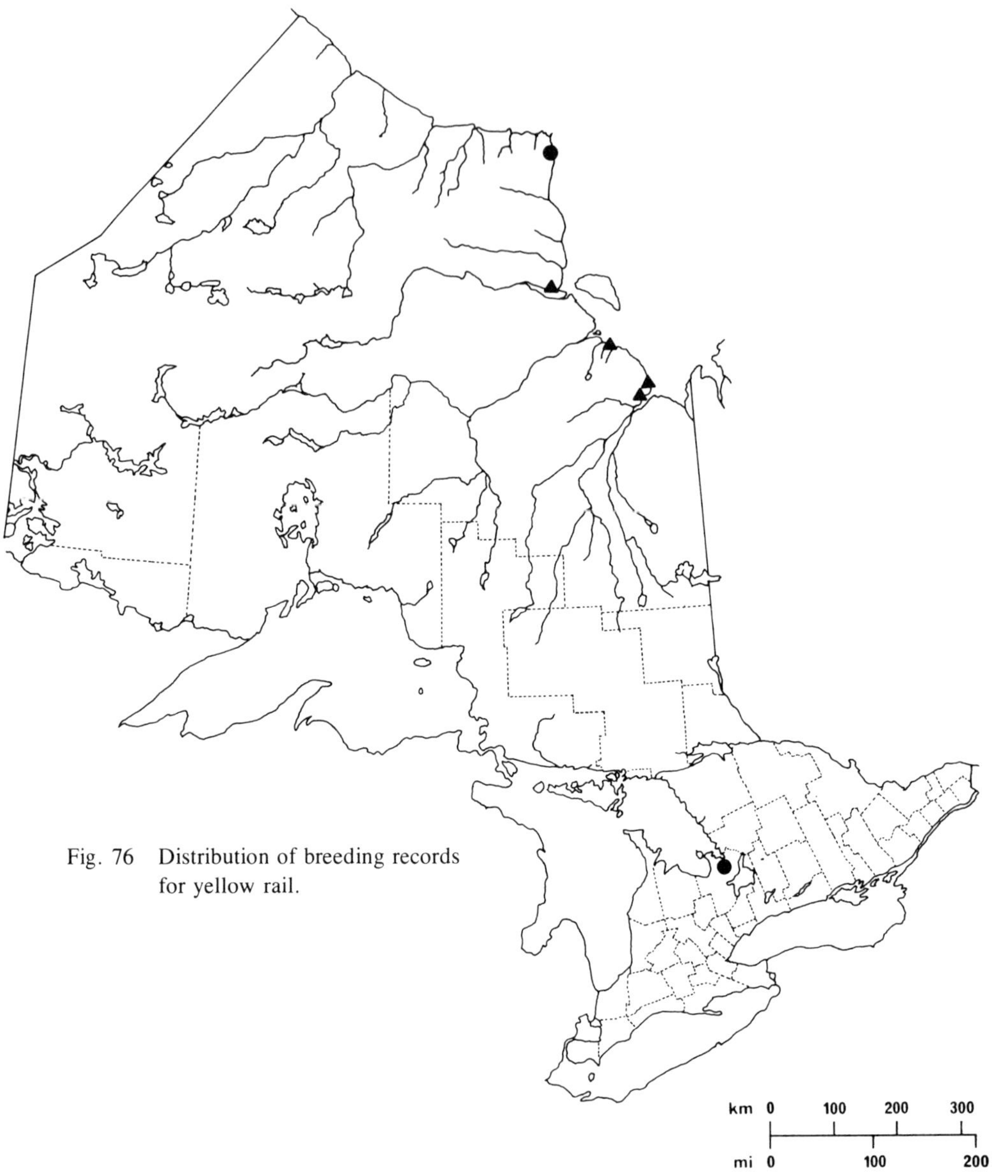

Fig. 76 Distribution of breeding records for yellow rail.

Yellow Rail, *Coturnicops noveboracensis (Gmelin)*

Nidiology

RECORDS 3 nests representing 3 provincial regions.
This rarely observed rail undoubtedly is a more common provincial breeding species than is suggested by the 3 known nest records (Cochrane and Kenora districts and York County) (Devitt, 1939, Elliot and Morrison, 1979). A fourth record described the finding of hatched egg shells on 8 July 1948, Kenora District (Peck, 1972).

It breeds in freshwater marshes, which support a fairly heavy cover of sedges, rushes, bulrushes, horsetails, grasses, and other marsh vegetation.

All 3 known nests were positioned in grasses and sedges under dead, standing vegetation, which formed a canopy over at least 1 nest and hid it completely. One nest was supported at a height of 5 cm (2 inches) above water that was 5 to 10 cm (2 to 4 inches) deep. The base of a second nest was just resting on the ground, and its rim was 7.5 cm (3 inches) above ground.

One nest was composed of fine blades of dead marsh grasses, coiled in the form of a cup. Its outside diameter was 9.5 cm (3.75 inches); outside depth, 8 cm (3 inches); inside diameter, 7 cm (2.75 inches); inside depth, 6 cm (2.4 inches). Inside diameter of a second nest was also 7 cm (2.75 inches).

EGGS 3 nests with ca 6 eggs, 7 eggs, 10 eggs.

INCUBATION PERIOD 1 nest, ca 17 to 18 days.

EGG DATES 3 nests, 12 June, 12 June to 30 June, 20 June (nest found deserted).

Breeding Distribution

The only area of Ontario where the yellow rail occurs regularly and in substantial numbers is in the north, along the west coast of James Bay and probably along the Hudson Bay coast (Fig. 165). Outside this region, the species has been recorded sporadically in summer at a few widely scattered marshes in southern Ontario. While potentially it may breed wherever suitable habitat exists, it appears to be absent from much of the province.

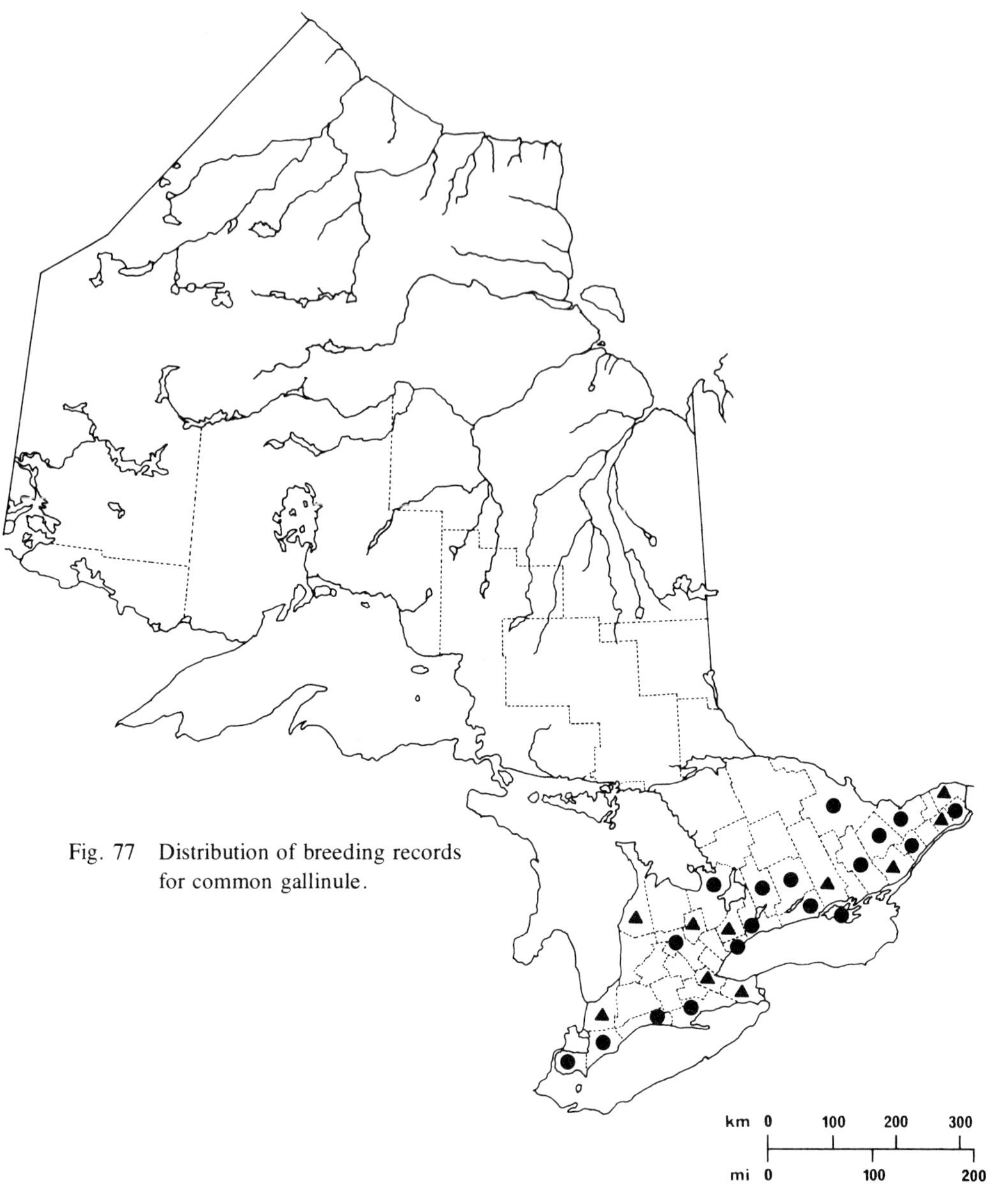

Fig. 77 Distribution of breeding records for common gallinule.

Common Gallinule, *Gallinula chloropus* (Linnaeus)

Nidiology

RECORDS 406 (413 nests) representing 23 provincial regions.
In Ontario this species usually breeds in freshwater cattail marshes, and often in marshes of bulrush, sedge, horsetail, bur-reed, sweet gale, arrowhead, and common reed grass, with associated growths of grass, loosestrife, leatherleaf, alder, and willow. Other breeding habitats include the marshy borders of lakes, rivers, ponds, sewage ponds, and willow swales and swamps.

Nests were invariably positioned over water or over wet masses of vegetation and were usually in stands of dead and living cattails and less often in grasses and sedges, bulrushes and bur-reed, willow and dogwood clumps, loosestrife, common reed grass, sweet gale, horsetail, and arrowhead. One nest was in the roots of a dead tree, and another was on a floating islet of mud and dead vegetation. Nest situations ranged from near shore to as far as 0.8 km (0.5 mi) out from shore, and from the edge of open water to as far as 9 m (30 ft) distant from it, in stands of vegetation. Water depths near 20 nests ranged from 20.5 to 91.5 cm (8 to 36 inches). Heights above water of 121 nests ranged from 2.5 to 91.5 cm (1 to 36 inches), with 61 averaging 12.5 to 30.5 cm (5 to 12 inches). The bottoms of many nests were close to or at water level.

Nests were flat to well-hollowed, bulky platforms and were loosely woven, usually of pieces of dead cattail and less often of grasses, sedges, weeds, and other dead marsh vegetation. They were occasionally lined with grasses and plant down. Many nest platforms were described as being floating, but they were maintained in position, anchored to living and dead emergent vegetation. Nests were usually well concealed and were often canopied by living and dead cattails and other marsh plants. Ramps of flattened vegetation led to many nests. Outside diameters of 16 nests ranged from 18 to 30.5 cm (7 to 12 inches); outside depths of 10 nests ranged from 10 to 30.5 cm (4 to 12 inches); inside diameters of 4 nests ranged from 7.5 to 20.5 cm (3 to 8 inches); inside depths of 3 nests ranged from 2.5 to 4 cm (1 to 1.5 inches).

EGGS 387 nests with 1 to 18 eggs; 1E (13N), 2E (15N), 3E (19N), 4E (18N), 5E (26N), **6E** (29N), **7E** (52N), **8E** (68N), **9E** (69N), **10E** (39N), 11E (15N), 12E (13N), 13E (5N), 14E (1N), 15E (3N), 18E (2N).
Average clutch range 6 to 9 eggs (218 nests).

INCUBATION PERIOD 5 nests: 1 of at least 19 days, 1 of ca 19 to 22 days, 1 of ca 21 days, 1 of ca 22 days, 1 of at least 22 days. Incubation commenced before clutches were complete. There were 3 records of second broods. One nest with 7 eggs of gallinule also contained 2 eggs of coot; another nest with 6 eggs of gallinule also contained 1 egg of coot.

EGG DATES 401 nests, 4 May to 21 July (453 dates); 201 nests, 31 May to 15 June.

Breeding Distribution

The common gallinule breeds throughout the agricultural part of southern Ontario, but is less common in the Canadian Shield where there are few marshes to serve as nesting habitat.

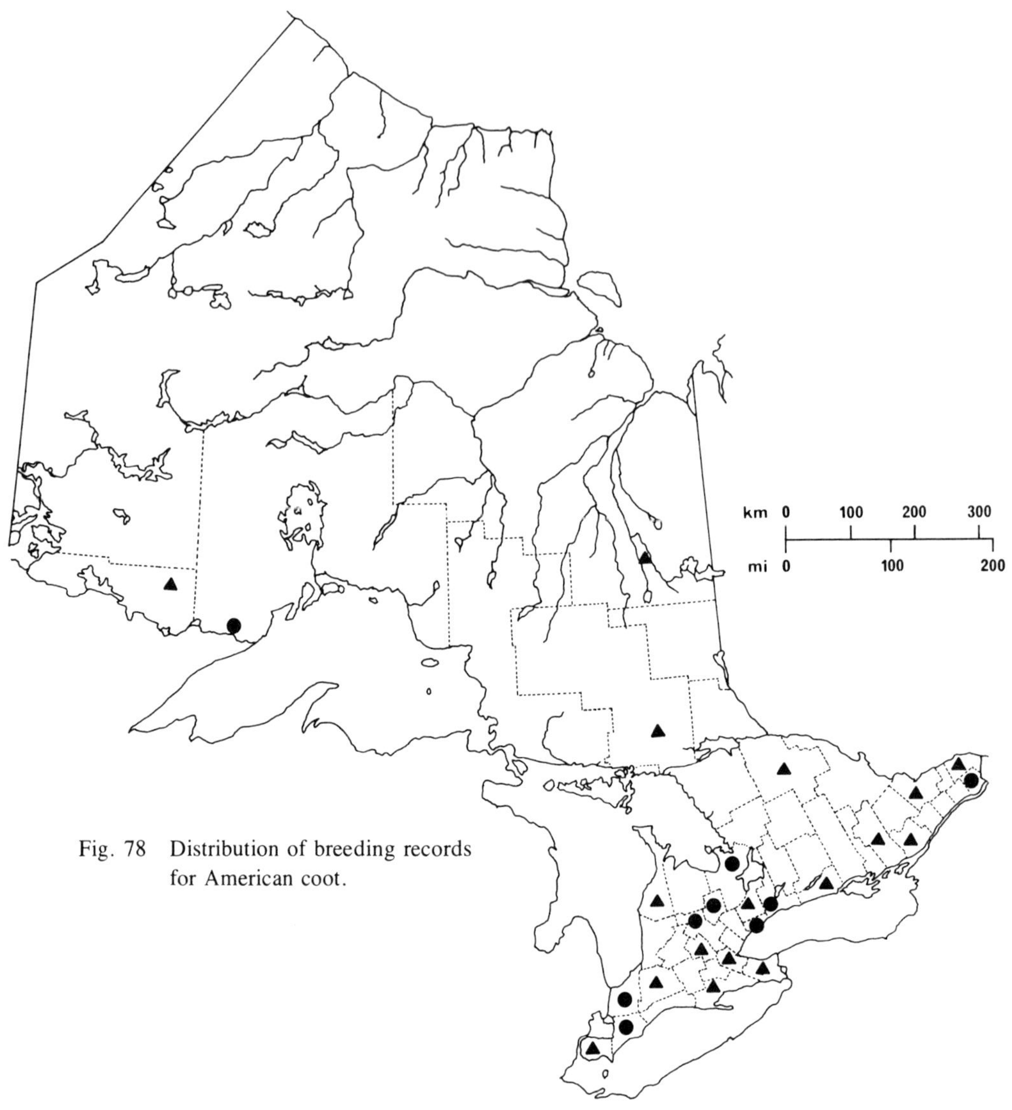

Fig. 78 Distribution of breeding records for American coot.

American Coot, *Fulica americana* Gmelin

Nidiology

RECORDS 376 (387 nests) representing 16 provincial regions.

The breeding habitats and nest sites of American coot are similar to those of common gallinule. Coots breed most often in freshwater cattail marshes and less frequently in stands of bulrush, sedge, bur-reed, and common reed grass, and in willow swales and bogs, with associated vegetation of grasses, nightshade, horsetails, heaths, and alder. Other breeding habitats include marshy shores of lakes, rivers, sewage ponds, ponds, and canals.

With 1 exception, nests were positioned over water, usually in stands of dead and living cattails and less often in bulrushes, sedges, bur-reeds, horsetails, common reed grasses, and other emergent aquatic vegetation. Three nests were attached to willow branches protruding from the water; 2 nests were in clumps of nightshade; 1 nest was attached to a water lily; 1 nest was in a clump of speckled alder; 1 nest was in a heath plant. Occasionally nests were recorded on floating islets of mud and vegetation. Coots built floating or semi-floating nests more frequently than did gallinules. Almost all nests were anchored to emergent vegetation. Nests were variously situated from 3.5 to 400 m (12 to 1320 ft) distant from shore, and were sometimes in open water but usually in stands of marsh vegetation from 1.5 to 30.5 m (5 to 100 ft) distant from open water. Water depths near 113 nests ranged from 10 to 152.5 cm (4 to 60 inches). Heights above water of 82 nests ranged from 5 to 61 cm (2 to 24 inches), with 41 averaging 9 to 30.5 cm (3.5 to 12 inches). The bottoms of most nests were at or below water level. Open-water nests tended to be built up higher than nests in dense growths of vegetation.

Nests were flat-topped to well-hollowed bulky mounds, loosely constructed of pieces of dead cattail leaves and stalks and less often of bulrushes, sedges, grasses, and other dead marsh vegetation. Some nests contained mud; some were lined with reeds and finer material, grasses, and green cattails. Some nests were canopied by nearby standing vegetation; many nests had ramps of dead vegetation leading to them. Outside diameters of 5 nests ranged from 25.5 to 61 cm (10 to 24 inches); outside depths of 9 nests ranged from 10 to 25.5 cm (4 to 10 inches); inside diameters of 2 nests were 18 and 30.5 cm (7 and 12 inches); inside depth of 1 nest was 5 cm (2 inches).

EGGS 359 nests with 1 to 18 eggs; 1E (19N), 2E (16N), 3E (21N), 4E (14N), **5E** (29N), **6E** (30N), **7E** (53N), **8E** (66N), **9E** (55N), **10E** (28N), **11E** (11N), 12E (11N), 13E (3N), 14E (2N), 15E (1N), 16E (1N), 18E (1N).
Average clutch range 5 to 9 eggs (233 nests).

INCUBATION PERIOD 3 nests: 1 of at least 21 days, 1 of more than 22 days, 1 of ca 24 days. Eggs were laid on consecutive days and incubation commenced before clutches were complete. Coot eggs were laid in 2 nests of common gallinule.

EGG DATES 373 nests, 4 May to 16 July (415 dates); 187 nests, 30 May to 16 June.

Breeding Distribution

The range of the American coot is limited more by lack of suitable habitat than by northern latitude. It breeds throughout the agricultural area of southern Ontario but is scarce in the Canadian Shield. In northern Ontario there are only a few breeding records available from the southern part of the region; the species is probably absent from most of the northern part of the province, yet in Manitoba it is known to breed as far north as Hudson Bay coast, at Churchill.

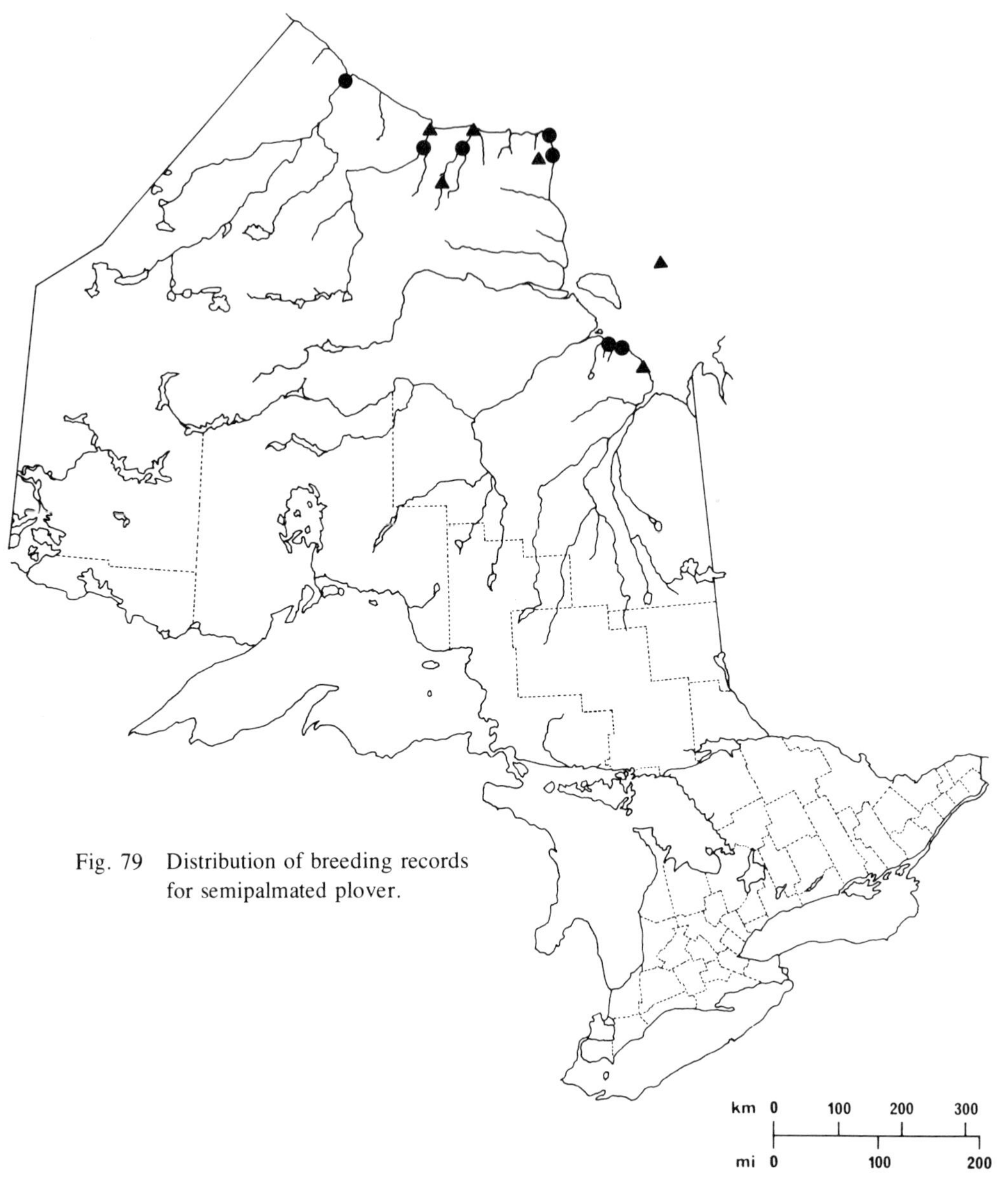

Fig. 79 Distribution of breeding records for semipalmated plover.

Semipalmated Plover, *Charadrius semipalmatus* Bonaparte

Nidiology

RECORDS 23 nests representing 2 provincial regions (and an adjacent island in James Bay, NWT).

Breeds in subarctic and boreal regions near the James Bay and Hudson Bay coasts. One breeding record was on the shore of a large island in James Bay, near Akimiski Island (NWT).

Nest sites selected were usually gravel and sometimes sandy areas of tidal flats, shores or rivers and islands, tundra ridges, and old beach lines; a gravel airstrip in an area of heath tundra was used for 4 nests.

Nests were slight depressions in open areas of gravel or sand, in sparsely vegetated areas, in mats of tundra vegetation, and once beneath a dwarf willow that was 15 cm (6 inches) high. Some nests had no lining; others were sparsely lined with dead grasses, sedges, leaves of willow and various heaths, and tiny twigs.

EGGS 22 nests with 1 to 4 eggs; 1E (1N), 2E (1N), **3E** (4N), **4E** (16N).
Average clutch range 4 eggs (16 nests).

INCUBATION PERIOD No information.

EGG DATES 23 nests, 5 June to 30 July (30 dates); 12 nests, 23 June to 30 June.

Breeding Distribution

The semipalmated plover breeds along the James Bay and Hudson Bay coasts and for some distance inland along the major rivers, where suitable habitat is available. Breeding was reported as early as 1904 (Macoun and Macoun, 1909) but not substantiated until 1947 when downy young were collected (Manning, 1952), and no nests were found until 1948 (Peck, 1972).

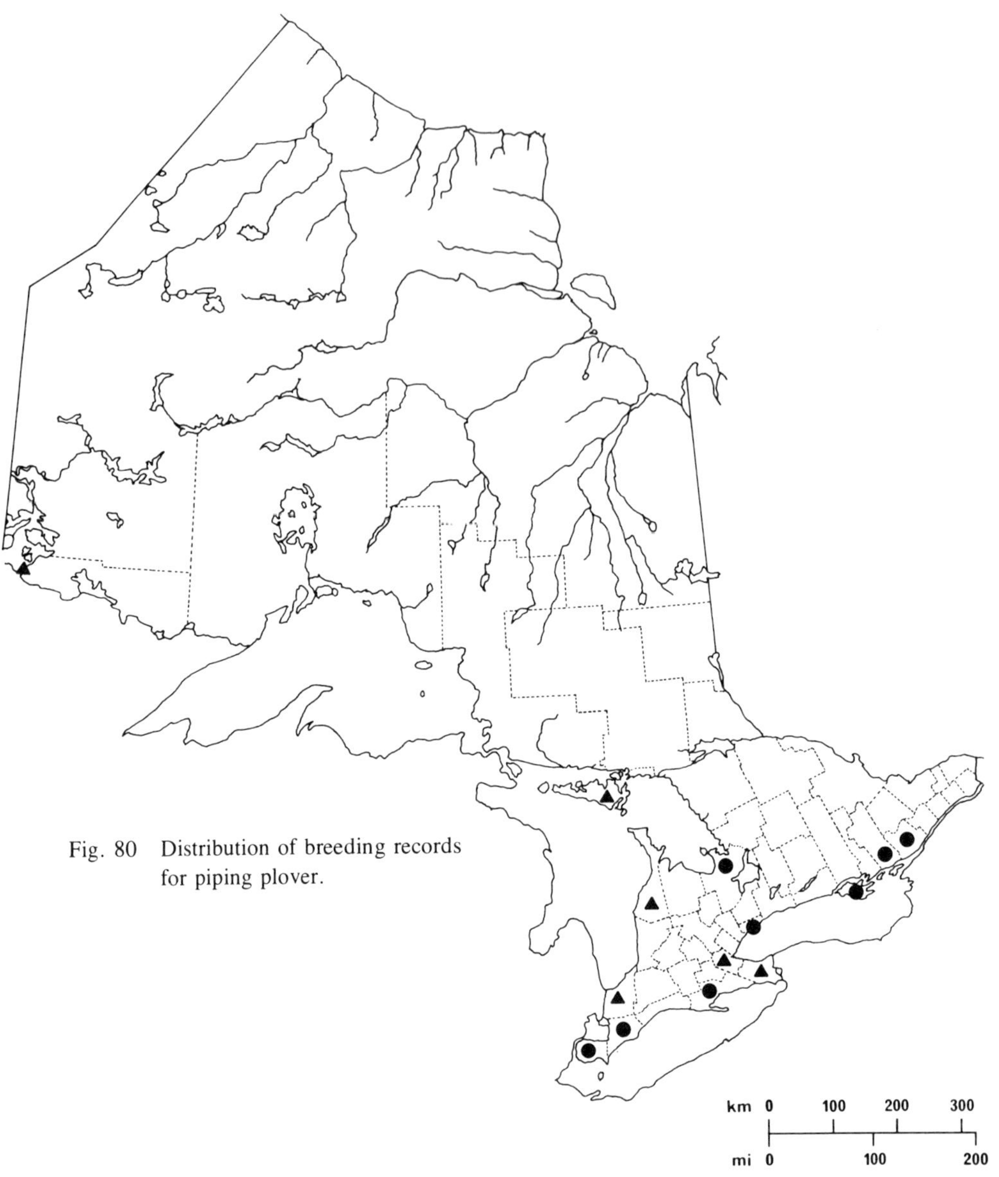

Fig. 80 Distribution of breeding records for piping plover.

Piping Plover, *Charadrius melodus* Ord

Nidiology

RECORDS 87 (88 nests) representing 11 provincial regions.
As sand beaches have succumbed to human pressures, this shorebird has disappeared from all but a few of its former nesting areas and appears to be in a precarious position at the present time.

Recorded breeding habitats were sand and gravel beaches on the shores of large lakes and rivers and on islands in lakes. Beaches were usually of sand with various amounts of gravel, and were sometimes sparsely vegetated with grass clumps and cottonwood saplings. Preferred nesting sites were large, flat expanses, but some sites had knolls and dunes nearby.

Nests were shallow scrapes or depressions in sand and gravel, and were in open situations or beside grass clumps, stones, and cottonwood and willow saplings. Nest locations ranged from 7.5 to 90 m (25 to 300 ft) from the shoreline. Some nests were unlined; some were lined with grasses and vegetation; most were lined with small, flat pebbles. Small clam-shell fragments were included in the lining of 1 nest.

EGGS 76 nests with 1 to 8 eggs; 1E (5N), 2E (3N), **3E** (11N), **4E** (54N), 5E (1N), 6E (1N), 8E (1N).
Average clutch range 4 eggs (54 nests).
Clutches in excess of 4 eggs may have been the result of laying by 2 or more females.

INCUBATION PERIOD 5 nests, 22 to 26 days.

EGG DATES 83 nests, 8 May to 29 July (122 dates); 41 nests, 28 May to 10 June.

Breeding Distribution

In recent years, nesting by the piping plover (Fig. 150A) has been almost entirely restricted to the relatively undisturbed beaches of Long Point on Lake Erie. It is also known to breed in at least one location at Lake of the Woods. However, in former years its range included the shores of lakes Ontario, Erie, and Huron, and parts of the St. Lawrence River. Although a few birds may persist, the species has virtually ceased to breed in the province.

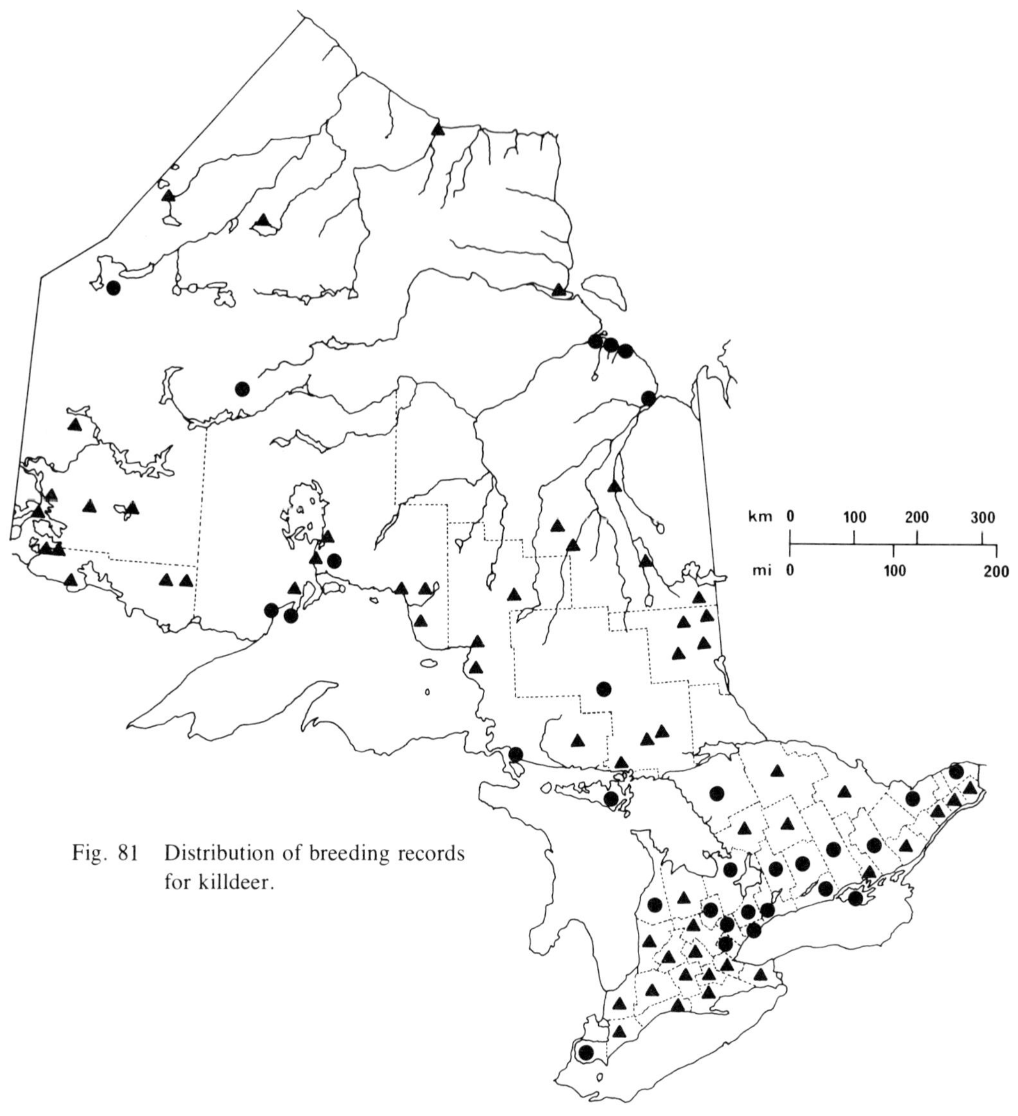

Fig. 81 Distribution of breeding records for killdeer.

Killdeer, *Charadrius vociferus* Linnaeus

Nidiology

RECORDS 946 (947 nests) representing 48 provincial regions.
Nest-habitat selection varies widely and includes, in order of preference, fields, pastures, and savannahs; sand dunes, sand and gravel beaches and shorelines; gravel roads, road shoulders, and railroad right-of-ways; lawns, schoolyards, nurseries, and residential and

market gardens; rock and gravel areas (including parking lots, gravel roofs, airstrips, and campgrounds); orchards and woodland clearings; gravel pits and quarries; mud and tidal flats; mine clearings; marshes; dykes; and barnyards.

Almost all nests were on the ground (4 nests were on gravel roofs from 3 to 9 m (10 to 30 ft) in height). Most nests were in open situations, but a few were partly concealed. Nests were usually on flat surfaces, but 13 nests were on small hummocks. Nests were usually on gravel or sand (370 nests) and occasionally on or among wood chips, sawdust, and burnt embers, and in burnt or decayed stumps (9 nests); in coal cinders (8 nests); in weeds or grass clumps (6 nests); under plants, rocks, bushes, and trees, and once under a hydro standard (6 nests); on cow chips or horse dung (5 nests); on rock surfaces and once on a horizontal tombstone (5 nests); in mosses and lichens (3 nests); on earthen furrows (2 nests); on marsh vegetation (1 nest was floating) (2 nests); on a mat of dead bracken fern (1 nest); and in dead leaves (1 nest). In 1 record, eggs were laid in a fallen nest of common grackle in a field. Two nests were within a few metres of each other; 2 other nests were 27.5 cm (90 ft) apart; another nest was near active nests of ring-billed gull.

Nests were almost invariably hollows or depressions, usually lined (402 nests) but occasionally unlined (38 nests); a few nests were on solid surfaces, such as gravel roofs and large rocks, and were unlined. Lining materials included, in order of preference, twigs, wood chips, and bark; pebbles and small stones; grasses and straw; weeds, plant stalks, skins and seeds of fruit, and buds; leaves; algae, lichens, and mosses; dead rushes; shell fragments; corn stalks; rootlets; and plastic and cigarette filters. Diameters of 7 nests ranged from 7 to 15 cm (2.8 to 6 inches), with 4 averaging 10 to 12.5 cm (3.9 to 5 inches); depths of 6 nests ranged from 2 to 4.5 cm (0.8 to 1.8 inches).

EGGS 365 nests with 1 to 7 eggs; 1E (4N)—1.1%, 2E (15N)—4.1%, **3E** (45N)—12.3%, **4E** (292N)—80%, **5E** (8N)—2.2%, 7E (1N)—0.3%.
Average clutch range 4 eggs (292 nests).
Six young hatched in the 7-egg nest. Some pairs were double brooded, and in some second nestings, clutches were laid in the original scrape. Second nestings and late nestings sometimes involved clutches of 3 or fewer eggs. In a few nests 1 egg in the clutch was infertile. Eggs were usually laid daily, but in 1 nest a 4-day gap occurred and incubation commenced before the last egg was laid, and in another nest at least 6 days elapsed between the third and fourth eggs.

INCUBATION PERIOD 28 nests, 24 to 30 days: 7 of 24 days, 10 of 25 days, 8 of 26 days, 2 of 27 days, 1 of 30 days. Although incubation usually commenced upon clutch completion, it was perhaps delayed in nests with longer incubation periods. Eggs remained in 1 nest for at least 33 days before hatching. Simultaneous hatching during a 24-hour period was usual, but hatching occasionally took several days.

EGG DATES 365 nests, 1 April to 21 August (567 dates); 183 nests, 18 May to 11 June.

Breeding Distribution

The killdeer breeds throughout the province.

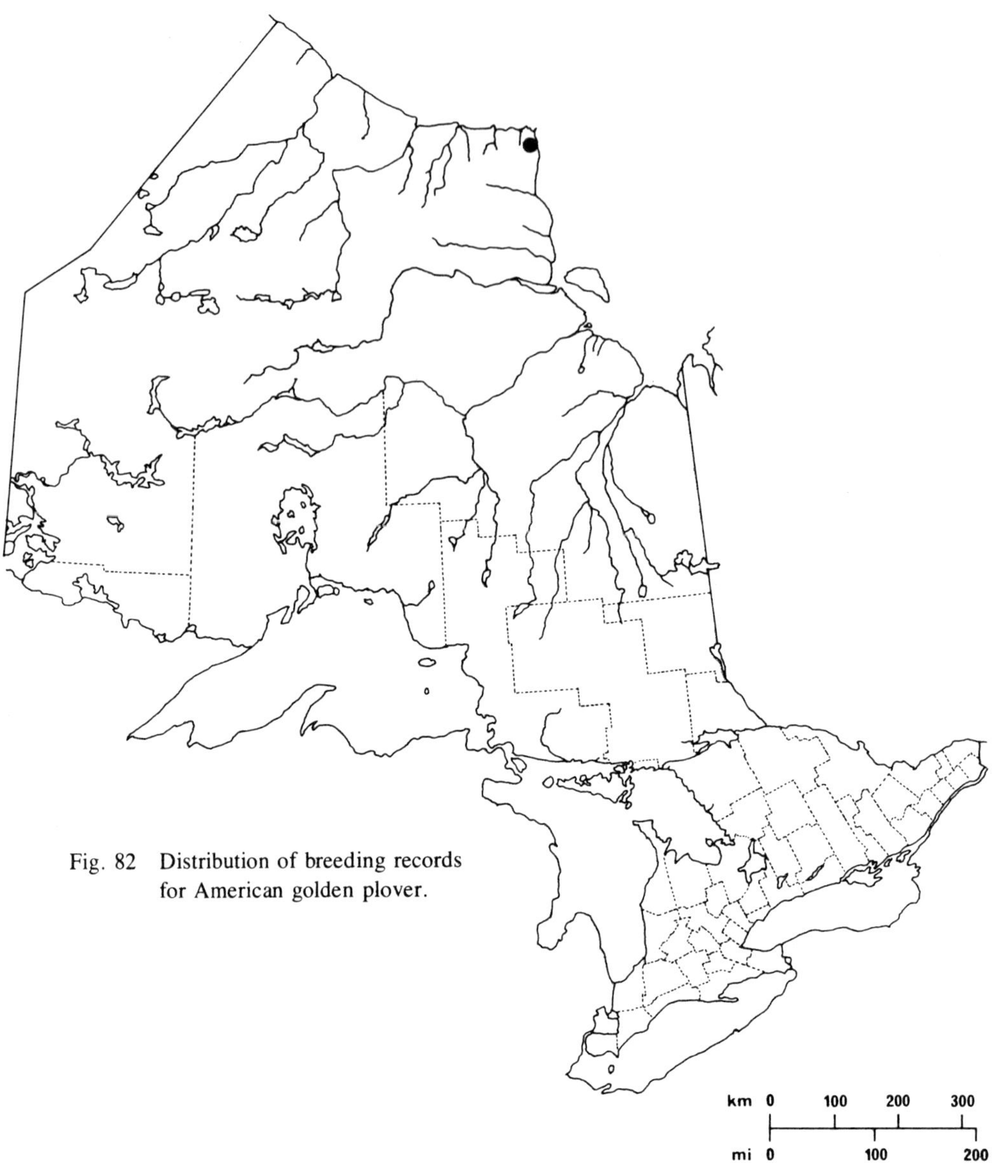

Fig. 82 Distribution of breeding records for American golden plover.

American Golden Plover, *Pluvialis dominica* (Müller)

Nidiology

RECORDS 2 nests representing 1 provincial region.
Although this northern breeding species probably nests regularly in the province's narrow strip of tundra, to date only 2 nests have been found.

Both nests were on relatively high, dry tundra areas, which supported a heath-lichen complex of vegetation.

The 2 nests were depressions surrounded by reindeer lichens, grasses, mountain avens, dwarf willow, and various heaths. They were lined with pieces of lichen, dead leaves, and grasses. Inside diameter of 1 nest was 12.5 cm (5 inches); its inside depth was 5 cm (2 inches).

EGGS 2 nests, each with 4 eggs.

INCUBATION PERIOD No information.

EGG DATES 2 nests, 23 June to 1 July (3 dates).

Breeding Distribution

The American golden plover (Fig. 168A) is a breeding species of the tundra (Fig. 167) and was recently found nesting in Ontario near the abandoned radar site 415 in the Cape Henrietta Maria region (Peck, 1972). This nest and a second nest found subsequently near the same radar site seem to indicate a small breeding population in the extreme northeast of the province.

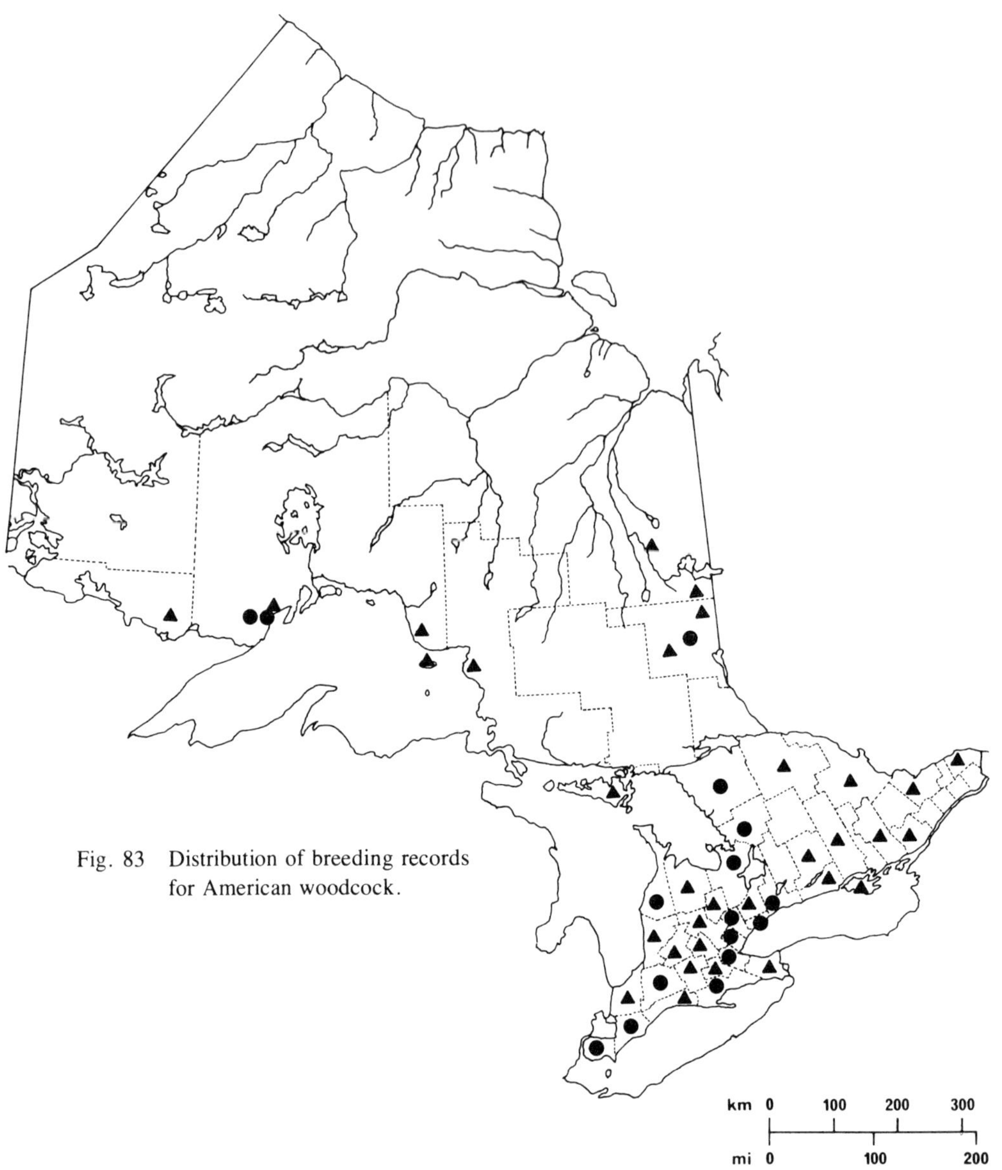

Fig. 83 Distribution of breeding records for American woodcock.

American Woodcock, *Philohela minor* (Gmelin)

Nidiology

RECORDS 194 nests representing 37 provincial regions.
This species breeds in a wide variety of habitats with nest sites in both open and closed surroundings. Open habitats recorded included fields and shrubby pastures, clearings, marshes and bogs, swales, beaches and sand-dune ridges, roadsides and ditches, fence rows, and burnt-over areas. Closed habitats included deciduous woods and swamps, coniferous woods and plantations, and mixed tree stands. In all closed habitats, forest edges and openings were much more frequently used than were dense, central wooded areas. In both habitat types, dry situations (85 nests) were preferred to wet situations (28 nests). Various trees and shrubs were recorded at many nest sites; the most common of these were poplar spp. (21 nests), cedar and juniper spp. (19 nests), pine spp. (10 nests), birch spp. (8 nests), and hawthorn spp. (7 nests).

Nests were invariably positioned on the ground or on vegetation and debris (1 nest was placed on a decayed log overgrown with grass). Some nests were placed on hummocks or small knolls up to 20.5 cm (8 inches) high (16 nests); 1 nest on a hummock was at water level. Nests were most often at the bases of or under trees and shrubs (52 nests), in tall grasses (23 nests), and in dead leaves (17 nests). They were less often in mosses, weeds and other plants, and sedge clumps and other marsh vegetation; in clumps of cedar, dogwood, aspen, and willow; in raspberry canes; in poison ivy; on pine needles and twigs; and under boughs and brush.

Usually nests were shallow depressions and occasionally they were flat structures; nests were variously lined, in order of preference, with leaves, grasses, weed and plant stems, sedge and marsh vegetation, twigs, pine needles, bracken fern, mosses, and feathers. Diameters of 3 nests ranged from 12.5 to 15 cm (5 to 6 inches). Five clutches of eggs were deposited on the existing vegetation without a nest.

EGGS 162 nests with 1 to 4 eggs; 1E (6N), 2E (12N), **3E** (35N), **4E** (109N).
Average clutch range 4 eggs (109 nests).

INCUBATION PERIOD 2 nests, ca 20 to 21 days.

EGG DATES 161 nests, 2 April to 11 July (228 dates); 80 nests, 2 May to 21 May.

Breeding Distribution

The American woodcock breeds throughout southern Ontario, and as far north as the north shore of Lake Superior and southern Cochrane District. However, its numbers are small in the Canadian Shield.

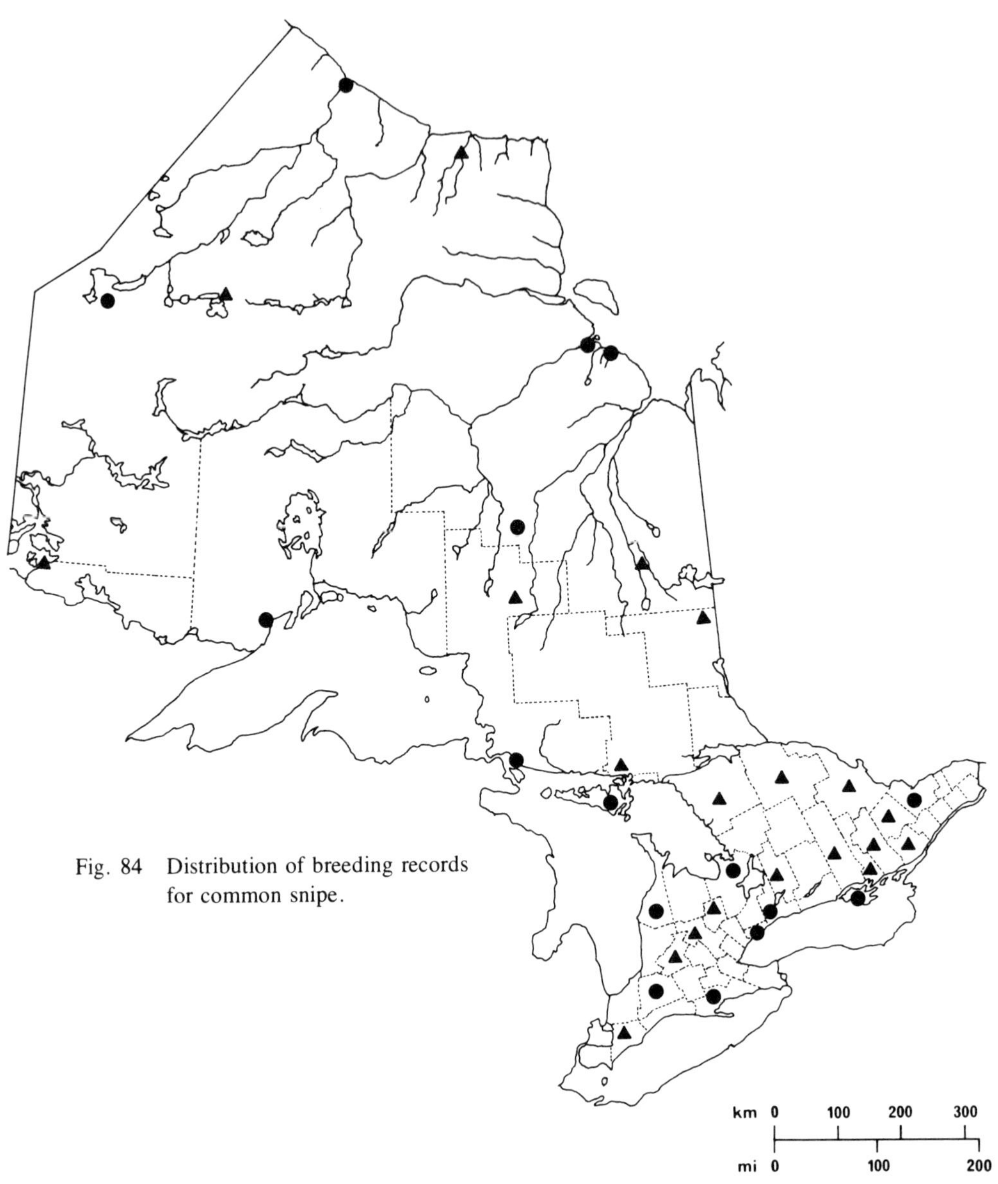

Fig. 84 Distribution of breeding records for common snipe.

Common Snipe, *Capella gallinago* (Linnaeus)

Nidiology

RECORDS 48 nests representing 26 provincial regions.
Breeding habitats recorded were both wet (31 nests) and dry (6 nests). Wet habitats included freshwater marshes, wet meadows and pastures, and bogs and swamps; dry habitats were fields, pastures, and an area of second-growth willows.

Nests were either on boggy or on dry ground, and were situated in tufts of grass 15 cm (6 inches) tall; in cattails and sedges; on mossy hummocks; among or under bracken fern, Labrador tea, leatherleaf, bog laurel, cinquefoil, and small red cedars; or in relatively bare, exposed areas.

Nests ranged from lined cups to depressions in mosses, grasses, and other plants; at times no nest was built. Nests were lined with grasses, cattails, mosses, horsetails, dead cedar, and bracken fern leaves; 1 nest was canopied by overhanging grasses.

EGGS 45 nests with 2 to 4 eggs; 2E (2N), **3E** (5N), **4E** (38N).
Average clutch range 4 eggs (38 nests).

INCUBATION PERIOD No information.

EGG DATES 42 nests, 24 April to 19 July (47 dates); 21 nests, 15 May to 8 June.

Breeding Distribution

With the possible exception of a few regions in the extreme south, the common snipe nests throughout Ontario.

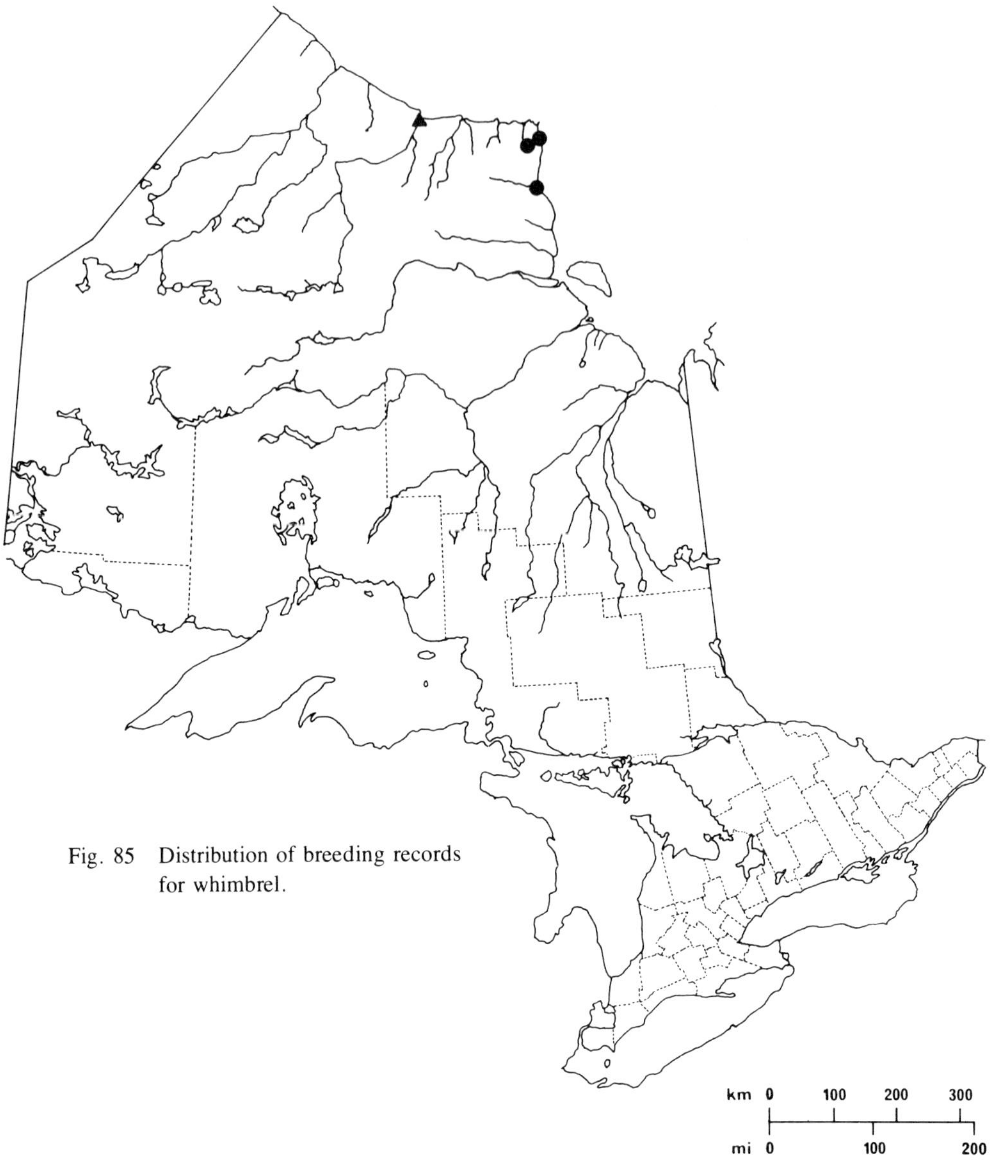

Fig. 85 Distribution of breeding records for whimbrel.

Whimbrel, *Numenius phaeopus* (Linnaeus)

Nidiology

RECORDS 4 nests representing 1 provincial region.
This species breeds in Ontario on the narrow strip of subarctic tundra on the north coast. Nests are on both dry and wet tundra. One nest recorded in a dry situation was on a ridge among some scattered dwarf birch ca 1 m (3 ft) high, and another was in a gravel area with sparse grass; 1 nest in a wet situation was in tussock tundra on a raised mound.

Nests were depressions on the ground, usually in grasses, and were either in the open or among dwarf birch and willow. Nests were composed of dead sedge and grass stalks and were sparsely lined with fine grass stems and a few leaves.

EGGS 4 nests with 3 to 4 eggs; 3E (1N), 4E (3N).

INCUBATION PERIOD No information.

EGG DATES 4 nests, 25 June to 17 July (6 dates); 2 nests, 27 June to 28 June.

Breeding Distribution

The first documented record of breeding of the whimbrel (Fig. 166A) in the province was obtained in 1947 at Lake River south of Cape Henrietta Maria, Kenora District (Manning, 1952). Subsequent records have mainly been in the Cape region, but adults are known to summer all along the Hudson Bay coast and they undoubtedly nest there as well.

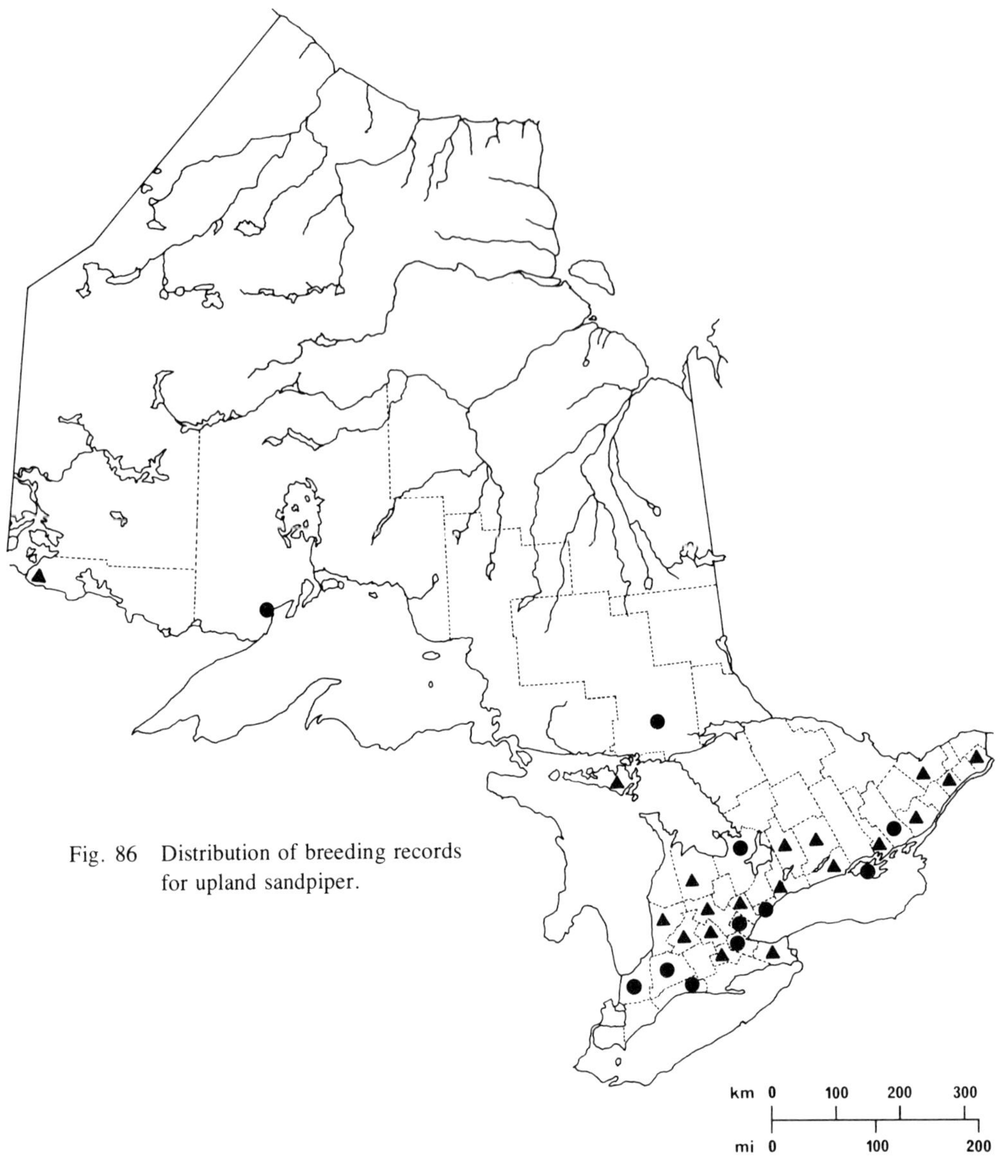

Fig. 86 Distribution of breeding records for upland sandpiper.

Upland Sandpiper, *Bartramia longicauda* (Bechstein)

Nidiology

RECORDS 44 nests representing 20 provincial regions.
Breeds on agricultural land in hay fields, unused pastures, and hawthorn meadows; breeds occasionally on airport grasslands.

Nests were on the ground in standing grasses, which were sometimes short but more often at least 15 to 25 cm (6 to 10 inches) tall. Nests were usually in open grasslands, but 2 nests were at the bases of small trees, and 1 nest was within 9 m (30 ft) of a fence. Two nests were 9 and 61 m (30 and 200 ft) away from other occupied nests of the same species.

Nests scrapes ranged from shallow to fairly deep depressions in the ground and were lined, often sparsely and loosely, with grasses and weed stalks, and sometimes with a few feathers. Some nests were canopied by tall grasses surrounding the site.

EGGS 42 nests with 2 to 4 eggs; 2E (2N), **3E** (5N), **4E** (35N).
Average clutch range 4 eggs (35 nests).

INCUBATION PERIOD No information.

EGG DATES 38 nests, 12 May to 9 July (48 dates); 19 nests, 27 May to 8 June.

Breeding Distribution

The upland sandpiper (Fig. 152A) breeds throughout the nonforested portions of southern Ontario and as far north as Sudbury. In the west it is observed and probably breeds sparingly between Thunder Bay and western Rainy River District.

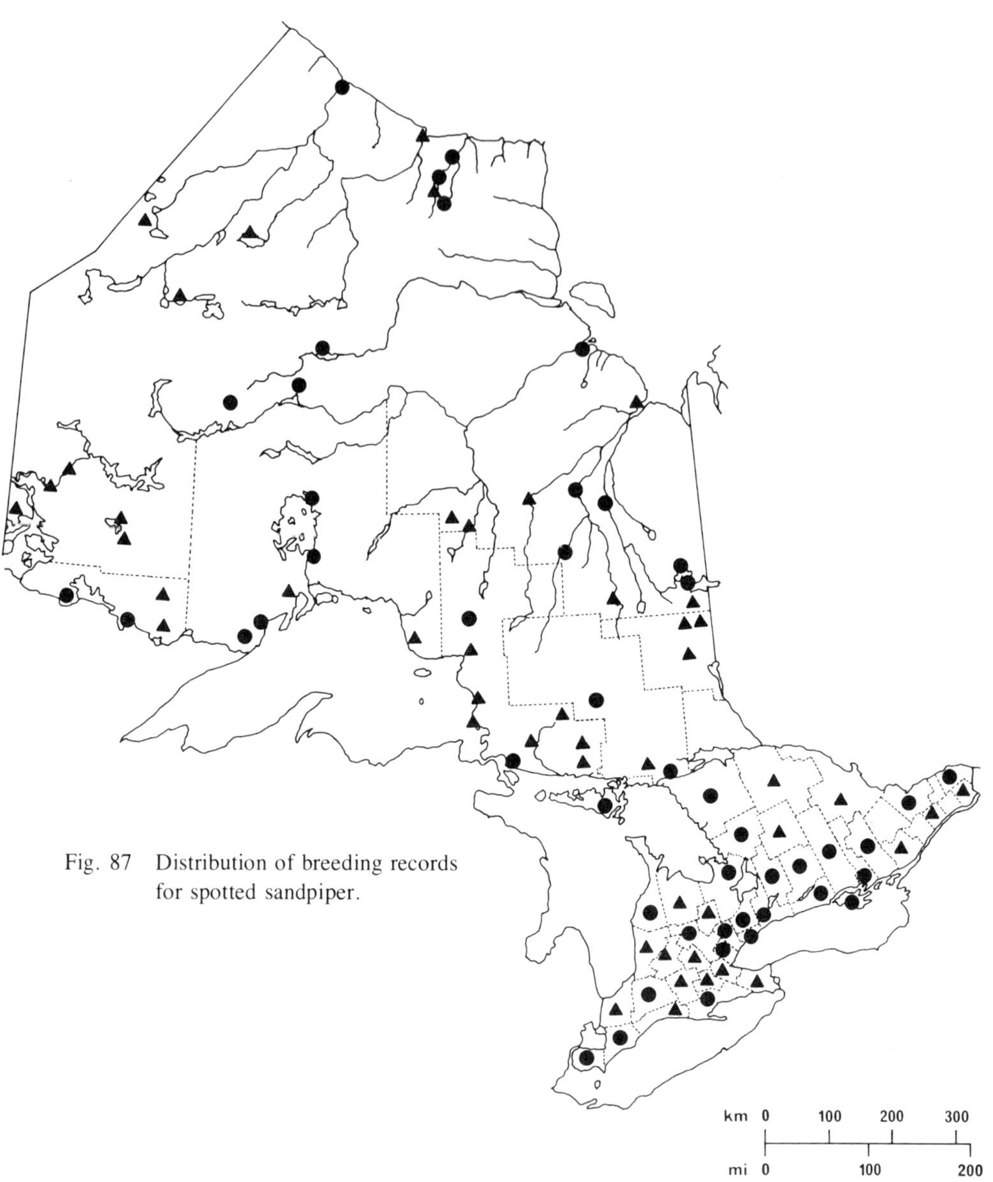

Fig. 87 Distribution of breeding records for spotted sandpiper.

Spotted Sandpiper, *Actitis macularia* (Linnaeus)

Nidiology

RECORDS 520 (523 nests) representing 46 provincial regions.
This species breeds in a variety of habitats. Nest sites were selected most commonly on mainland areas but often on islands (54 nests), where the species appeared to be semi-colonial. Nests were often near water, but many were at a considerable distance from it, and more dry sites were chosen than wet. In order of preference, nesting habitats included sandy beaches and dunes; clearings, woodland edges, and burns; lakes and river banks; rock and gravel areas and gravel pits; fields, roadsides, and ditches; marshes, bogs, beaver meadows, mudflats, and sewage lagoons; gardens and orchards; and schoolyards and parking lots.

Nests were invariably on the ground and were usually hidden but occasionally in exposed situations. Nests were slightly more often in nongrassy areas (84 nests) than in grassy areas (64 nests). They were positioned in grasses or grass clumps (71 nests); among, in, or at the bases of trees, bushes, weeds, flowers, and other plants (66 nests); on stones and on or under rocks (11 nests); under sticks, branches, roots, and driftwood (5 nests); in mosses (3 nests); and under a cattail clump (1 nest). Three unusual nest positions were as follows: on a rocky ledge at an elevation of 46.5 cm (18 inches) (1 nest), on a ledge in a pile of cinders (1 nest), and in a pocket 7.5 cm (3 inches) deep in a rock (1 nest). Occasionally nests were located 15 to 76 m (60 to 300 ft) from other nests of the same species (13 records); in 2 records they were near nests of mallard and killdeer.

Nests were cuplike depressions or hollows in sand or gravel (147 nests), in grasses (55 nests), and occasionally in other vegetation. One nest was located 20.5 cm (8 inches) above ground. Nests were woven and variously lined, homogeneously or in combination, with the following materials, listed in order of preference: grasses, leaves, weed stalks, sticks and twigs, straw, dried bulrushes, plant down, pine needles and cedar sprigs, fibres and rootlets, small stones, mosses, feathers, pieces of wood, and bracken fern. In 4 nests there was no lining. Diameters of 2 nests were 10 and 12.5 cm (4 and 5 inches); inside depths of 2 nests were 2.5 and 3 cm (1 and 1.25 inches).

EGGS 500 nests with 1 to 5 eggs; 1E (11N)—2.2%, 2E (15N)—3%, **3E** (57N)—11.4%, **4E** (414N)—82.8%, 5E (3N)—0.6%.
Average clutch range 4 eggs (414 nests).
One nest of 4 eggs also contained 1 egg of brown-headed cowbird.

INCUBATION PERIOD 10 nests: 5 of at least 19 days, 4 of at least 20 days, 1 of 20 days. Eggs were laid at 1 day intervals and hatching was usually simultaneous and occurred within 24 hours, although in 1 nest it took 4 days.

EGG DATES 485 nests, 14 May to 23 July (602 dates); 243 nests, 7 June to 21 June.

Breeding Distribution

The spotted sandpiper nests throughout the province.

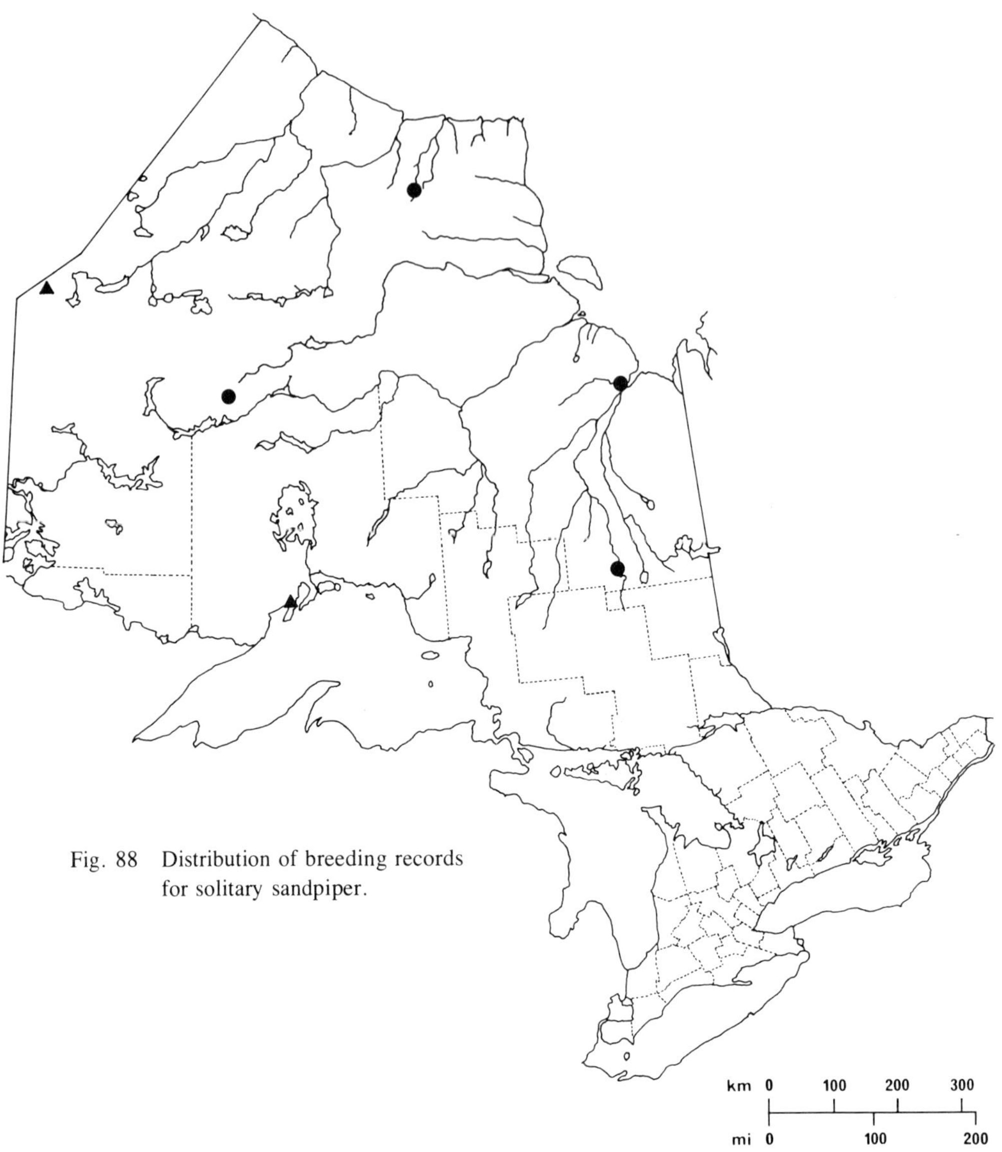

Fig. 88 Distribution of breeding records for solitary sandpiper.

Solitary Sandpiper, *Tringa solitaria* Wilson

Nidiology

RECORDS 1 nest representing 1 provincial region.
To date only 1 nest of this species has been found in Ontario. It contained 4 eggs and was discovered on 28 June 1964 at Sutton Lake, Kenora District. The nest was near a pond in a black spruce forest located on a hill above the southern end of Sutton Lake. It was situated in a 4.5 m (15 ft) black spruce, and was positioned against the trunk, at a height of 2 m (6.5 ft).

This nest was an old nest of American robin, composed of mosses, lichens, mud, grasses, and spruce twigs.

Both nest and eggs were collected (ROM 9479), and the eggs contained 15 mm (0.5 inch) embryos with formed eyes.

Breeding Distribution

Records of the solitary sandpiper observed with downy young extend back to 1906. However, it was not until 1964 that the first nest of this species was found in Ontario (Schueler et al., 1974). Although there are very few records of breeding, the species probably nests in wetland (Fig. 163) throughout the forested regions of northern Ontario, where it has been observed exhibiting territorial behaviour during the breeding season. Several published records of nesting in southern Ontario were never substantiated and are now believed to be erroneous.

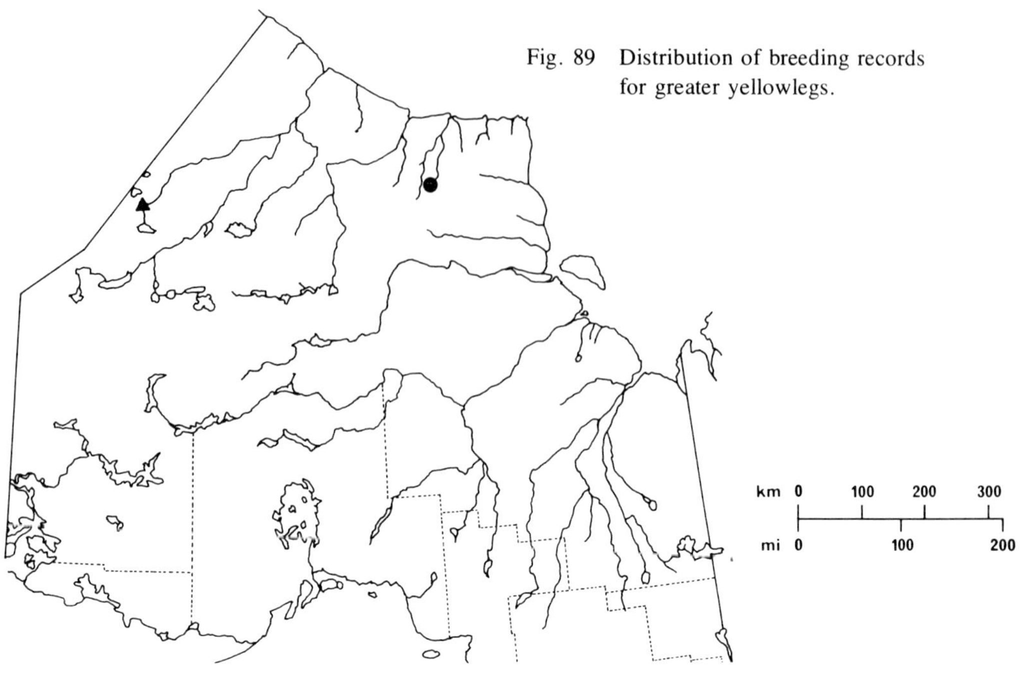

Fig. 89 Distribution of breeding records for greater yellowlegs.

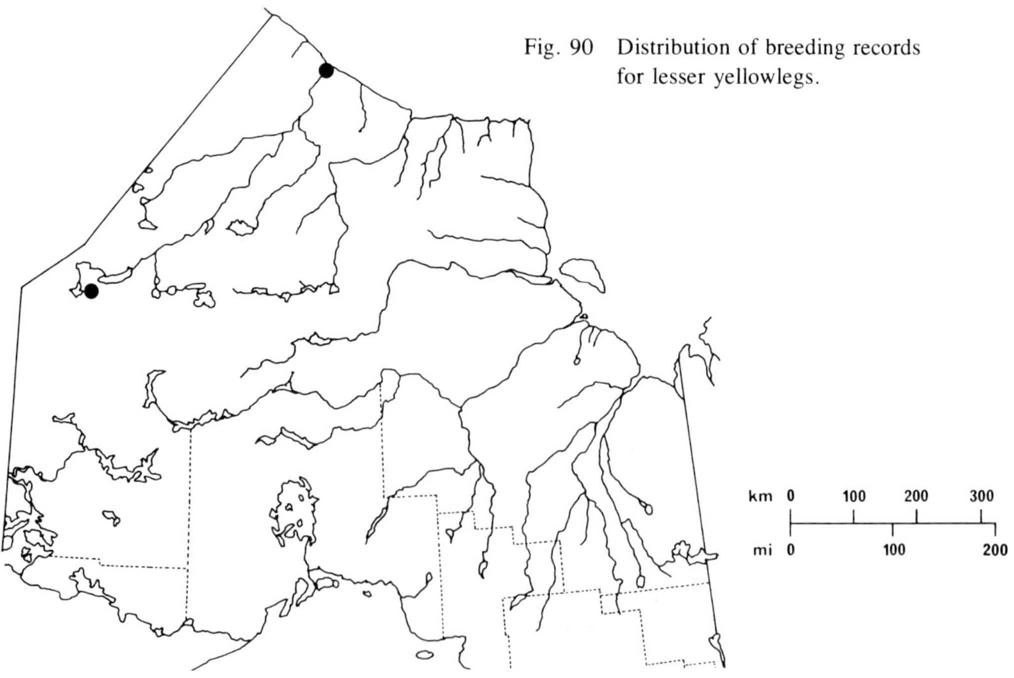

Fig. 90 Distribution of breeding records for lesser yellowlegs.

Greater Yellowlegs, *Tringa melanoleuca* (Gmelin)

Nidiology

RECORDS 2 nests representing 1 provincial region.
Until recently this shorebird had only hypothetical breeding status in the province—based on the collection of an adult female with enlarged ova (ROM 94540) on 17 June 1964 at the north end of Hawley Lake, Kenora District.

On 12 June 1977 a nest with 2 eggs was found at Little Sachigo Lake, Kenora District, but no material evidence was obtained. This nest was in a tamarack bog on a sphagnum-moss hummock, at the base of 3 tamaracks that were 1.5 m (5 ft) in height. The nest depression was lined sparingly and haphazardly with a few strands of sedge.

On 1 July 1980 a nest with 4 eggs was found and collected (ROM 12654) near Aquatuk Lake, Kenora District. This nest was in an open, treed bog with sloughs; the nest was situated on a raised area that was surrounded by reindeer lichen and was at the base of a black spruce ca 2 m (7 ft) in height. The nest depression was lined with dried spruce twigs 2.5 to 5 cm (1 to 2 inches) in length.

Breeding Distribution

Although the greater yellowlegs (Fig. 164A) occurs throughout much of the Hudson Bay Lowland and the Boreal Forest region across northern Ontario, only the above two nests have ever been reported. The difficulty in locating nests is undoubtedly the major reason for such a paucity of information regarding its breeding.

Lesser Yellowlegs, *Tringa flavipes* (Gmelin)

Nidiology

RECORDS 1 nest representing 1 provincial region.
The first documented nest record of this species in Ontario was at Favourable Lake, Kenora District, when a nest with 2 eggs was found on 4 June 1938, then collected on 12 June 1938, by C. E. Hope after the bird had deserted it (ROM 5337). The nest was in a mine clearing in the midst of bog pools and stumps.

The only other breeding record was from Fort Severn, where downy young were collected on 9 and 10 July 1940 (ROM 66330, 66331), also by C. E. Hope.

Breeding Distribution

The lesser yellowlegs probably breeds over most of the northern third of the province, but because of the difficulty in differentiating the two yellowleg species, and because of the lack of nesting records, its range has not been precisely determined. It is regularly seen along the Hudson Bay and northern James Bay coasts in summer.

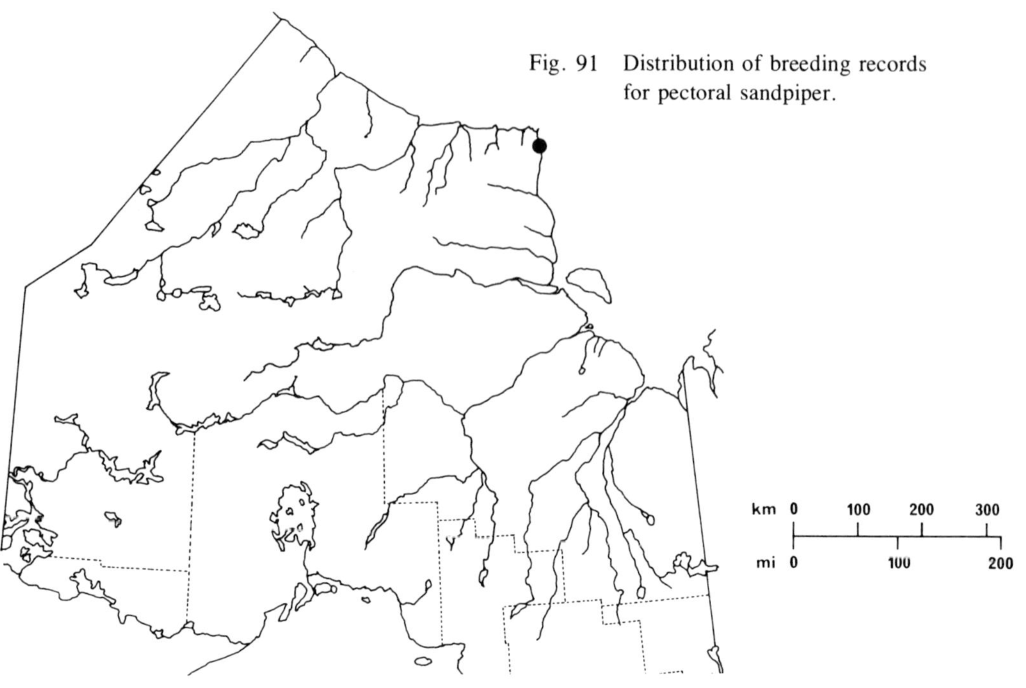

Fig. 91 Distribution of breeding records for pectoral sandpiper.

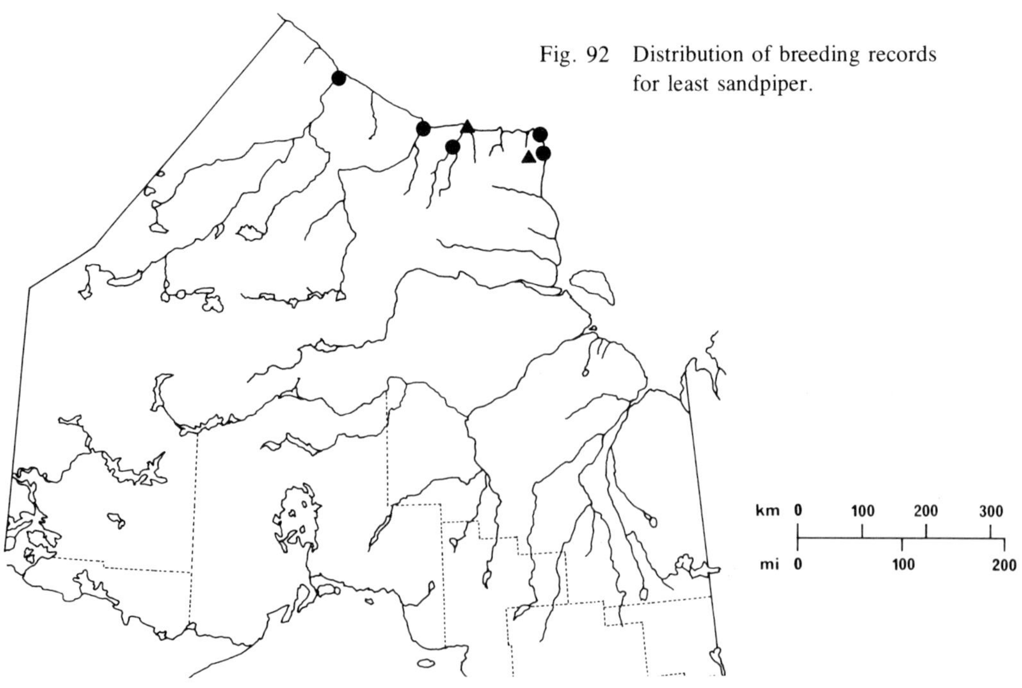

Fig. 92 Distribution of breeding records for least sandpiper.

Pectoral Sandpiper, *Calidris melanotos* (Vieillot)

Nidiology

This tundra-breeding shorebird was established as an Ontario breeding species on 5 July 1948, in the Cape Henrietta Maria region, Kenora District, when 2 downy young (ROM 75680, 75681) were collected (Peck, 1972). To date no nests have been discovered.

Breeding Distribution

Although the pectoral sandpiper was believed to breed all along the Hudson Bay coast (Godfrey, 1966), there is little evidence to support this assumption. Migrant birds have been encountered all along the Hudson Bay and James Bay coasts (Manning, 1952), but the only indications of breeding are on the tundra of the Cape Henrietta Maria region, where adults that were apparently nesting have been observed.

Least Sandpiper, *Calidris minutilla* (Vieillot)

Nidiology

RECORDS 5 nests representing 1 provincial region.
Breeds on subarctic tundra along the north coast. The 5 known nest records all described a wet breeding habitat in grassy tussock tundra. There were a few small tamaracks near 1 nest.

All nests were in tufts of grass, and 1 nest was bordered by living dwarf willow. Nest depressions were lined with dead leaves of dwarf willow and dwarf birch and with dead grass stems.

EGGS 5 nests with 3 to 4 eggs; 3E (2N), 4E (3N).

INCUBATION PERIOD No information.

EGG DATES 5 nests, 22 June to 21 July; 3 nests, 29 June to 2 July.

Breeding Distribution

The least sandpiper is another species with a breeding range in Ontario that is largely restricted to the tundra areas of the northern coasts (Fig. 169). Its range undoubtedly extends both for some distance south along the James Bay coast and for a short way inland along the major rivers in the northern coastal regions. Few nests have been found since the first one in 1940, and most records are based on the collections of downy young (Baillie, 1961; Peck, 1972).

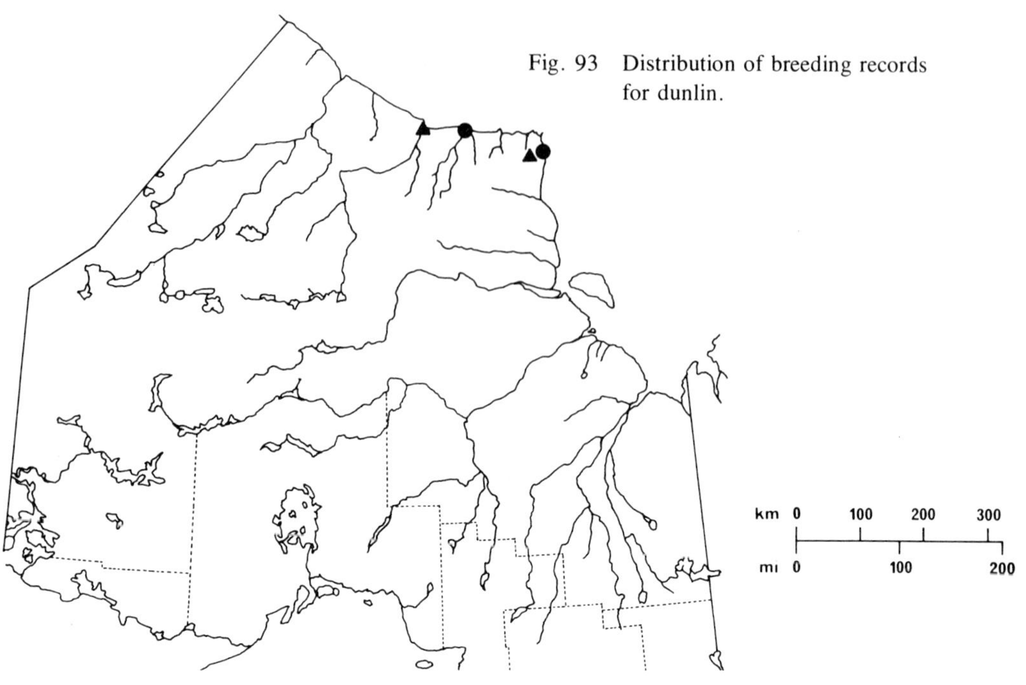

Fig. 93 Distribution of breeding records for dunlin.

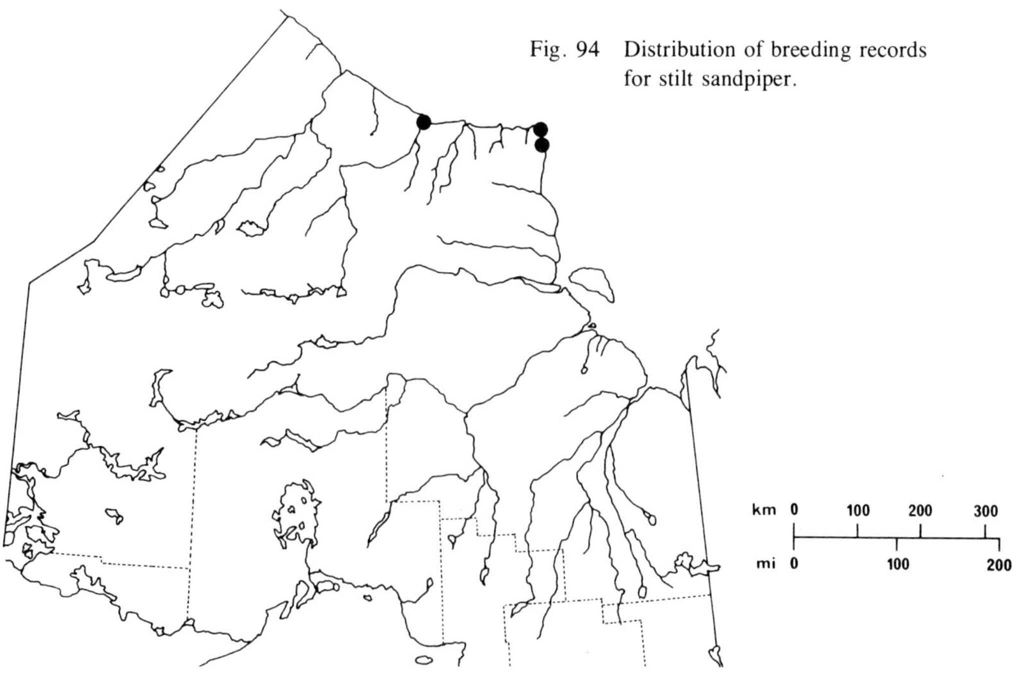

Fig. 94 Distribution of breeding records for stilt sandpiper.

Dunlin, *Calidris alpina* (Linnaeus)

Nidiology

RECORDS 5 nests representing 1 provincial region.
This shorebird breeds in Ontario only in the northern tundra region. All recorded nests were in wet tundra areas and most were in grassy tussock tundra (Fig. 169). One nest was on a low lichen-covered ridge in a marsh area; another was at the edge of a large slough.

Most nests were in grass tufts on hummocks; 1 was in a clump of sedge near some small willows. Nest depressions were lined with dead leaves of dwarf birch and willow and with dried stems of grass and sedge.

EGGS 5 nests with 2 to 4 eggs; 2E (1N), 4E (4N).

INCUBATION PERIOD No information.

EGG DATES 5 nests, 23 June to 20 July (7 dates); 3 nests, 27 June to 28 June.

Breeding Distribution

The first confirmation of breeding in Ontario by the dunlin came from Cape Henrietta Maria in 1947 (Manning, 1952). The few other records from the Cape region and along the Hudson Bay coast indicate that this species probably breeds throughout the coastal tundra.

Stilt Sandpiper, *Micropalama himantopus* (Bonaparte)

Nidiology

This northern breeding shorebird was first established as a provincial breeding species with the collection, between 20 and 23 July 1947, of a half-fledged juvenile found near the limestone ridges of Cape Henrietta Maria, Kenora District (Manning, 1952).

On 9 July 1948, 2 downy young (ROM 75764, 75765) were collected on the tundra, a few miles south of Cape Henrietta Maria (Peck, 1972). On 24 June 1962 another downy young (ROM 92998) was collected from a brood of 4 found at the mouth of the Sutton River, Kenora District (Baillie, 1962).

Breeding Distribution

The stilt sandpiper may actually breed throughout the tundra in northern Ontario (Baillie, 1962) in spite of the scarcity of documented breeding records.

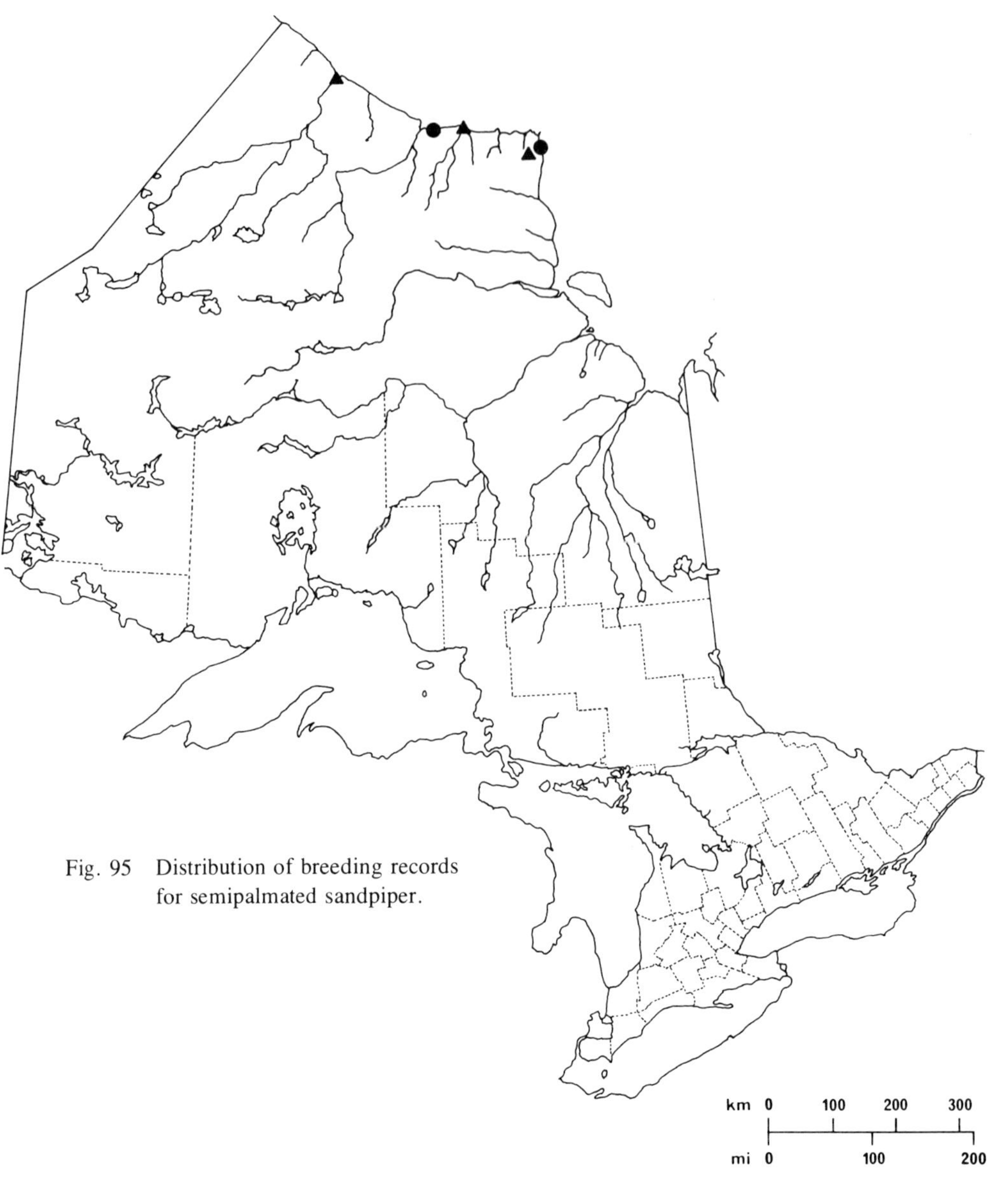

Fig. 95 Distribution of breeding records for semipalmated sandpiper.

Semipalmated Sandpiper, *Calidris pusilla* (Linnaeus)

Nidiology

RECORDS 11 nests representing 1 provincial region.
Breeds only on subarctic tundra in both dry and wet sites. Dry sites are heath-lichen tundra (Fig. 167) on ridges and in raised beach-line areas; wet sites are wet, grassy tussock tundra (Fig. 169) and marshy areas.

Nests were depressions in the ground in open scrapes or, more often, in grasses and lichens, where they were among dwarf willows 8 cm (3 inches) in height, and grass in tufts or tussocks. Nests were frequently lined with small, dead leaves, such as those of dwarf willow, and with dead grass stems.

One nest was placed within 46 m (150 ft) of an active nest of American golden plover.

EGGS 11 nests, with 3 to 4 eggs; 3E (3N), 4E (8N).

INCUBATION PERIOD No information.

EGG DATES 11 nests, 24 June to 30 June (13 dates); 6 nests, 25 June to 29 June.

Breeding Distribution

The semipalmated sandpiper occurs all along the Hudson Bay coast in summer, but the first evidence of breeding was not secured until 14 July 1948, when a juvenile was collected in the Cape Henrietta Maria region (Baillie, 1962). Subsequently, two nests were found in 1962 at the Sutton River; one nest in 1965 east of Winisk (Schueler et al., 1974); and five nests in 1970 (Peck, 1972), one nest in 1975, and two nests in 1980, all near the Cape.

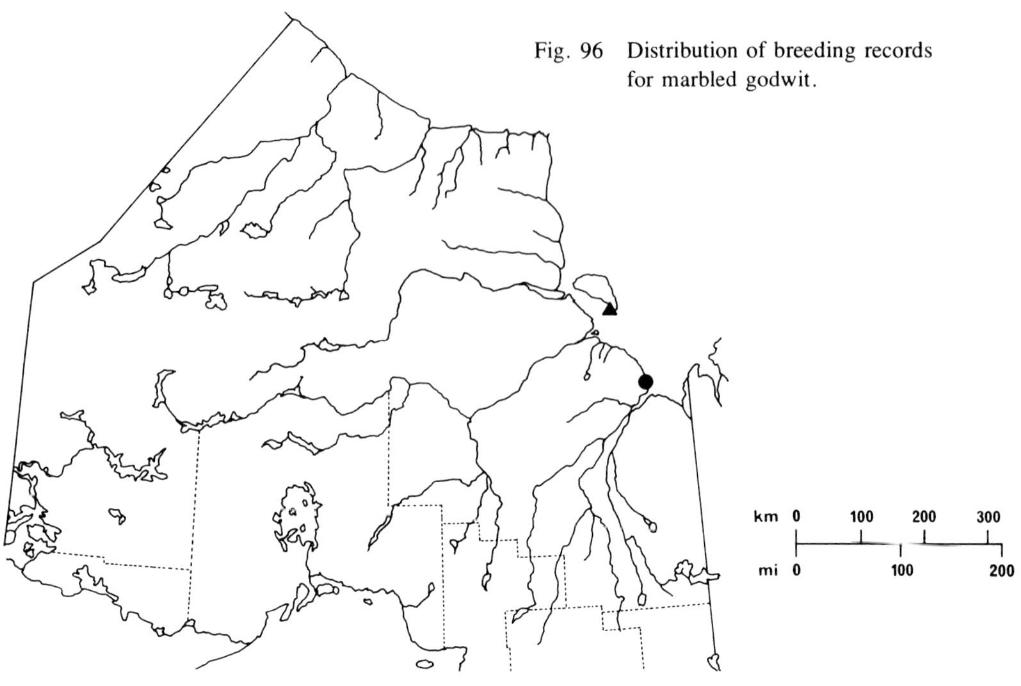

Fig. 96 Distribution of breeding records for marbled godwit.

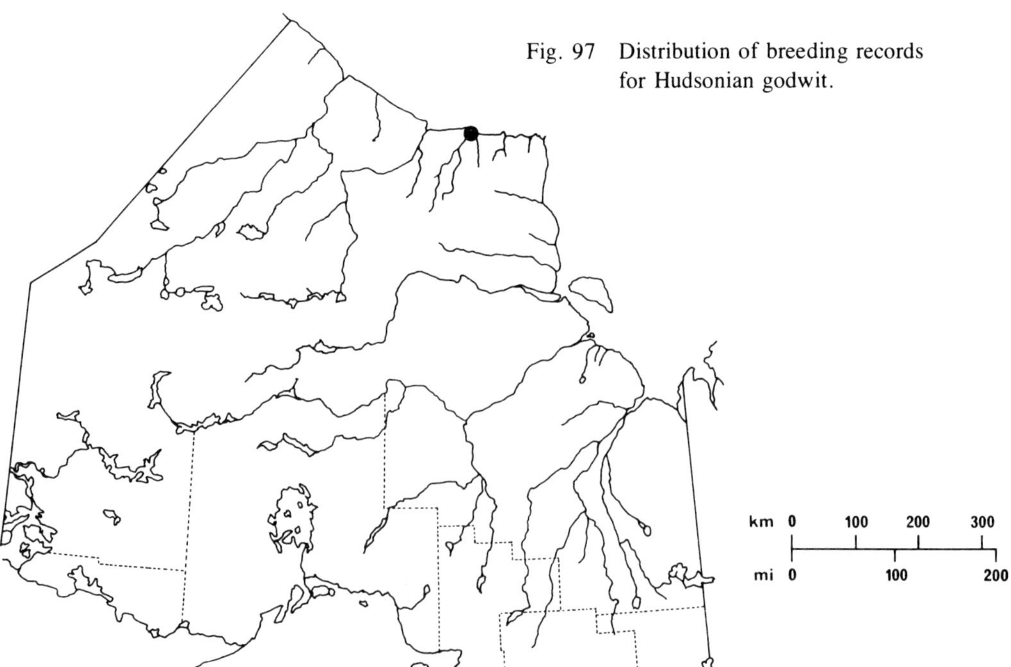

Fig. 97 Distribution of breeding records for Hudsonian godwit.

Marbled Godwit, *Limosa fedoa* (Linnaeus)

Nidiology

RECORDS 1 nest representing 1 provincial region.
This prairie-breeding species became a recent addition to the provincial breeding-bird list with the collection of a 10-day-old bird on 4 July 1975 near North Point, on James Bay, Cochrane District (NMC 63284).

In 1976 in the same locality, a nest with 4 pipped eggs was found on 1 July. By 2 July, 3 of the eggs had hatched. This nest was found in an open, brackish, coastal marsh. It was situated on a small, slightly raised, grassy ridge between 2 ponds and was ca 10 cm (4 inches) above the water surface. The nest was a shallow cup, scantily lined with dry grasses and pieces of sedge. Its diameter was 13 by 20 cm (5 by 8 inches); its inside depth was 7.5 cm (3 inches).

Breeding Distribution

Recorded occurrences of marbled godwit on the James Bay coast in summer date back to 1750 (Manning, 1952). But for many years, the only known nesting area of the species was on the central Great Plains, and it was believed to be only a migrant on the northeast coast of Ontario. Manning (1952) noted pairs that were apparently breeding in the province in 1947, but not until 1975 was he able to collect a downy young, which he found at North Point (Morrison et al., 1976). The discovery of a nest in 1976 and the observation of a downy, flightless young on the southwest coast of Akimiski Island in 1978 seem to indicate at least a small breeding population on the James Bay coast (R.I.G. Morrison, pers. comm.).

Hudsonian Godwit, *Limosa haemastica* (Linnaeus)

Nidiology

This treeline-nesting shorebird was added to the provincial breeding-bird list on 31 July 1962, when a flightless young was collected (ROM 93127) at the mouth of the Sutton River on Hudson Bay, Kenora District.

Breeding Distribution

Breeding in Ontario by the Hudsonian godwit is substantiated by a single record, the collection of a flightless young in 1962 (Baillie, 1963). To date no nests have been found in the province. Although summer records of birds that were apparently breeding are scarce (Peck, 1972), it probably breeds at several locations along the Hudson Bay coast.

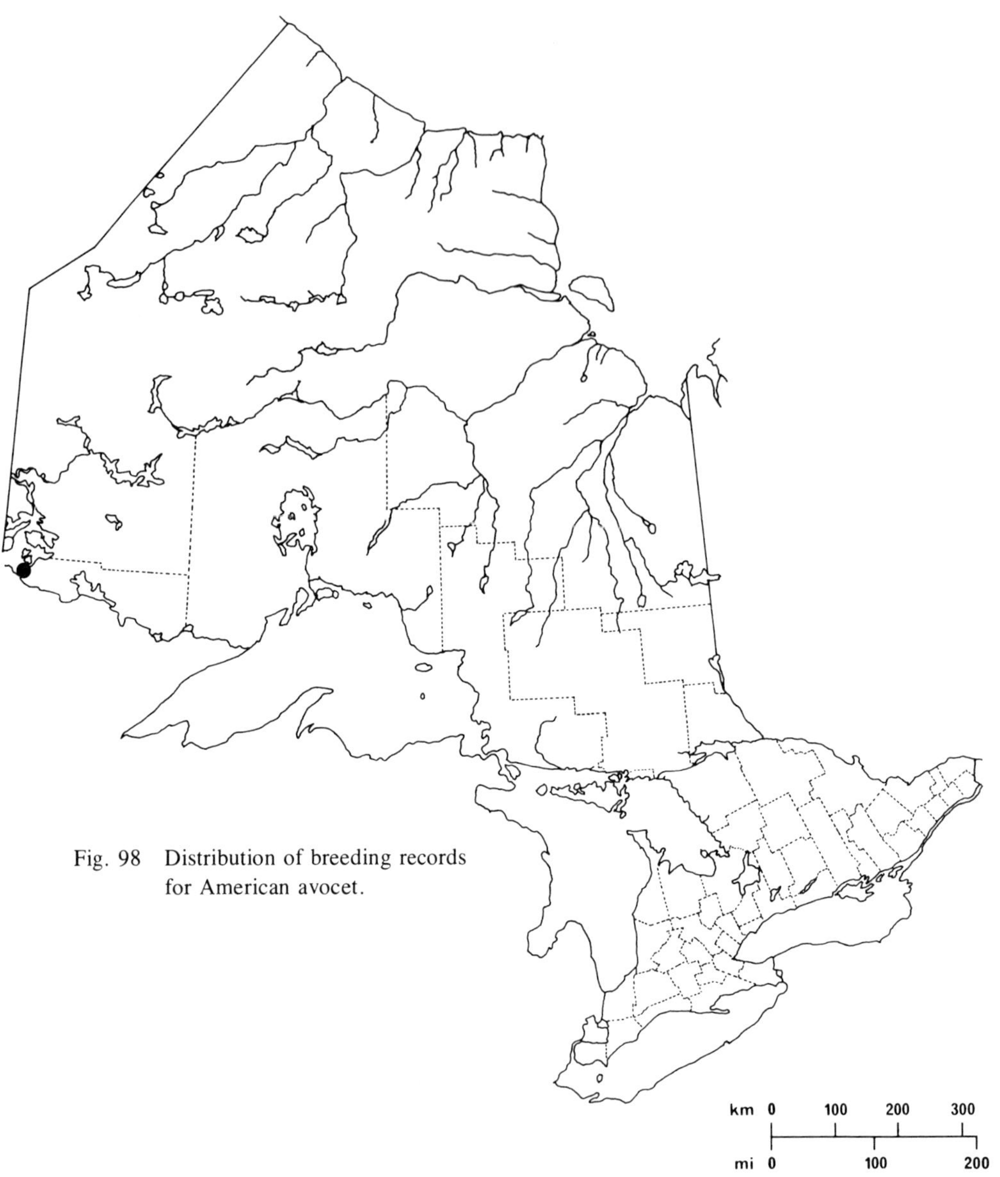

Fig. 98 Distribution of breeding records for American avocet.

American Avocet, *Recurvirostra americana* Gmelin

Nidiology

Avocets ordinarily breed along the sparsely vegetated shores of lakes and sloughs in the western prairies. However, this species was added to the list of provincial breeding birds on 6 July 1980, when photographs were taken by W. Wilson of an adult and a flightless young on Sable Island, Lake of the Woods, Rainy River District (ROM PR 1154–1160). A pair of adults and 3 flightless young were seen on the beach on that island, however, no nest was found.

Sable Island, in the southeastern part of Lake of the Woods, is a low, sandy island with wide beaches (Fig. 149) and sparsely vegetated dunes in the centre. It is similar to typical avocet habitats.

Breeding Distribution

Before 1980 the American avocet had been reported in the province about two dozen times, with some reports dating as far back as 1860. All occurrences were in the spring and autumn during the migration period. Breeding was never suspected, even though the species nests in Manitoba within 225 km (140 mi) of Ontario. The record mentioned above is the only evidence to date of breeding of this species in Ontario.

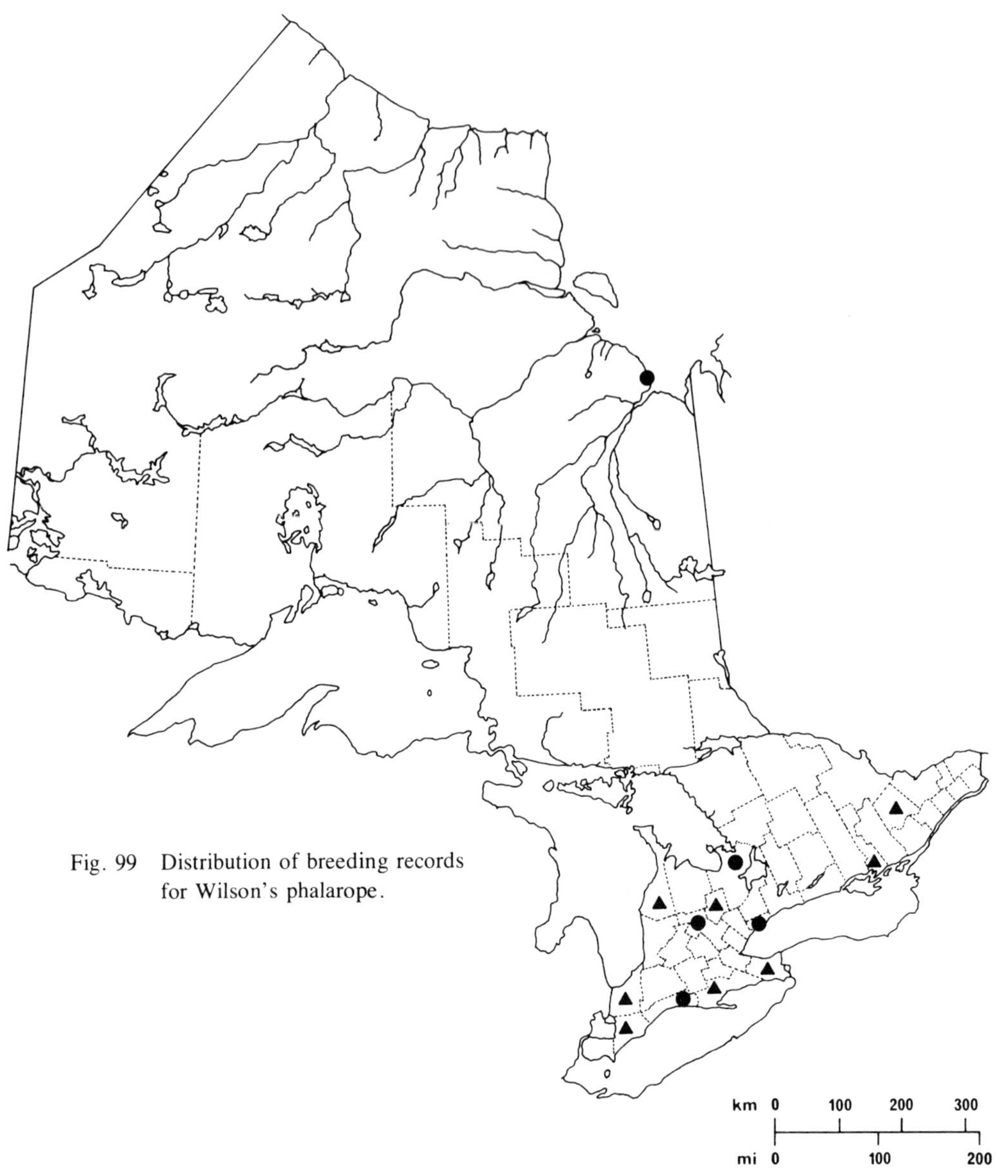

Fig. 99 Distribution of breeding records for Wilson's phalarope.

Wilson's Phalarope, *Steganopus tricolor* Vieillot

Nidiology

RECORDS 16 nests representing 10 provincial regions.
Breeds in grass and sedge areas usually in or near freshwater marshes, on islands, lakeshores, and river banks, and in swales and muskeg. Nests are close to water or as far as 45 m (150 ft) distant from it. At times several pairs nested in close proximity, as a small, loose colony.

Nests were invariably in grass and sedge areas, which were often in association with cattails and various other plants such as hawkweed and buckbean; some nests were close to dogwood and willow clumps. Nests were depressions in the ground, and despite the close proximity of water, they were in dry situations. Vegetation around the nests ranged from short to tall, and nests were sometimes canopied by tall grasses and weeds. The depressions were shallow and were lined with dead grass and weed stems and with feathers. Inside diameter of 1 nest ranged from 9 to 10 cm (3.5 to 4 inches); its inside depth was 3 cm (1.25 inches).

EGGS 16 nests with 3 to 4 eggs; **3E** (4N), **4E** (12N).
Average clutch range 4 eggs (12 nests).

INCUBATION PERIOD No information.

EGG DATES 15 nests, 18 May to 24 June (20 dates); 8 nests, 6 June to 10 June.

Breeding Distribution

The Wilson's phalarope (Fig. 150B) is usually a breeding bird of the western prairies, and by 1936 this species was known to have nested at only two places in Ontario, in the extreme southwest of the province (Baillie and Harrington, 1936). Since then, isolated nests and breeding records have been reported from a number of scattered localities covering the area south of the Canadian Shield, including the Ottawa area.

In 1975, the species was recorded nesting in northern Ontario, on the James Bay coast. It was found at North Point, Cochrane District (Morrison and Manning, 1976). The only other possible breeding area for the species is near Lake of the Woods, where summering birds have been observed.

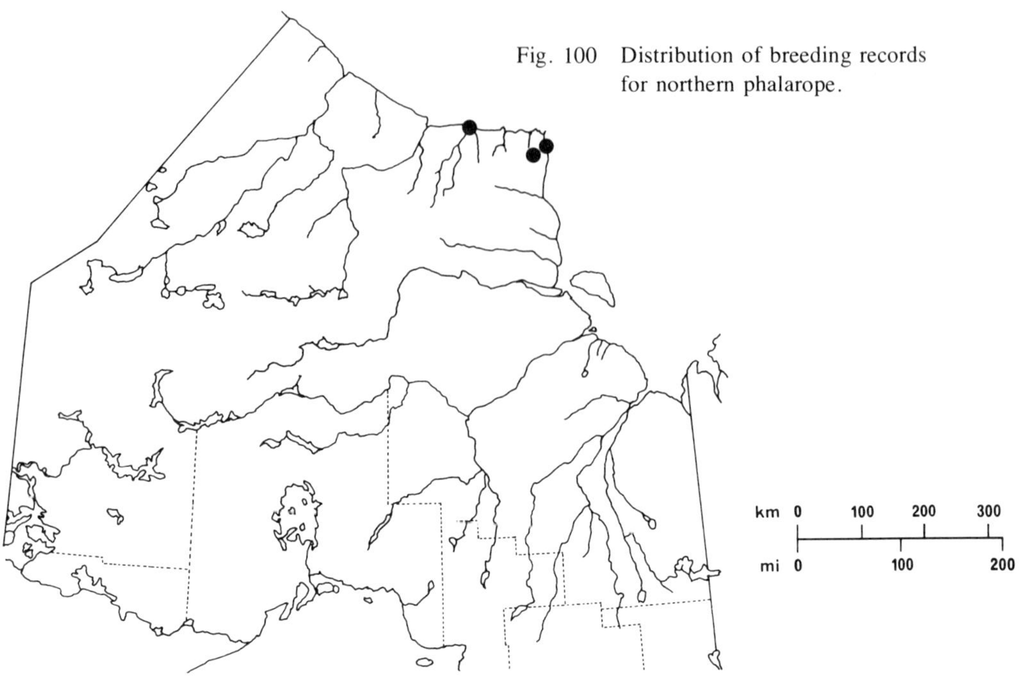

Fig. 100 Distribution of breeding records for northern phalarope.

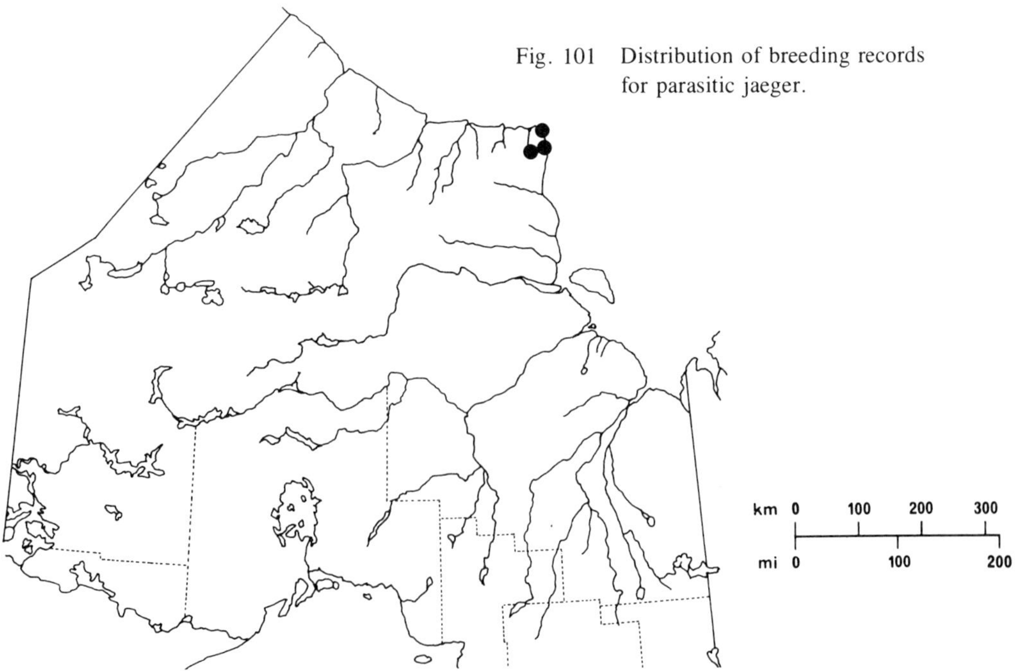

Fig. 101 Distribution of breeding records for parasitic jaeger.

Northern Phalarope, *Lobipes lobatus* (Linnaeus)

Nidiology

RECORDS 4 nests representing 1 provincial region.
Breeds only on subarctic tundra along the north coast and prefers wet tussock tundra. Nests were usually placed beside pools or sloughs (Fig. 169).

Nests were depressions hidden in long grasses or sedges and were lined with dead stalks of grass and sedge and with a few dwarf willow leaves. One nest was within 30 m (100 ft) of an active nest of whimbrel.

EGGS 4 nests, each with 4 eggs.

INCUBATION PERIOD No information.

EGG DATES 4 nests, 26 June to 4 July (7 dates).

Breeding Distribution

Ontario breeding of the northern phalarope was first documented in 1947 (Manning, 1952), when downy young were collected. The first nest was found in 1948 (Peck, 1972). The few nests that have been found were all in the Cape Henrietta Maria region, where the species is most abundant; however, evidence of breeding is available from scattered locations all along the Hudson Bay coast (Baillie, 1962).

Parasitic Jaeger, *Stercorarius parasiticus* (Linnaeus)

Nidiology

RECORDS 2 nests representing 1 provincial region.
Breeds only on subarctic tundra along the north coast. To date only 2 nests have been found, and one of these was on a raised mound in an area of wet, grassy tussock tundra (Fig. 169). This nest was a slight depression in mosses, grasses, heaths, and dwarf willow, and was lined sparsely with dead grass stems and a few dead leaves of dwarf willow.

One of the nests contained 2 eggs being incubated by a light-phase adult on 25 and 26 June, and there was a dead young near the other nest on 22 July.

Breeding Distribution

Although our only nesting and breeding records have come from the Cape Henrietta Maria region, the parasitic jaeger (Fig. 170A) probably breeds in small numbers all along the Hudson Bay coast. This is still another species first documented as breeding in Ontario in 1947 by Manning, when he collected downy young (Manning, 1952). The first of the two nests thus far discovered was found in 1948 (Baillie, 1962).

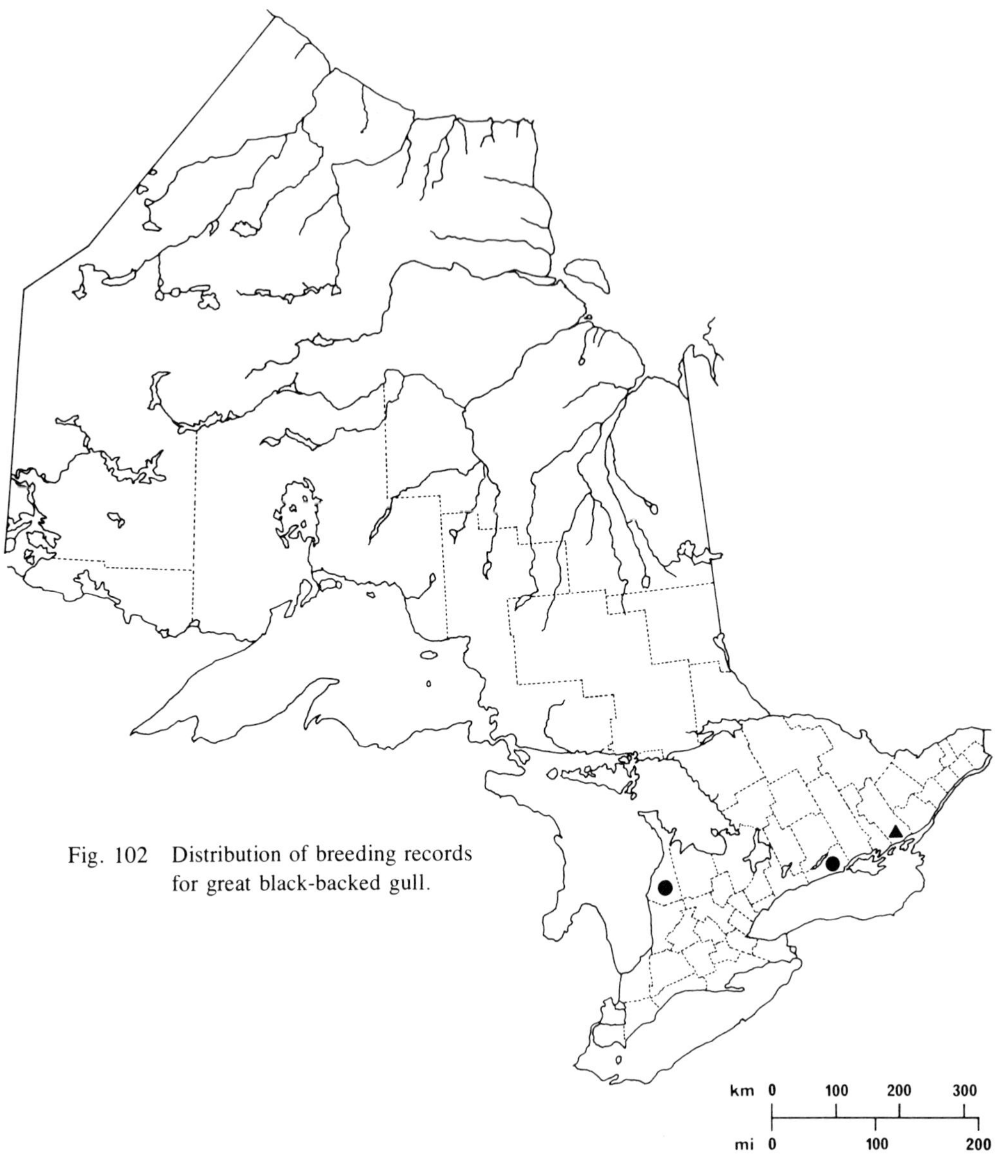

Fig. 102 Distribution of breeding records for great black-backed gull.

Great Black-backed Gull, *Larus marinus* Linnaeus

Nidiology

RECORDS 4 nests representing 2 provincial regions.
This is mainly a marine breeding species. In Ontario we have nest records at only 2 locations: 3 records on a small island in Lake Ontario, with colonies of herring gull, ring-billed gull, and common tern; and 1 record on an island in Lake Ontario near the source of the St. Lawrence River, on the outer edge of a colony of ring-billed gull.

One of the 2 nesting sites was on a rocky island that was covered with nettles, remnants of old buildings, and some mature lilac bushes. The other site was on a long, narrow island that had a cover of weeds and grasses. Both islands had stony beach areas where the nests were located. The nests were isolated from those of the gulls and terns. One nest was placed among short growths of nettles.

One nest was a scrape with little or no vegetation, and the other nests were mounds of both fresh and dead vegetation that resembled nests of herring gull. One nest was flat, with most of the egg cavity in a scrape below ground level; its rim was formed mostly of dead vegetation and feathers, with some fresh stonecrop and nettles. Its outside diameter was 56 cm (22 inches); outside depth, 5 cm (2 inches); inside diameter, 25.5 cm (10 inches); inside depth, 10 cm (4 inches). A second nest was a deeply hollowed mound of fresh and dead vegetation. Its outside diameter was 53.5 cm (21 inches); outside depth, 12.5 cm (5 inches); inside diameter, 33 cm (13 inches); inside depth, 9 cm (3.5 inches).

EGGS 4 nests with 1 to 3 eggs; 1E (1N), 2E (2N), 3E (1N).

INCUBATION PERIOD No information.

EGG DATES 4 nests, 7 June to 19 July.

Breeding Distribution

For many years the great black-backed gull has ventured inland to the lower Great Lakes in the fall and winter. It was first recorded as breeding in Ontario in 1954, when three large, flightless young were found in a colony of herring gulls on an island off the Bruce Peninsula (Krug, 1956). The species later began nesting on two islands in eastern Lake Ontario (Baillie, 1963; Quilliam, 1973). It has continued to nest in gull colonies on these islands, but no great increase in numbers has occurred on the Great Lakes (Angehrn et al., 1979).

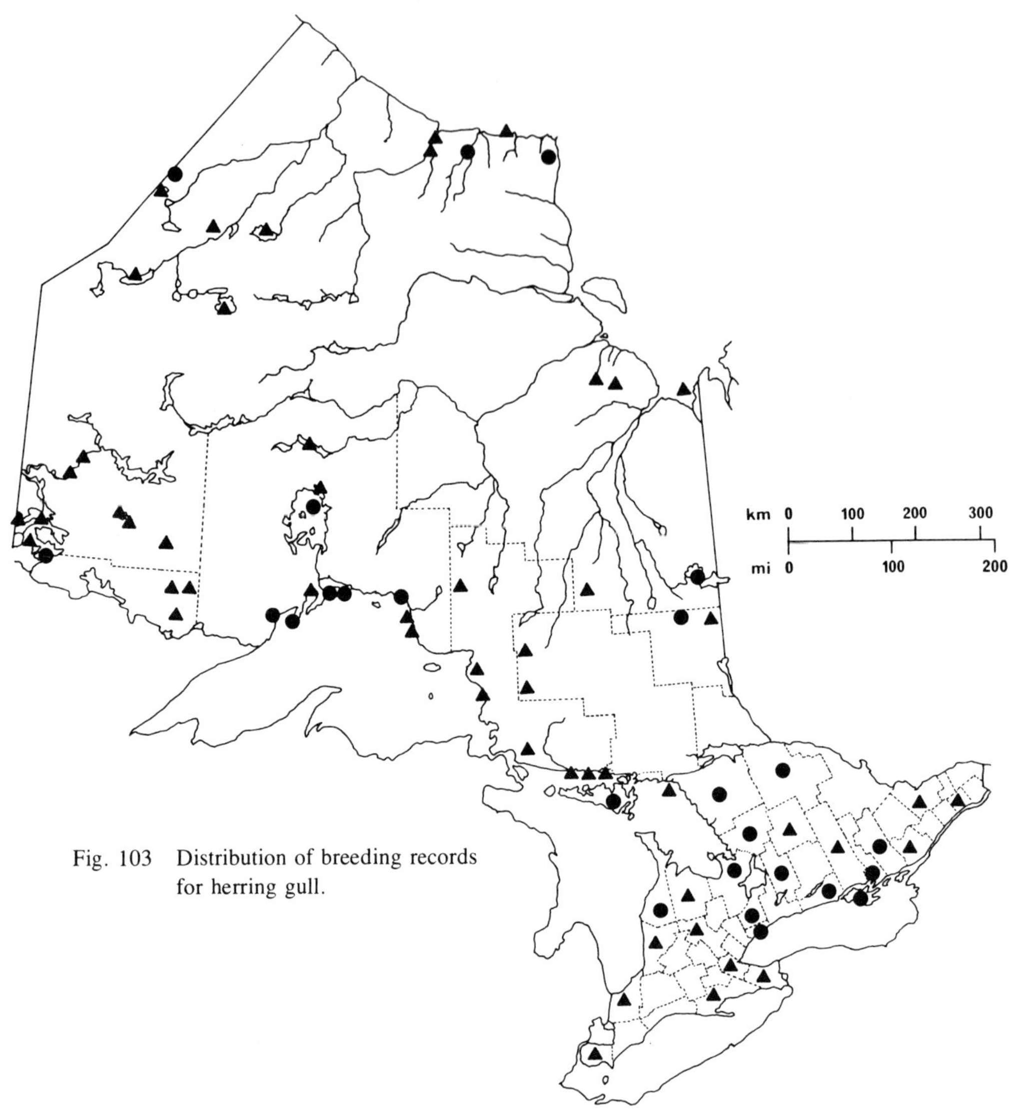

Fig. 103 Distribution of breeding records for herring gull.

Herring Gull, *Larus argentatus* Pontoppidan

Nidiology

RECORDS 512 (235 colonies, 148 isolated nestings, ca 16 811 nests) representing 32 provincial regions.

The herring gull is a much less colonial species than the ring-billed gull. It often nests singly; its colonies are more loosely knit, and usually its colonial nests are more widely spaced than those of the ring-billed gull. Most herring gull nests were on islands or rock islets (362 records); some were on the mainland (14 records); and a few were former nests of great blue heron, bald eagle, and osprey, situated in trees standing in water (6 records). Nest sites were located throughout the province, usually on freshwater lakes and rivers, with a few records from marine island and mainland locations in James and Hudson bays. Of 200 colonies, totalling 16 512 nests, the average size was 83 nests per colony. Some nestings were in mixed colonies with ring-billed gull; with common tern, arctic tern, and Caspian tern; and occasionally with great black-backed gull. Nest sites were also sometimes shared with other ground-nesting colonial species such as white pelican and double-crested cormorant. North of the Great Lakes, the numbers of isolated nestings increased, and nest sites were located on smaller inland lakes and on rivers with numerous shoals and small rock islets.

Island nests were usually on rock surfaces, with or without vegetation; sometimes on sand and gravel and in grassy and weedy areas; and occasionally elevated on lighthouses, shed roofs, and driftwood, and once on a leaning bush. Some nests were on islets in muskeg ponds and on sedge islets on tundra sloughs. Mainland nest sites were in landfill areas; on shorelines of rock, sand, and gravel; on cliffs; and, in 2 records, on a marshy pond edge and on a fertilizer elevator. Most nests were in the open but some were in vegetation, and a few were under bushes or at the bases of trees. Nests were usually on level surfaces and less often in rock crevices and indentations. Heights of elevated nests ranged from 1 m (3 ft) in a bush, to 12 m (40 ft) in a bald eagle nest, to 18 m (60 ft) on a fertilizer elevator. One nest was positioned 2 m (7 ft) from the nearest ring-billed gull nest.

Nests were usually slight to bulky raised mounds or platforms, with shallow to deep cups, and often with large bases; nests were less commonly depressions in sand, soil, or vegetative debris. They were variously composed of grasses, sedges and other aquatic vegetation, twigs, mosses, weed stalks, mud, feathers, bark, bones, string, and leaves. Nests were variously lined with grasses, feathers, conifer needles, plant fibre, lichen, bark, pebbles, string, and mosses. One nest was lined with fresh leaves. In 3 records no nest was built. Outside diameters of 5 nests ranged from 36 to 91.5 cm (14 to 36 inches); inside diameters of 3 nests ranged from 25.5 to 35.5 cm (10 to 14 inches); outside depths ranged from a few centimetres to 38 cm (15 inches); bowl depths ranged from almost flat to 20.5 cm (8 inches).

EGGS 5065 nests with 1 to 6 eggs; 1E (613N)—12.1%, **2E** (1096N)—21.6%, **3E** (3227N)—63.7%, **4E** (124N)—2.5%, 5E (2N)—0.04%, 6E (3N)—0.06%.
Average clutch range 3 eggs (3227 nests).
Because a number of the 1- and 2-egg clutches recorded were incomplete or partially hatched, the above percentages may be somewhat biased. Records for colonies visited before hatching show a larger percentage of 3-egg clutches. Single nests of this species have been recorded in both ring-billed gull and common tern colonies.

INCUBATION PERIOD No information.

EGG DATES 352 records, 14 April to 27 July (440 dates); 176 records, 26 May to 9 June. Because egg dates were often recorded late in the breeding season, on colony visits for the banding of young, the average egg dates may be somewhat biased.

Breeding Distribution

The herring gull nests throughout the province at any suitable body of water.

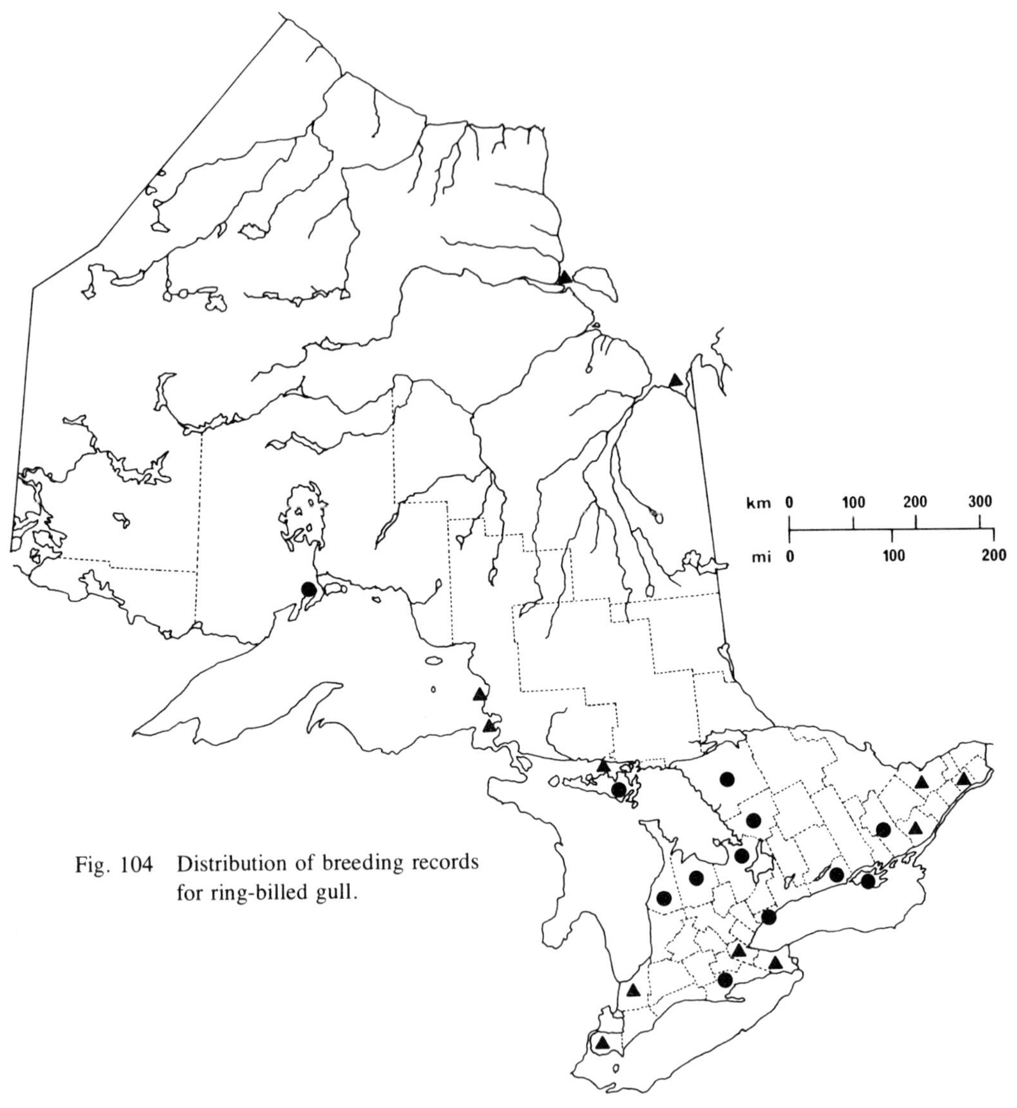

Fig. 104 Distribution of breeding records for ring-billed gull.

Ring-billed Gull, *Larus delawarensis* Ord

Nidiology

RECORDS 247 (101 colonies, 8 isolated nestings, ca 151 830 nests) representing 22 provincial regions.

A highly colonial species, the ring-billed gull usually breeds on islands, and only in the Great Lakes, the St. Lawrence and Ottawa rivers, and James Bay (100 records). Only occasionally does it breed on mainland sites, on the shores of the Great Lakes (7 records). The average size of 65 estimated colonies totalling 150 505 nests, was 2316 nests. This species sometimes nested in mixed colonies with herring gull, common tern, Caspian tern, and once with great black-backed gull.

Island nest sites were most often on solid rock surfaces, usually vegetated but sometimes barren, and less often in gravel and grassy areas. Mainland nest sites were on sand beaches or in gravel and rocky areas. They were usually vegetated. One record described the use of a rock log-boom crib as a nest site.

Nests were positioned most frequently in grasses, weeds, shrubs, and other vegetation; sometimes under small trees and bushes; occasionally on bare rock; and in 3 records, on driftwood.

Nests were usually raised mounds or platforms, but at times they were depressions or scrapes, variously formed of grasses, weed and plant stalks, aquatic plants, mosses, sticks, twigs, bits of wood, feathers, and fish bones. Nests were lined with grasses, weeds, feathers, bark chips, and aquatic and other vegetation. In a few records, no nest was built at all, and eggs were simply laid on the existing surface or in a scrape or hollow in soil or rock. Outside height of 1 nest was 3.8 cm (1.5 inches).

EGGS 38 919 nests with 1 to 6 eggs; 1E (526N)—1.4%, **2E** (3871N)—10%, **3E** (34 163N)—87.8%, **4E** (281N)—0.7%, 5E (68N)—0.2%, 6E (10N)—0.003%.
Average clutch range 3 eggs (34 163 nests).
Because a number of 1- and 2-egg clutches were undoubtedly incomplete, the above percentages are biased. Some 5- and 6-egg clutches may have involved laying by more than 1 female. An additional bias has been created because on some colony visits exact counts were recorded only of sets or 4 or more eggs. One egg was laid in a herring gull nest that contained 2 herring gull eggs; 2 isolated nests were situated in herring gull colonies.

INCUBATION PERIOD No information.

EGG DATES 98 records, 14 April to 30 July (118 dates); 49 records, 4 June to 18 June. Because egg dates were often recorded late in the breeding season on colony visits for banding of young, the average egg dates may be somewhat biased.

Breeding Distribution

The ring-billed gull has undergone a considerable increase in numbers in recent years. It breeds mainly on islands in all the Great Lakes and the St. Lawrence River. Single colonies on southern James Bay and on the Ottawa River have also been reported. It has recently (1978) been reported as breeding on an island in Akimiski Strait, James Bay (R.I.G. Morrison, pers. comm.).

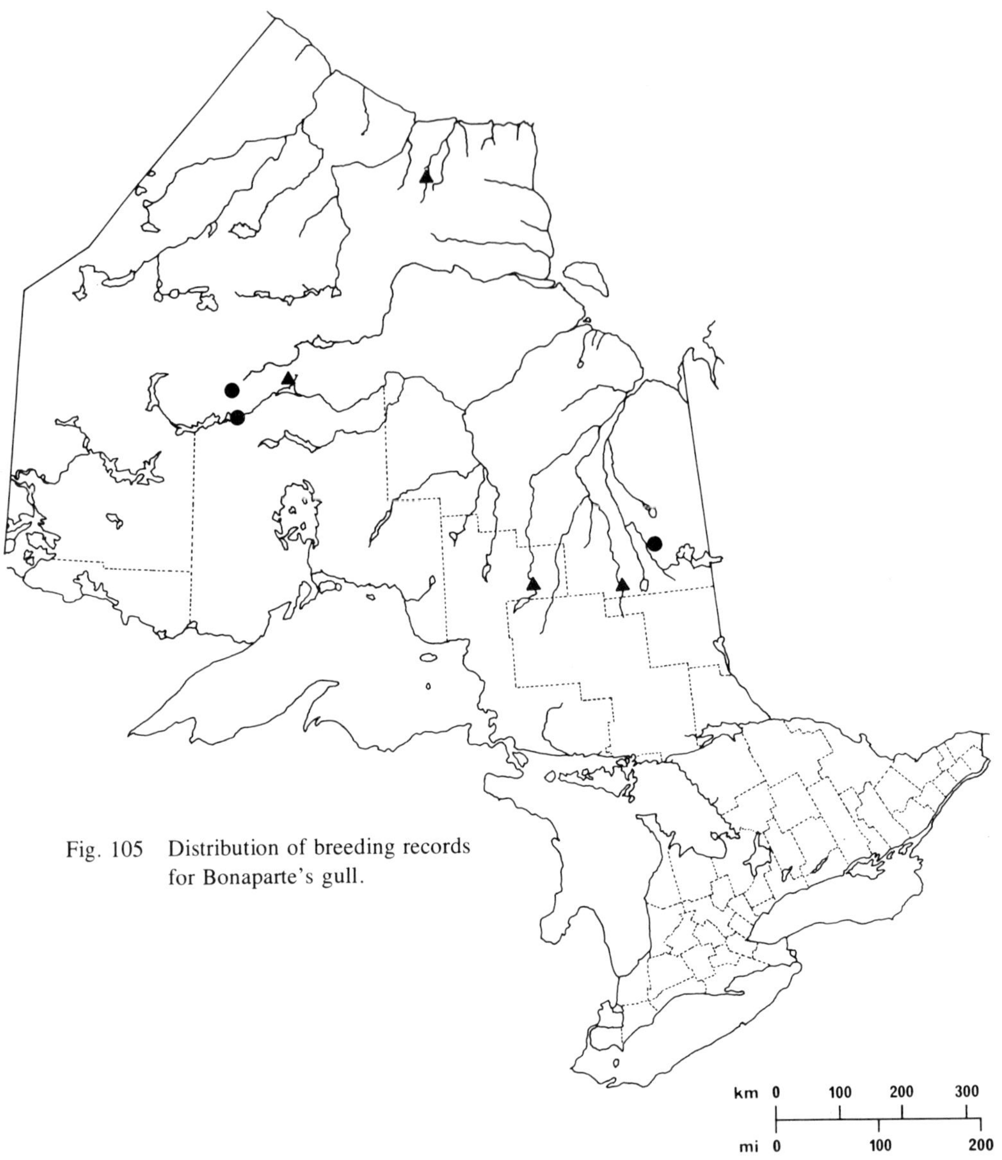

Fig. 105 Distribution of breeding records for Bonaparte's gull.

Bonaparte's Gull, *Larus philadelphia* (Ord)

Nidiology

RECORDS 6 nests representing 2 provincial regions.
This tree-nesting larid breeds in open black spruce muskeg, cedar-spruce bogs, and strips of black spruce along water edges. Nest locations ranged from the shores of lakes, sloughs, and rivers, to as far as 180 m (600 ft) distant from shore.

Most nests trees were black spruce, with 1 record of white cedar. One nest was on a leaning branch and another on small branches against the trunk; 1 nest was situated near the crown and another at ca two-thirds the height of the tree. Heights of 6 nests ranged from 3 to 7 m (10 to 23 ft), with 4 averaging 3.5 to 5 m (11 to 16 ft).

Nests were crudely made, with very shallow bowls. Outside diameter of 1 nest was 25.5 cm (10 inches).

Nests were composed of small twigs, mosses, lichens, and reeds and other marsh vegetation.

EGGS 6 nests with 2 to 3 eggs; 2E (2N), 3E (4N).

INCUBATION PERIOD No information.

EGG DATES 6 nests, 29 May to 19 July.

Breeding Distribution

Although few nest records exist for Bonaparte's gull in Ontario, this species appears to be widely distributed and breeds throughout the forested portions of the north, extending south as far as northern Thunder Bay, northern Algoma, and southern Cochrane Districts. The first flightless young was collected in 1937 (Baillie, 1939, 1960), and established the species as a breeding bird in the province. However, the inaccessibility of much of the northern forest and the nomadic habits of this gull have combined to prevent a clearer picture of its breeding distribution and nest habits in Ontario.

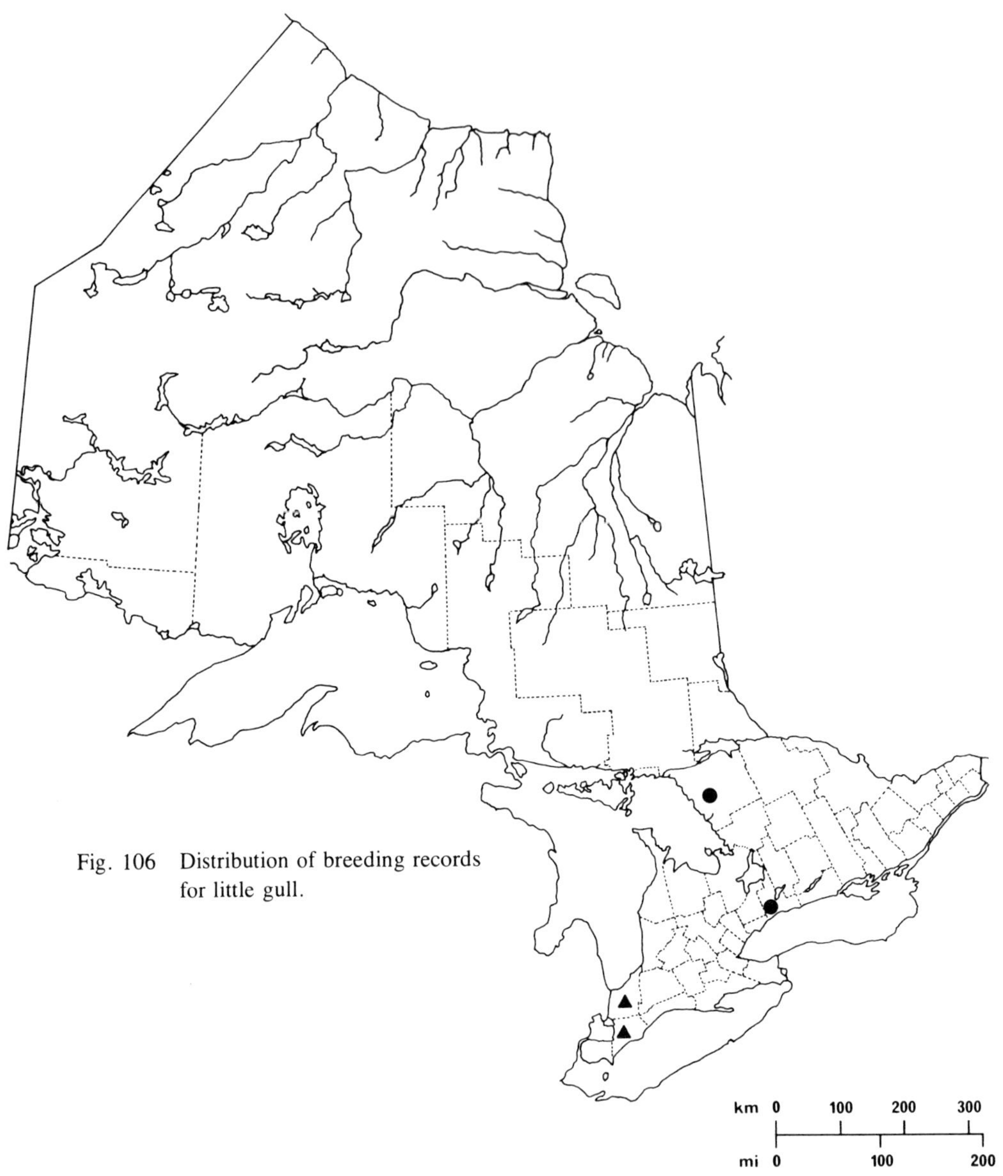

Fig. 106 Distribution of breeding records for little gull.

Little Gull, *Larus minutus* Pallas

Nidiology

RECORDS 15 (4 colonies, 2 isolated nestings, 15 nests) representing 3 provincial regions. This Eurasian straggler winters regularly in small numbers on the maritime coast and on the Great Lakes. The first record of nesting in the western hemisphere was in 1962 in Ontario, at Oshawa, in what is locally named "Second Marsh". Since then, the species has nested on at least 6 other occasions in Ontario, sometimes with black tern, once with Forster's tern, and once with common tern.

Breeds in large freshwater marshes of cattail, bulrush, and bur-reed. Nests are usually situated in open areas, but sometimes in dense cattail borders. In 1 recent record, nesting occurred on a vegetated, flat limestone island near a common tern colony.

Nests were placed on floating mats of cattail, bur-reed, and other marsh vegetation (6 nests); on semi-floating islands of reeds, mud, and decaying vegetation that were 90 to 120 m (300 to 400 ft) in length (3 nests); on small mudflats (2 nests); on muskrat "push-ups" (2 nests); and on living bulrush "islets" formed at submerged tree branches and stumps (2 nests). Water depths near 5 nests ranged from 20 to 67 cm (8 to 30 inches).

Nests were composed mainly of dry plant material (cattails, bulrushes, bur-reeds, and grasses), occasionally with some green algae and decaying vegetation included; nests had shallow cups. Rim heights above water in 2 records were 7 and 9 cm (2.8 and 3.5 inches); outside diameters of 2 nests ranged from 16.5 to 20 cm (6.5 to 8 inches); inside diameter of 1 nest was 12 cm (4.75 inches); inside depths of 2 nests were 2 and 2.5 cm (0.8 and 1 inch).

EGGS 15 nests with 1 to 3 eggs; 1E (2N), **2E** (1N), **3E** (12N).
Average clutch range 3 eggs (12 nests).

INCUBATION PERIOD 1 nest, ca 21 days.

EGG DATES 15 nests, 27 May to 7 July (27 dates); 8 nests, 12 June to 15 June.

Breeding Distribution

The first record of a little gull in Ontario was in 1930. Since then its numbers slowly increased until it became a regular visitor in small numbers (Baillie, 1963). The first known nesting of the species in North America occurred in 1962 at Oshawa, in Ontario (Scott, 1963). Since then nesting has not been reported annually, but this gull may well be nesting in one or more parts of southern Ontario each year. Details of the first several years of nesting in Durham RM (Fig. 148B) and in Kent County are to be found in Tozer and Richards (1974). A third nesting region was reported in 1979, in Lambton County, and a fourth in 1980, on North Limestone Island, Parry Sound District (ROM PR 1164).

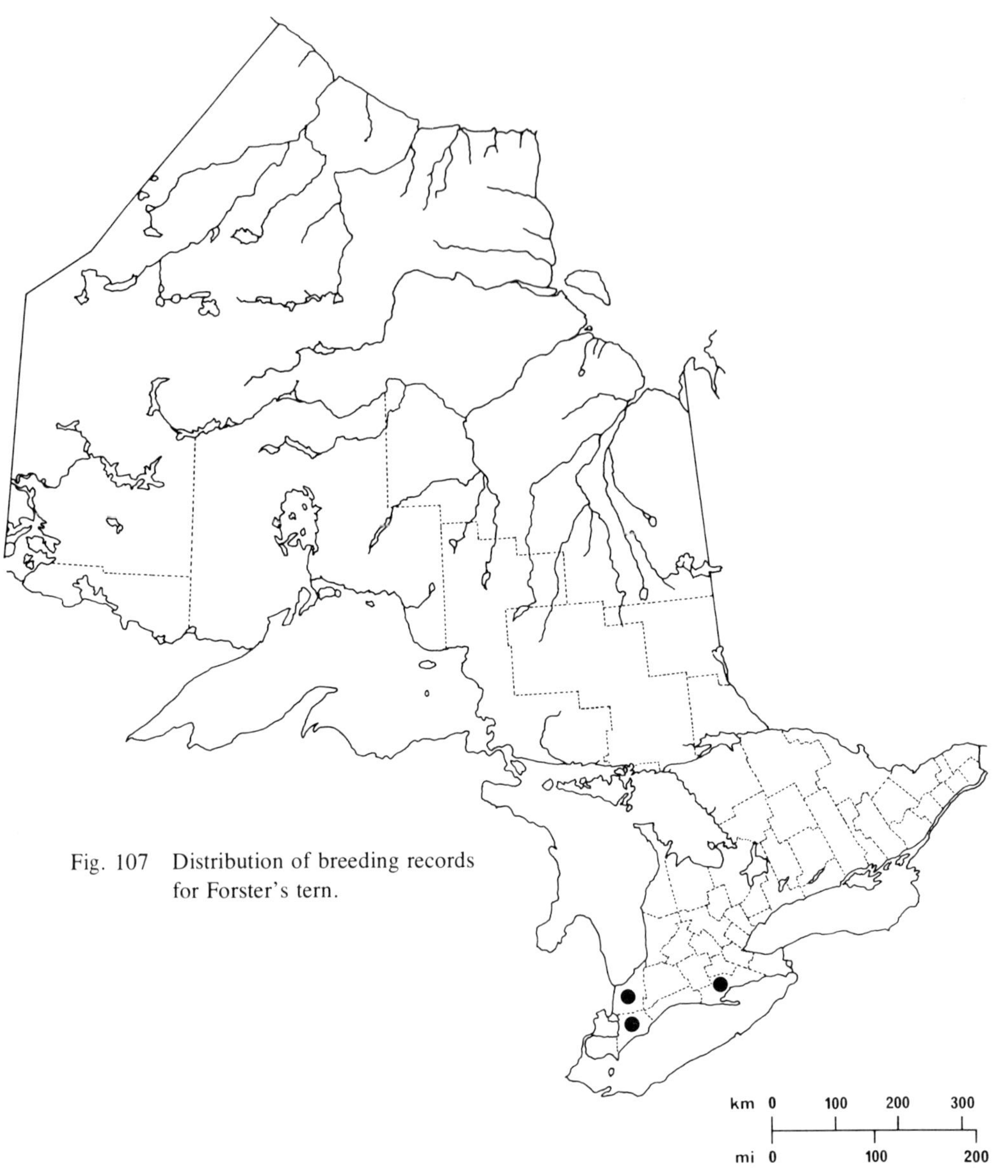

Fig. 107 Distribution of breeding records for Forster's tern.

Forster's Tern, *Sterna forsteri* Nuttall

Nidiology

RECORDS 25 (6 colonies, 34 nests) representing 3 provincial regions.
Since 1976 this western, marsh-nesting, colonial species has been documented at 3 locations in Ontario, in 6 colonies, and the average colony size was ca 6 nests.

Nests were in large cattail marshes situated near the shores of islands, peninsulas, and bays of large freshwater lakes. In at least 3 records this tern nested near nests of black tern and once near a single nest of little gull. In 1 location 2 colonies were 274 m (900 ft) apart. Distances between individual nests in a colony ranged from 1.8 to 27 m (6 to 90 ft).

Nests were placed on floating cattail mats (18 nests) and on muskrat houses (5 nests). One nest was on a floating grass mat 7.5 cm (3 inches) thick. Heights of 6 nests ranged from 5 to 25 cm (2 to 10 inches) above water.

Nests were depressions in the accumulated vegetation, with dead cattail stems and grasses loosely arranged around the depressions. Outside diameters of 2 nests were 53 by 66 cm (21 by 26 inches); inside diameters of 2 nests were 15 by 15 cm (6 by 6 inches) and 7.5 by 12.5 cm (3 by 5 inches).

EGGS 33 nests, with 1 to 3 eggs; 1E (2N), **2E** (6N), **3E** (25N).
Average clutch range 3 eggs (25 nests).

INCUBATION PERIOD No information.

EGG DATES 6 colonies (33 records), 22 May to 6 June (7 dates); 3 colonies, 29 May.

Breeding Distribution

The Forster's tern (Fig. 148A) was reported as breeding commonly in the Lake St. Clair marshes in the late 1800s (Baillie, 1958), although no material evidence was secured at that time. In this century no reports of nesting were secured for more than 75 years. In 1976 a colony was located at Long Point, Lake Erie (Peck, 1976), and subsequently two more small colonies were found at Lake St. Clair. As far as we are aware the species is restricted to the extreme south of the province.

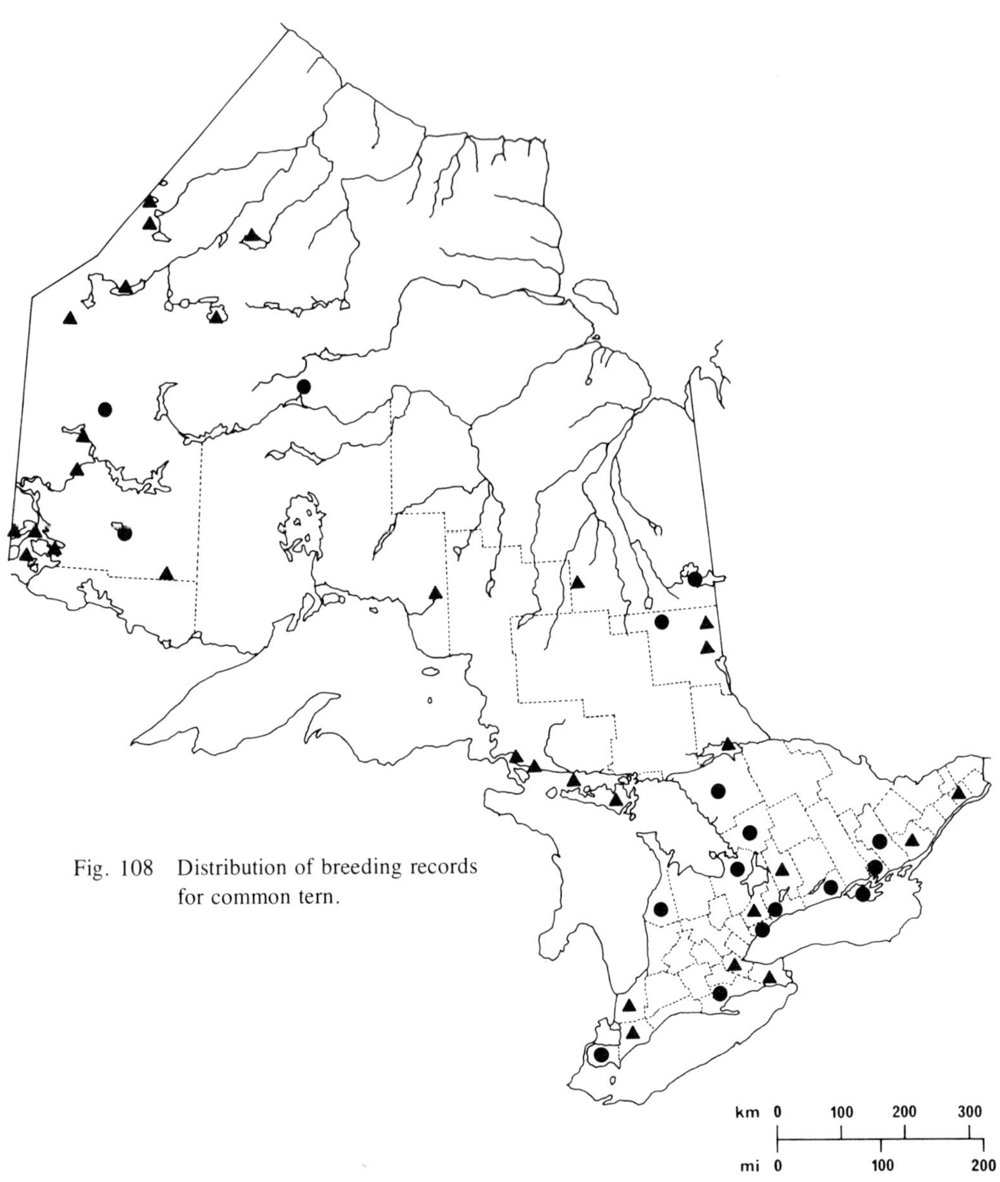

Fig. 108 Distribution of breeding records for common tern.

Common Tern, *Sterna hirundo* Linnaeus

Nidiology

RECORDS 432 (86 colonies, ca 18 553 nests) representing 26 provincial regions.
This colonial species breeds on islands and shores and in marshes of lakes and rivers. Breeding sites recorded included rocky, grassy/weedy, and sandy areas of barren or vegetated islands (68 records); areas of sand, rock/gravel, landfill, and grass on mainland shores (42 records); growths of cattail, marsh islets, and muddy or boggy edges of marshes (22 records).

The average size of all estimated colonies was ca 46 nests. Nest sites were often shared with herring gull, ring-billed gull, and Caspian tern, and less often with little gull, Forster's tern, and black tern.

Nests were variously situated on or in rocks; on sand and gravel areas; in grasses, weeds, stonecrop, mosses, and nettles; among debris (driftwood, dead fish, and dead vegetation); on muskrat houses; in cattails; on floating vegetation and logs; and on mud banks. One nest was on a stump 0.6 m (2 ft) above water.

Nests were usually depressions, but sometimes were flat, and occasionally were raised mounds. At times no nest was built. Nest structures ranged from simple to elaborate bowls of vegetation. Nests were both lined and unlined; nest materials were, in order of frequency, fine grasses and weeds, small twigs and debris, rushes/cattails and other marsh vegetation, pebbles, straw, feathers, stonecrop, mosses, lichen, and leaves. In 1 record, a double nest with eggs in each bowl was reported.

EGGS 5094 nests with 1 to 7 eggs; 1E (871N)—17.1%, **2E** (1367N)—26.8%, **3E** (2798N)—54.9%, **4E** (50N)—1%, 5E (7N)—0.1%, 7E (1N)—0.02%.
Average clutch range 3 eggs (2798 nests).
A number of 1- and 2-egg clutches were undoubtedly incomplete and this biases the above percentages. Sets of 5 and 7 eggs may have involved laying by more than 1 female.

INCUBATION PERIOD 1 nest, 22 to 23 days.

EGG DATES 220 records, 8 May to 17 August (275 dates); 110 records, 5 June to 24 June.

Breeding Distribution

The common tern breeds throughout the province except in a strip along the Hudson Bay coast.

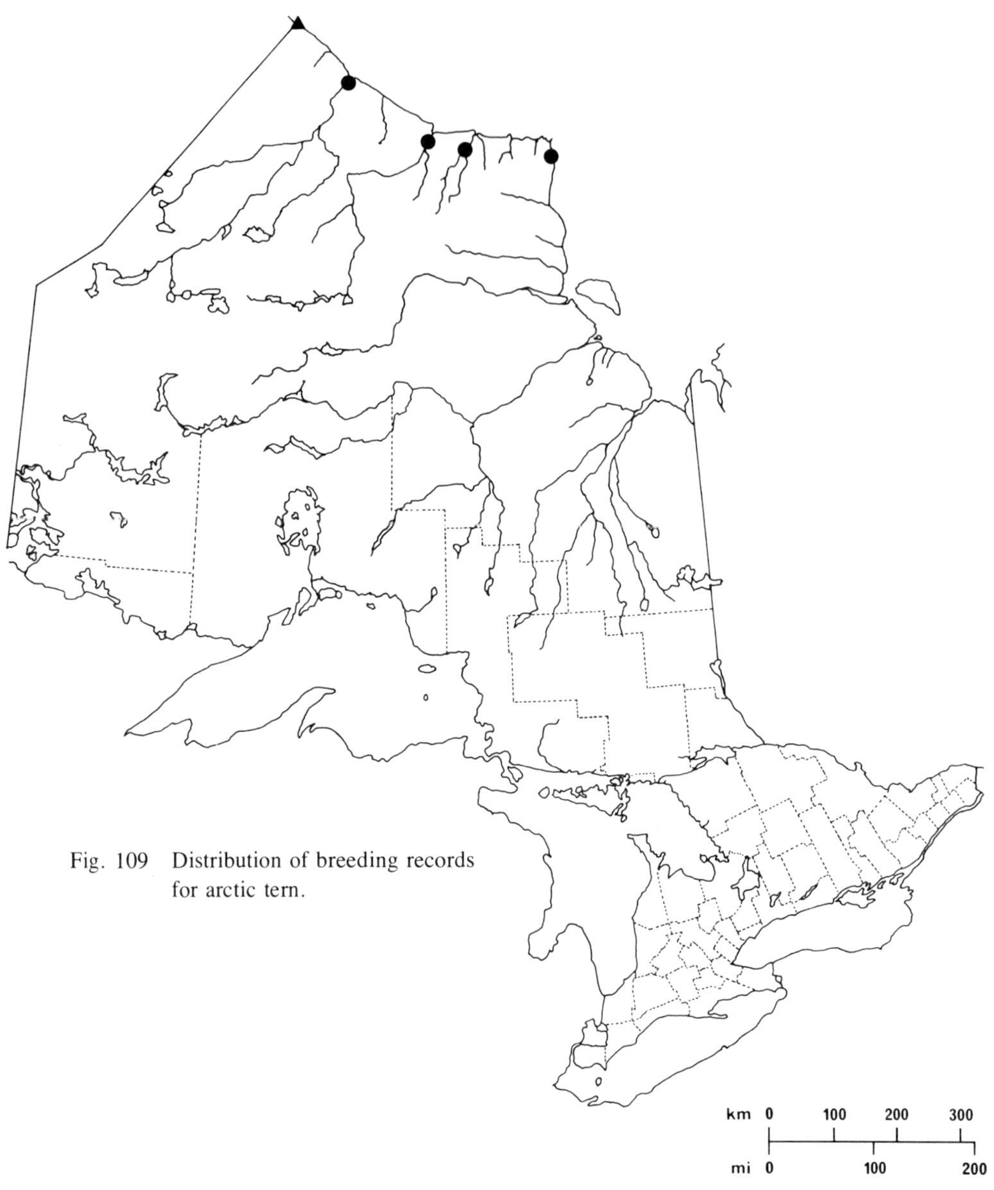

Fig. 109 Distribution of breeding records for arctic tern.

Arctic Tern, *Sterna paradisaea* Pontoppidan

Nidiology

RECORDS 6 (4 colonies, 13 nests) representing 1 provincial region.
The arctic tern is usually a colonial species and breeds on gravel bars, tidal flats, islands in coastal river estuaries, and islands close to the coasts. Recorded nest locations were sand and gravel areas, which in some cases were overgrown with grasses, scrub willow, and poplar trees, with a maximum height of 1.5 m (5 ft).

Nests were scrapes or depressions in gravelly ground, sometimes lined with a few tiny twigs.

EGGS 13 nests, with 1 to 2 eggs; 1E (4N), 2E (9N).
Average clutch range 2 eggs (9 nests).

INCUBATION PERIOD No information.

EGG DATES 5 records (3 colonies, 1 isolated), 16 June to 16 July (6 dates).

Breeding Distribution

The arctic tern breeds along the tundra portion of the James Bay and Hudson Bay coasts and inland for a short distance along some of the major rivers. The first evidence of breeding in Ontario was obtained in 1940 at Fort Severn on Hudson Bay (Baillie, 1961), but only a few nesting localities have been found subsequently. The first of these was on the James Bay coast, 20 miles south of Cape Henrietta Maria, where three nests were discovered in 1948 (Peck, 1972). Nesting was noted at the Sutton River in 1962, at East Penn Island in 1963, and at the Winisk River in 1965, all localities near the Hudson Bay coast in Kenora District. Even more recently, the species was reported in 1978 as breeding on an island in Akimiski Strait, James Bay (R.I.G. Morrison, pers. comm.).

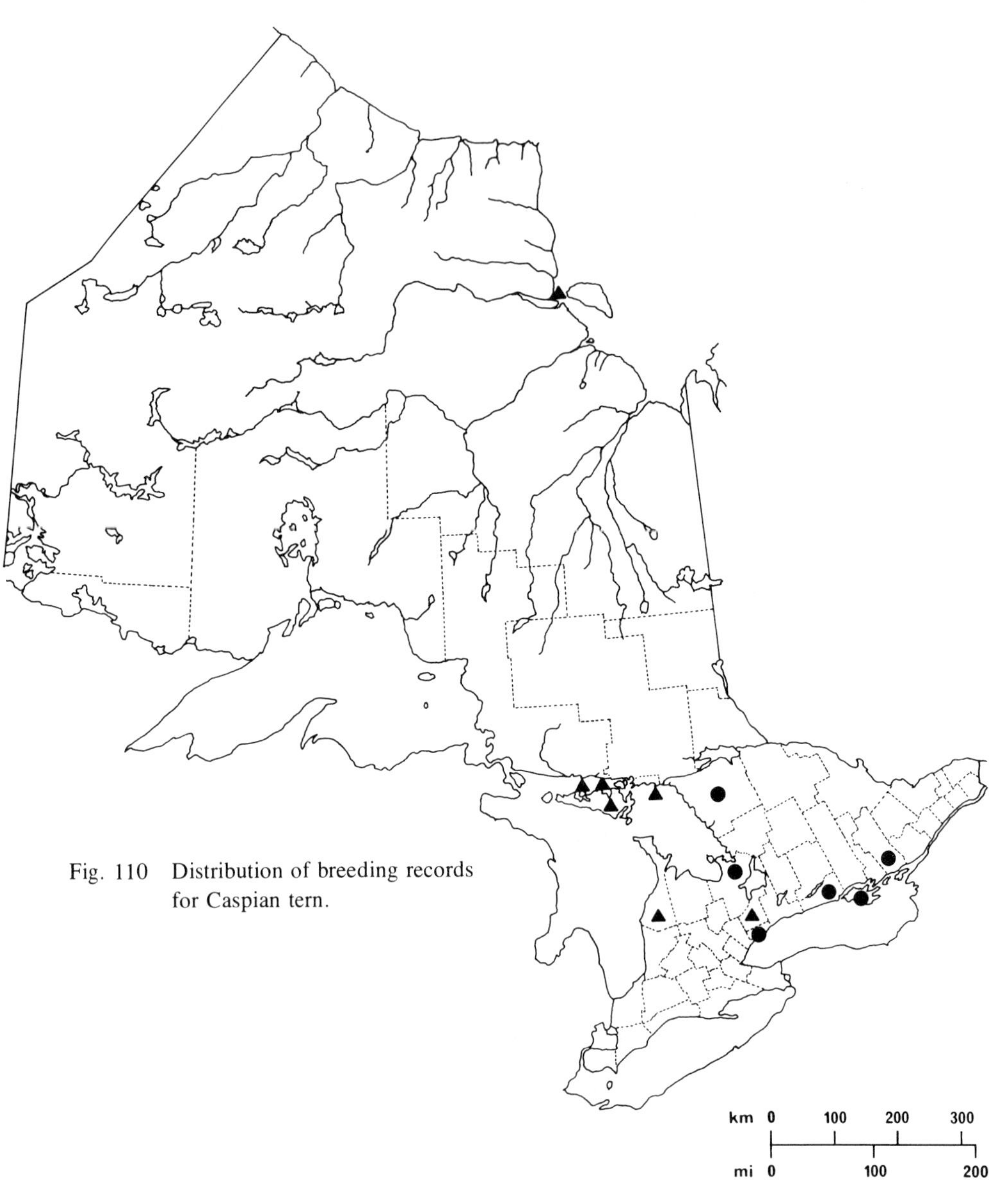

Fig. 110 Distribution of breeding records for Caspian tern.

Caspian Tern, *Sterna caspia* Pallas

Nidiology

RECORDS 71 (ca 3240 nests) representing 10 provincial regions.
The Caspian tern is primarily a colonial species and has been known to breed in Ontario since 1891. Nineteen nesting sites have been recorded (16 colonial, 3 isolated), most of them used again in successive years: Lake Huron, 11 sites; Lake Ontario, 7 sites; and Lake Simcoe, 1 site. The average size of 39 colonies was ca 83 nests. Nesting sites were often shared with herring gull, ring-billed gull, and common tern.

With 1 exception (a manmade landspit), all nesting sites were on shoals and islands in large lakes. These nest islands were small (up to 1.6 km [1 mi] in length) and had rock or gravel and sand surfaces. Several nest islands were formed of limestone. Most islands were covered with low vegetation and some trees.

Nests were placed on raised ridges or mounds if these were present, and were merely depressions in sand, soil, and pebbles, or among rocks. They were most often in open situations, but sometimes in or near vegetation. Some nests had little or no lining, but most were variously lined with grasses and dead vegetation, small pebbles, shell fragments, fish bones, twigs and driftwood, and feathers. At 1 colony, a number of nests were depressions in mounds or bowls of living stonecrop.

EGGS 799 nests with 1 to 3 eggs; **1E** (273N)—34.2%, **2E** (415N)—51.9%, **3E** (111N)—13.9%.
Average clutch range 2 eggs (415 nests).
A number of 1-egg sets may have been incomplete or replacement clutches.

INCUBATION PERIOD 1 nest, ca 21 days.

EGG DATES 34 records, 13 May to 19 July, (38 dates); 17 records, 6 June to 23 June.

Breeding Distribution

The Caspian tern has apparently not always been a resident of Ontario. The first colony was reported in 1891 (Baillie and Harrington, 1936). Since then, the species has slowly increased in numbers until today about 20 colonies are known on lakes Ontario, Huron, and Simcoe. It was recently reported as breeding on an island in Akimiski Strait, James Bay (R.I.G. Morrison, pers. comm.).

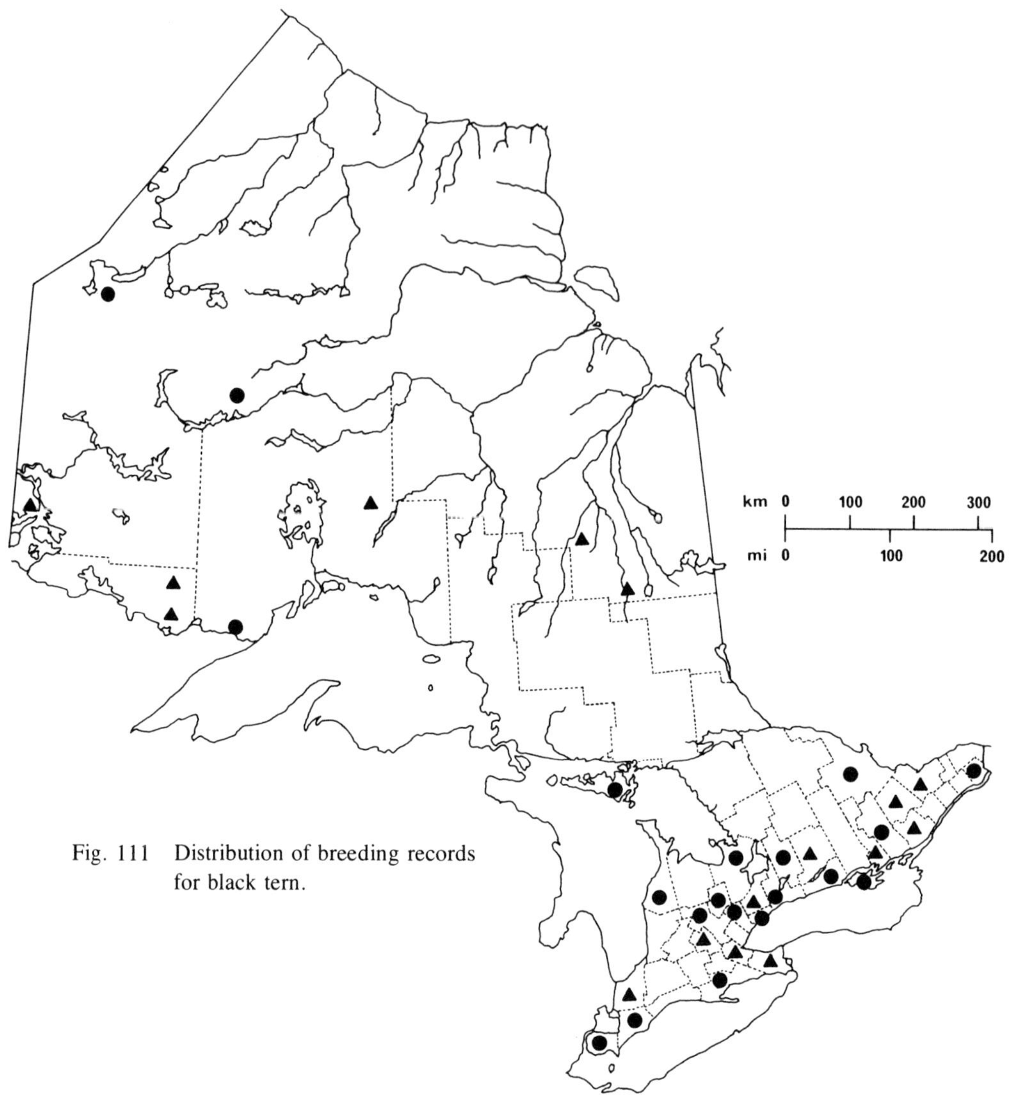

Fig. 111 Distribution of breeding records for black tern.

Black Tern, *Chlidonias niger* (Linnaeus)

Nidiology

RECORDS 722 (153 colonies, 17 probably isolated nestings, ca 1166 nests) representing 31 provincial regions.

The black tern is a colonial species that usually breeds in small, loosely knit colonies, but occasionally breeds singly. The average size of 102 estimated colonies was 9 nests, but this figure is undoubtedly too low owing to the loose nature of these aggregations and the consequent difficulty in locating all nests. A few nest sites were shared with little gull, Forster's tern, and common tern.

Breeds most commonly in wet cattails (Fig. 147), bulrushes and sedges found in marshes, on marshy borders of lakes, rivers, and ponds, and on marshy islands and islets. Sometimes it breeds in wet heath bogs and in flooded areas of willow, both with some marsh vegetation. On rare occasions, it breeds in drier sites among grasses and marsh vegetation. In some nest habitats, common reed grasses, wild rice, horsetails, leatherleaf, and sweet gale predominated.

Nests were most frequently positioned on either unanchored or anchored floating bases (405 records) and less frequently on more solid bases (143 records). The majority of floating nests were on mats of dead vegetation, but some were on pieces of wood such as boards and doors (26 records) and on logs (14 records). In 1 record eggs were laid on a lily pad without a nest. The solid-base nests were on raised mud patches (59 records), on piles of nonfloating vegetation (44 records), on muskrat houses (39 records), and in 1 record, among upturned tree roots with attached vegetation. Water depths ranged from 7.5 to 122 cm (3 to 48 inches); nest heights above water or ground ranged from 2.5 to 5 cm (1 to 2 inches) and were as much as 76 cm (30 inches) at muskrat-house nest sites.

Nests were most commonly depressions in nest-site material—in 4 records, they were in depressions in logs—and were less often built-up mounds of dead vegetation. Most nests were variously lined, in order of preference, with dry, and less often with wet, pieces of cattail, bulrush, sedge, common reed grass, grass, weeds, and water-lily roots, as well as with mud, peat, algae, sphagnum moss, and sticks. In 4 records no nest was built. One clutch was laid in a nest of pied-billed grebe. Outside diameters of 4 nest platforms ranged from 25.5 to 30.5 cm (10 to 12 inches); inside diameters of 5 nests ranged from 10 to 15 cm (4 to 6 inches); thicknesses of 4 nests ranged from 2.5 to 7.5 cm (1 to 3 inches).

EGGS 881 nests with 1 to 5 eggs; 1E (87N)—9.9%, **2E** (184N)—20.9%, **3E** (597N)—67.8%, **4E** (11N)—0.5%, 5E (2N)—0.2%.
Average clutch range 3 eggs (597 nests).

INCUBATION PERIOD 12 nests, 19 to 21 days: 3 of 19 days, 2 of ca 19 days, 1 of 20 days, 2 of ca 20 days, 3 of 21 days, 1 of ca 21 days. Incubation usually commenced with the last egg, but in 1 nest it began before the clutch was complete.

EGG DATES 145 records, 17 May to 24 July (209 dates); 73 records, 31 May to 21 June. Eggs were usually laid daily but lapses of 2 to 5 days were reported in 3 records. In 1 nest, 2 additional eggs were laid following the loss of 2 of the original 3 eggs.

Breeding Distribution

The black tern breeds throughout the agricultural part of southern Ontario, but becomes thinly scattered in the Canadian Shield, where suitable nesting habitat is scarce. There are records as far north as Fort Albany (Macoun and Macoun, 1909) and Favourable Lake, and it may nest even farther north.

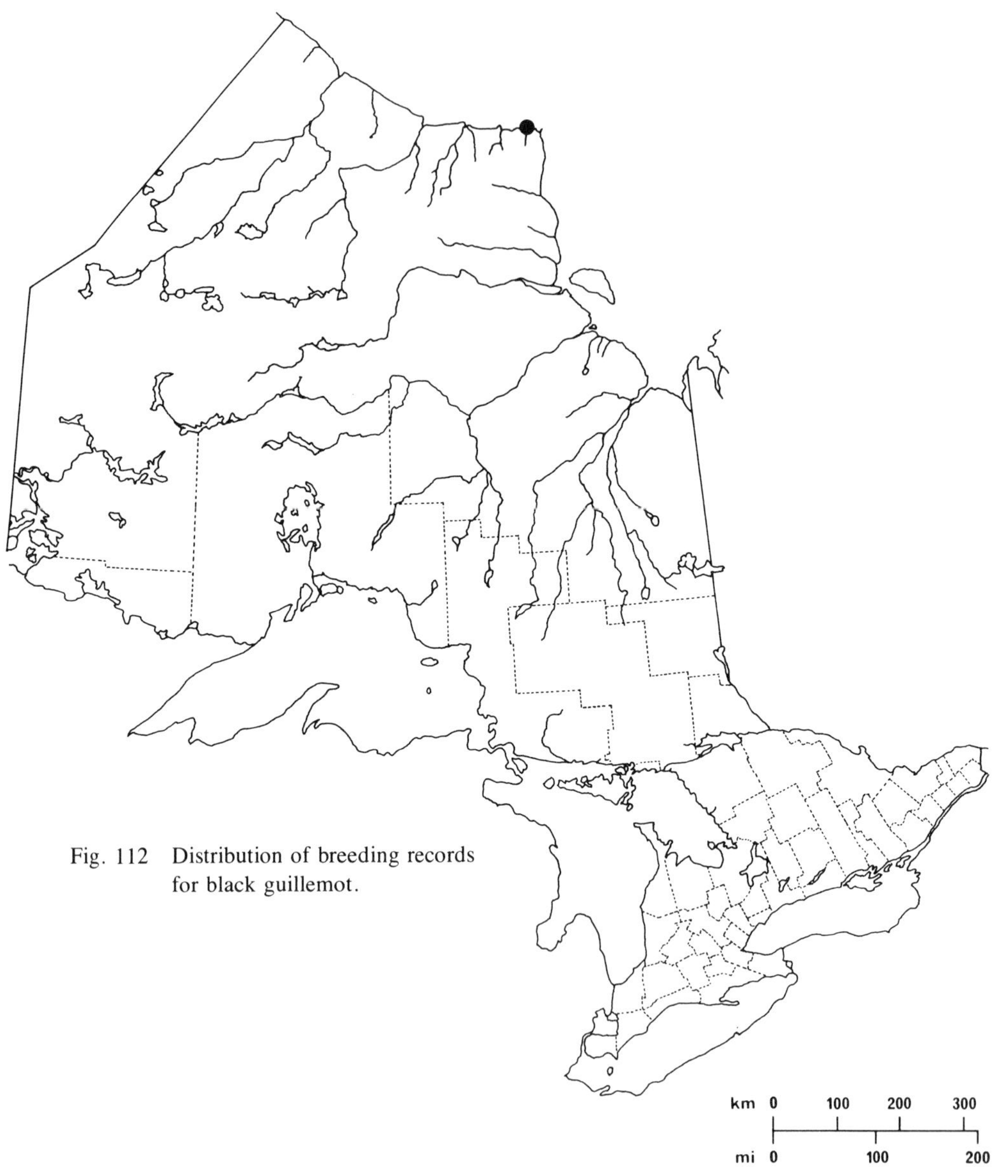

Fig. 112 Distribution of breeding records for black guillemot.

Black Guillemot, *Cepphus grylle* (Linnaeus)

Nidiology

This marine alcid has been observed in Ontario along the coasts of James and Hudson bays since 1942. It was added to the provincial breeding-bird list on 12 July 1957, when 80 adults were seen among the boulders on rocky Manchuinagush Island (which becomes a peninsula at low tide) on Hudson Bay, about 35.5 km (22 mi) west of Cape Henrietta Maria. Two adults (ROM 76854–76855) and the remains of a hatched egg (ROM 7512) were collected from the site (Lumsden, 1959b).

Breeding Distribution

The above record constitutes the only available evidence of nesting in Ontario by the black guillemot. Although this and earlier reports of groups of summering adults suggest that other nestings occurred, no recent records are available to indicate further breeding (H. G. Lumsden, pers. comm.).

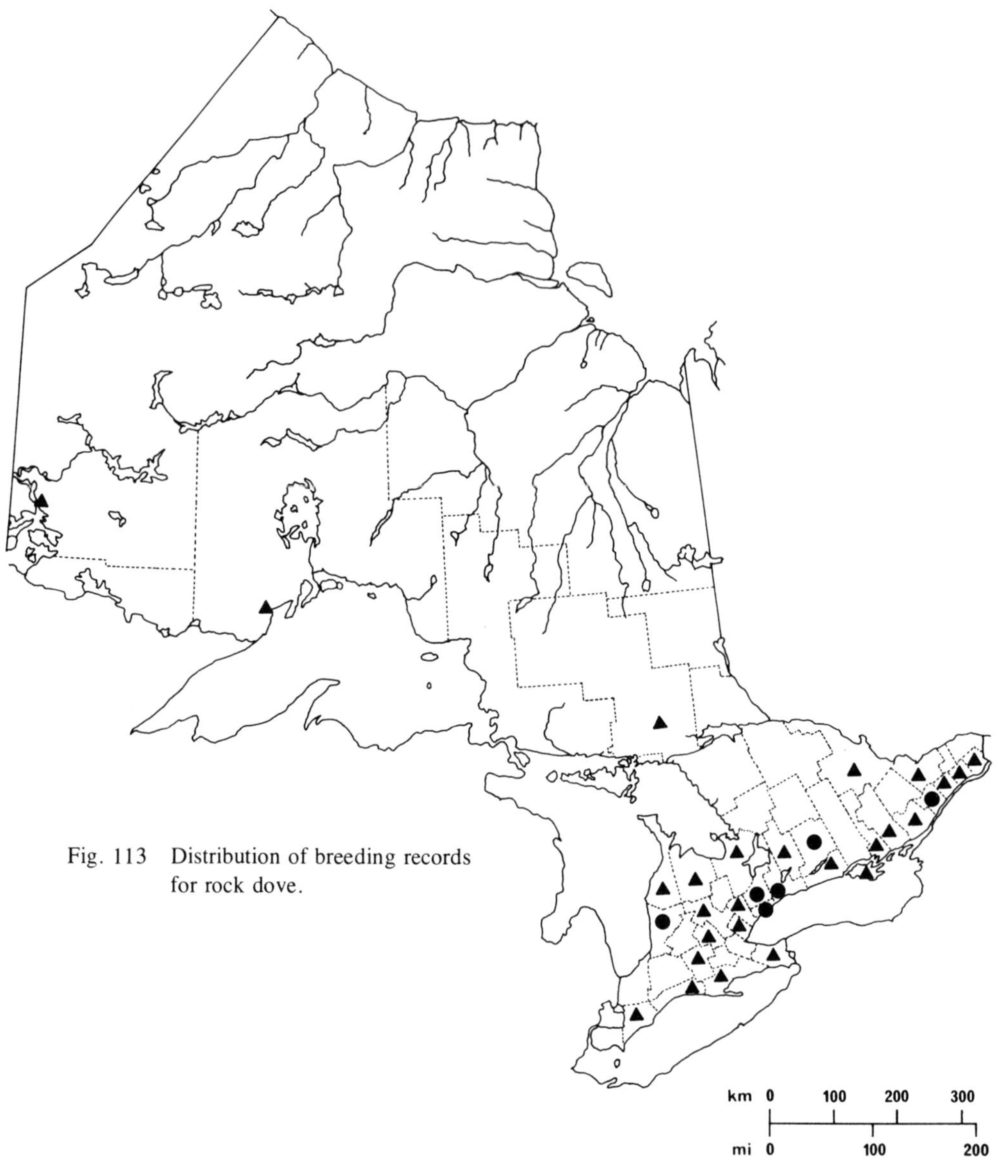

Fig. 113 Distribution of breeding records for rock dove.

Rock Dove, *Columba livia* Gmelin

Nidiology

RECORDS 151 (179 nests) representing 30 provincial regions.
Nest records for feral populations of this domesticated species were almost equally divided between those in rural areas (80 nests) and those in urban areas (76 nests).

All nests were in or on buildings and other manmade structures. In rural areas, 65 nests were in barns, silos, and pigeon coops, and 14 nests were under bridges. One nest was on a high ledge of a stone quarry, which closely approximated a natural cliff site. In urban areas, 59 nests were, in order of frequency, in or on houses, apartments, stores, unspecified buildings, church steeples, stadiums, bandstands, and garages, and 17 nests were under bridges.

Nests were variously positioned on barn beams, rafters, and metal tracks (51 nests); under eaves and cornices of roofs (26 nests); on window sills and various ledges (24 nests); on bridge girders (20 nests); on floors and platforms (5 nests); in nest boxes (4 nests); on eavestroughs (3 nests); on silo steps (2 nests); and on quarry ledges (1 nest). Heights of 122 nests ranged from 1 to 27.5 m (3 to 90 ft), with 61 averaging 4.5 to 9 m (15 to 30 ft). One nest in a barn was built on top of an old nest of barn swallow, situated on a metal hay-lift track.

Nests ranged from bulky, untidy platforms, with flat or slightly hollowed tops, to scant collections of straw and hay. Sometimes no nest was built. Most nests were accumulations of various materials, which included, in order of preference, straw, hay and grasses, feathers, weed stalks, twigs, twine and wire, paper, leaves, and mud. Many nests, particularly those containing young, were covered with excrement. Outside diameters of 7 nests ranged from 12.5 to 30.5 cm (5 to 12 inches).

EGGS 112 nests, with 1 to 3 eggs; **1E** (22N), **2E** (87N), 3E (3N).
Average clutch range 2 eggs (87 nests).
Some of the 1-egg clutches were incomplete or had suffered egg loss, and the 3-egg clutches may have been the product of 2 females.

INCUBATION PERIOD 2 nests: 1 of at least 17 days and 1 of at least 19 days. Three nests produced 2 broods, 3 broods, and 4 broods respectively, each in a single year.

EGG DATES 78 nests, 27 January to 24 December (90 dates); 39 nests, 16 April to 15 June. The height of the season is less significant for this species because, at least in southern Ontario, it nests throughout the year.

Breeding Distribution

The rock dove was introduced into North America probably with the arrival of the first settlers. Many are still maintained in captivity or in a semi-wild state, but truly feral birds have been breeding in Ontario for many years. The species breeds throughout most of southern Ontario, although it largely avoids forested areas. In northern Ontario, it occasionally nests near settlements and farms, as far north as Kenora, Thunder Bay, and Cochrane.

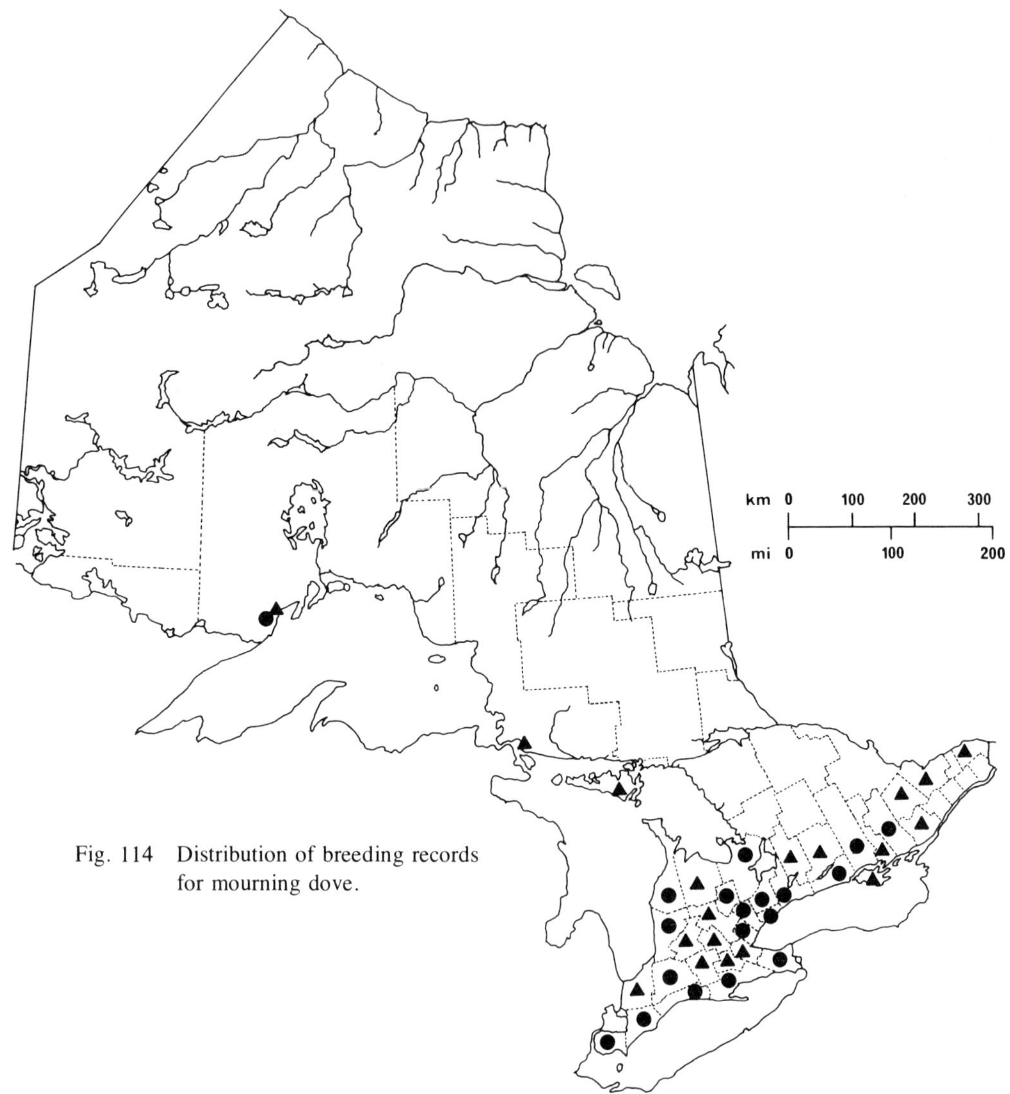

Fig. 114 Distribution of breeding records for mourning dove.

Mourning Dove, *Zenaida macroura* (Linnaeus)

Nidiology

RECORDS 1602 (1815 nests) representing 36 provincial regions.
Breeds in a wide variety of habitats. Areas recorded included shrubby fields and pastures, hawthorn meadows, and orchards (89 records); deciduous woods, hedgerows, and treed sand dunes (72 records); coniferous plantations, woods and hedgerows (51 records); mixed woods (28 records); residential gardens, lawns, cemeteries, and parks (62 records); and marshes, marsh edges, dykes, swamps, willow swales, river valleys, flood plains, and

beaver ponds (45 records). Occasionally this species appeared to be almost colonial, especially when it nested in small, dense coniferous groves; 1 group of more than 200 nests was reported in such a restricted area.

Most nests were elevated above ground, usually in trees, shrubs, and vines (121 records), and less often in or on fallen trees (10 records), on stumps (9 records), in cattails (3 records), on suspended pieces of bark (2 records), in tree and log cavities (2 records), and on rail fences (1 record). A few nests were on the ground (12 records). Most shrub and tree nests were in crotches and forks at or near the trunk, but many were situated 0.3 to 3 m (1 to 10 ft) from the trunk on branches that were usually horizontal. Coniferous trees (6 spp., 154 nests) were preferred to deciduous trees and shrubs (20 spp., 137 nests), and the types most commonly selected were spruce spp. (52 nests), pine spp. (49 nests), hawthorn spp. (46 nests), cedar and juniper spp. (45 nests), and apple spp. (30 nests). Some nests were in close proximity to other nests of the same species and to nests of other species such as American robin and common grackle. In 2 records the common grackle nests were in the same tree as the mourning dove nests. Heights of 325 nests ranged from on the ground to 12.2 m (40 ft), with 163 averaging 1.5 to 3 m (5 to 10 ft).

Nests ranged from flat to shallow-cupped platforms and were often frail and flimsy, but occasionally well built and substantial. Most nests were newly constructed but this species also used old nests of American robin (45 records), common grackle (13 records), and other species such as green heron, eastern kingbird, common crow, brown thrasher, gray catbird, cedar waxwing, and cardinal. Three nests were built on an old squirrel drey, a suspended basket, and a broken wren nest box respectively. Two nests were reused in successive years.

Nests were composed of small deciduous and coniferous sticks and twigs, grasses, weed stalks, rootlets, conifer needles, leaves, bark, feathers, and vine tendrils. Nests were variously lined with grasses, small twigs, rootlets, plant stalks, and once with bits of fine coloured wire. Outside diameters of 5 nests ranged from 12.5 to 20.5 cm (5 to 8 inches); inside diameters of 2 nests were 9 and 10 cm (3.5 and 4 inches); outside depths of 3 nests ranged from 5 to 10 cm (2 to 4 inches).

EGGS 341 nests, with 1 to 4 eggs; 1E (16N), **2E** (307N), 3E (16N), 4E (2N).
Average clutch range 2 eggs (307 nests).
Eggs were laid at daily intervals and incubation usually commenced with the laying of the second egg.

INCUBATION PERIOD 15 nests, 13 to 16 days; 1 of 13 days, 7 of 14 days, 6 of 15 days, 1 of 16 days. In 1 nest, because of an apparent delay in the onset of incubation, the eggs hatched on the 19th day after clutch completion.

EGG DATES 326 nests, 19 March to 28 September (457 dates); 163 nests, 9 May to 13 June. The extended breeding season, often involving second, third, and even fourth broods, renders the height of the season somewhat less significant for this species.

Breeding Distribution

The mourning dove breeds throughout the agricultural part of southern Ontario, but scarcely extends into the forested areas of the Canadian Shield. Limited numbers are found in a few localities as far north as Thunder Bay or Sudbury, and possibly in the agricultural areas of the Clay Belt of central Ontario (Smith, 1957).

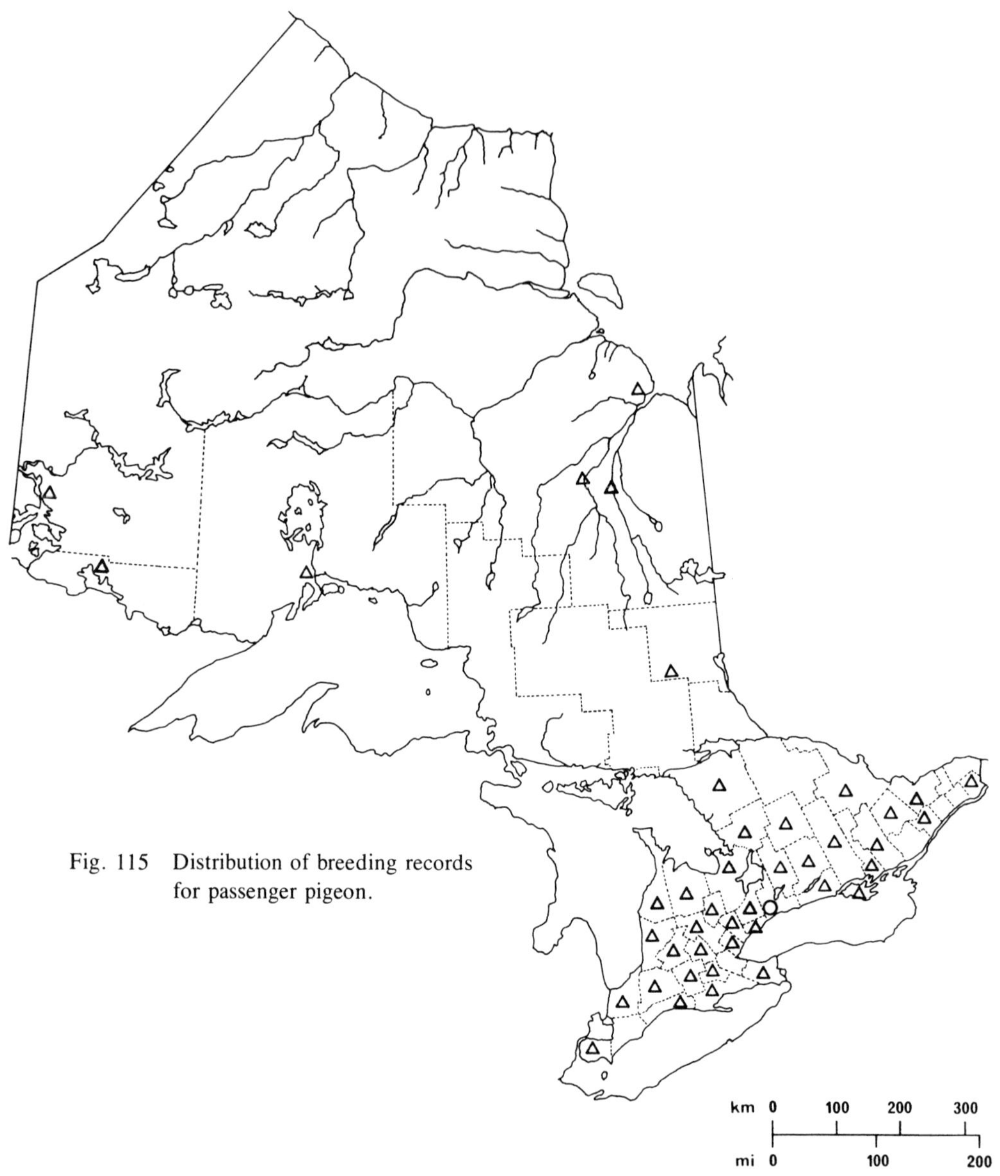

Fig. 115 Distribution of breeding records for passenger pigeon.

Passenger Pigeon, *Ectopistes migratorius* (Linnaeus)

Nidiology

RECORDS 3 representing 3 provincial regions.
Extinct. Formerly bred in deciduous, mixed, and coniferous wooded areas, in both wet and dry situations.

The passenger pigeon was a colonial species, and 2 records from Durham and Simcoe counties described huge nesting colonies of thousands of birds with from 4 or 5 up to 17 nests in a single tree. In 1 colony nest heights were 4.5 m (15 ft) or more. Mitchell (1935) quotes an observer who stated that most nest heights ranged from 2.5 to 3.5 m (8 to 12 ft) and some were as high as 7.5 to 9 m (25 to 30 ft).

Nests were placed on horizontal limbs of both deciduous and coniferous trees and were loosely built structures of broken twigs.

EGGS In 1 record, there were 2 eggs in most clutches, but only 1 egg in some clutches.

INCUBATION PERIOD According to Mitchell (1935), incubation periods were variously estimated to be ca 2 to 3 weeks (13 to 24 days).

EGG DATES The only egg date reported was June 1888, when a set of eggs was collected in Cochrane District.

Breeding Distribution

Apparently the passenger pigeon at one time nested commonly throughout southern Ontario, and at least locally as far north as Moose Factory (Mitchell, 1935). However, the full extent of its former range in northern Ontario will never be known. In 1891 the last provincial specimen was taken, and after 1902 no reliable sight records exist (Snyder, 1957).

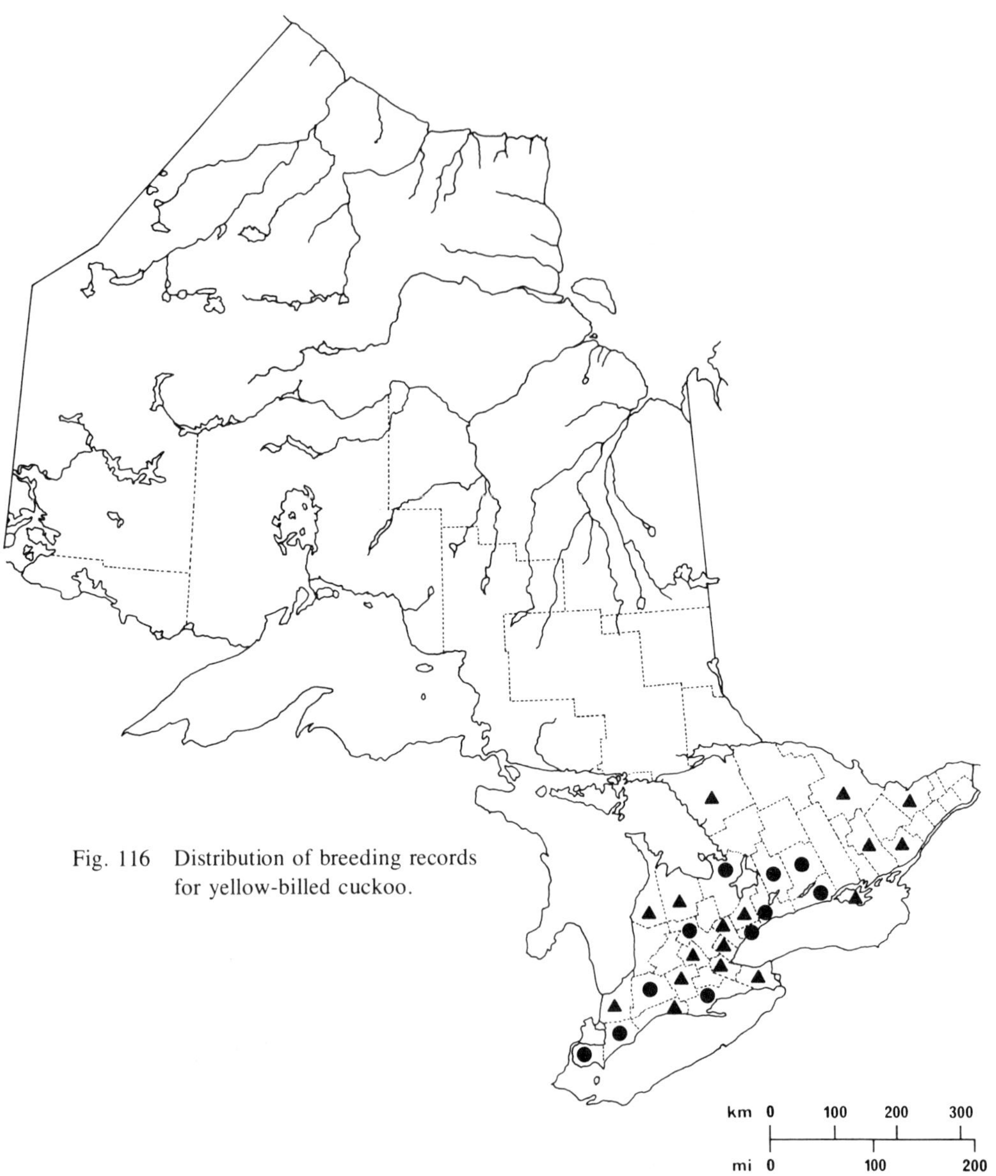

Fig. 116 Distribution of breeding records for yellow-billed cuckoo.

Yellow-billed Cuckoo, *Coccyzus americanus* (Linnaeus)

Nidiology

RECORDS 71 nests representing 23 provincial regions.
This species breeds on agricultural land in shrubby fields, hawthorn meadows, orchards, river flood plains, fence rows, and roadsides; in wooded areas, most often in deciduous and mixed wood lots and occasionally in coniferous plantations in urban parkland and cemeteries; and in overgrown swamps and marshes.

All nests were above ground, usually in shrubs, vines, and saplings, and less often in large trees. Deciduous saplings and trees (13 spp., 39 nests) were preferred to coniferous trees (3 spp., 5 nests). Grape and unspecified vine species (10 nests) and various shrubs and bushes (7 nests) were also used. Nests were usually saddled on horizontal limbs or in forks, at various distances from the central trunk and at various levels in the bush or tree. Heights of 55 nests ranged from 0.6 to 6 m (2 to 20 ft), with 28 averaging 1 to 2 m (3 to 7 ft).

Nests were frail, loosely constructed platforms of dead twigs, sticks, and grasses, with shallow cups, variously lined, in order of frequency, with green and dead leaves, grass stems, bark strips, willow and chestnut flowers, weed stems, rootlets, and cedar sprigs.

EGGS 63 nests with 1 to 5 eggs; 1E (6N), **2E** (26N), **3E** (25N), **4E** (5N), 5E (1N).
Average clutch range 2 to 3 eggs (51 nests).
One nest contained 3 eggs of yellow-billed cuckoo and 1 egg of black-billed cuckoo. Two nests, each with 4 eggs of yellow-billed cuckoo, contained single eggs of black-billed cuckoo. Similarly, this species occasionally laid in nests of black-billed cuckoo (see following species).

INCUBATION PERIOD 1 nest, ca 13 days.

EGG DATES 53 nests, 23 May to 7 August (61 dates); 27 nests, 9 June to 4 July.

Breeding Distribution

The yellow-billed cuckoo breeds throughout southern Ontario, most commonly south of the forested areas of the Canadian Shield. It possibly also nests as far north as Sudbury.

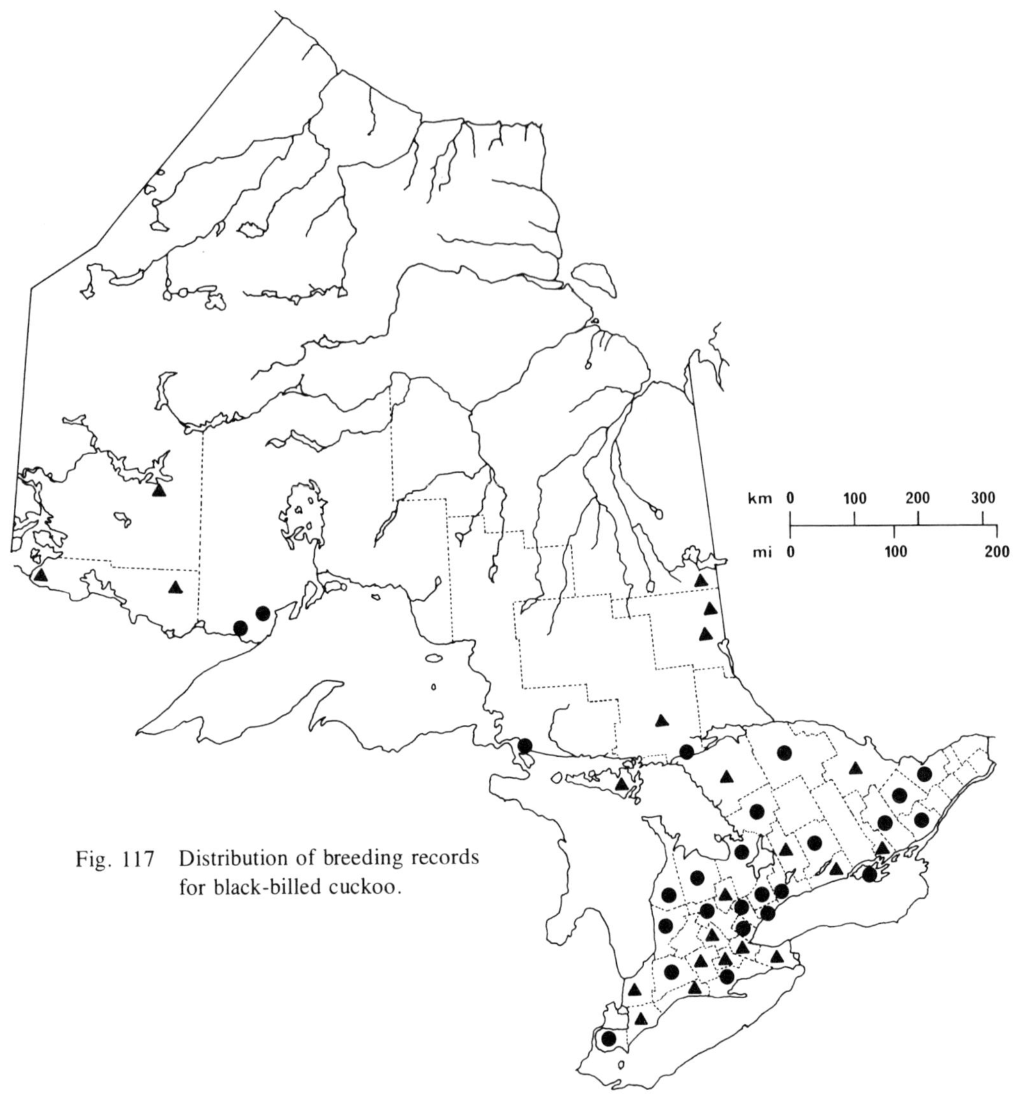

Fig. 117 Distribution of breeding records for black-billed cuckoo.

Black-billed Cuckoo, *Coccyzus erythropthalmus* (Wilson)

Nidiology

RECORDS 257 nests representing 40 provincial regions.
Breeds on agricultural land (shrubby fields and pastures, hawthorn meadows, roadsides and fence rows, orchards and berry patches, and abandoned gravel pits); on lake and river shores, on flood plains, and on small river islands (1 record); both at the edges and in clearings of

mixed (Fig. 155) and deciduous woods, and in coniferous plantations; in wet areas (most often among willows and sometimes in swamps and near the edges of bogs and marshes); and in urban areas (golf-course borders, urban parks, ravines, and residential gardens).

Nests were almost invariably in living trees, shrubs, bushes, and vines (232 records); there were only 2 records of nests in dead trees and 1 record of a nest on the ground at the base of a tree. Deciduous saplings and trees (11 spp., 119 nests) were preferred to coniferous trees (5 spp., 68 nests); shrubs, bushes (12 spp., 44 nests), and vines (2 spp., 4 nests) were also used. Hawthorn spp. (71 nests), pine spp. (41 nests), willow spp. (15 nests), and apple spp. (11 nests) were most often chosen as nest trees. Usually nests were saddled on, or placed between, 1 or more horizontal branches or were in crotches, either against the main trunk or up to 1.8 m (6 ft) from it. A few nests were over water, and 1 nest was between 2 upright trees, each 5 cm (2 inches) in diameter. Heights of 233 nests ranged from ground level to 13.5 m (45 ft), with 117 averaging 1 to 1.5 m (3 to 5 ft). One nest was 1.8 m (6 ft) from an active nest of American robin.

Nests ranged from flimsy, loosely woven platforms to more substantial, bulky structures and either were flat or had shallow cups. Exteriors were of twigs and grasses; twigs were usually of deciduous and occasionally of coniferous trees and were ca 0.3 cm (0.1 inch) in diameter. Nests were variously lined, in order of frequency, with dead and green leaves, pine and other evergreen needles, plant and weed stalks and fibres, plant down, poplar and willow catkins and bud scales, grasses, bark strips, cedar and tamarack sprigs, mosses, fine twigs, feathers, rootlets, bracken fern, lichen, earth, and spiders' webs. A few nests were unlined. Nests were sometimes improved and strengthened during the egg-laying period. Outside diameters of 3 nests ranged from 12.5 to 15 cm (5 to 6 inches); outside depths of 3 nests ranged from 5 to 6.5 cm (2 to 2.5 inches).

EGGS 218 nests with 1 to 6 eggs; 1E (16N), **2E** (76N), **3E** (87N), **4E** (32N), 5E (6N), 6E (1N).
Average clutch range 2 to 3 eggs (163 nests).
One nest of 3 eggs contained 1 cowbird egg (parasitism—0.4%). This species sometimes laid in nests of yellow-billed cuckoo (see preceding species). Two nests, with 2 and 4 eggs respectively, contained single eggs of the yellow-billed cuckoo.

INCUBATION PERIOD 10 nests, 10 to 13 days: 1 of 10 days, 3 of ca 11 days, 5 of ca 12 days, 1 of ca 13 days. Since this and the preceding species commenced incubation with the first egg, young of various ages and partially incubated eggs were often found together in the nest. Unless eggs are marked, exact incubation periods are difficult to determine.

EGG DATES 180 nests, 20 May to 8 September (229 dates); 90 nests, 10 June to 2 July.

Breeding Distribution

The black-billed cuckoo (Fig. 158A) breeds throughout southern Ontario and extends into northern Ontario at least as far as Sioux Lookout and Cochrane. It breeds particularly where forests have been disturbed, as in agricultural areas (Fig. 157).

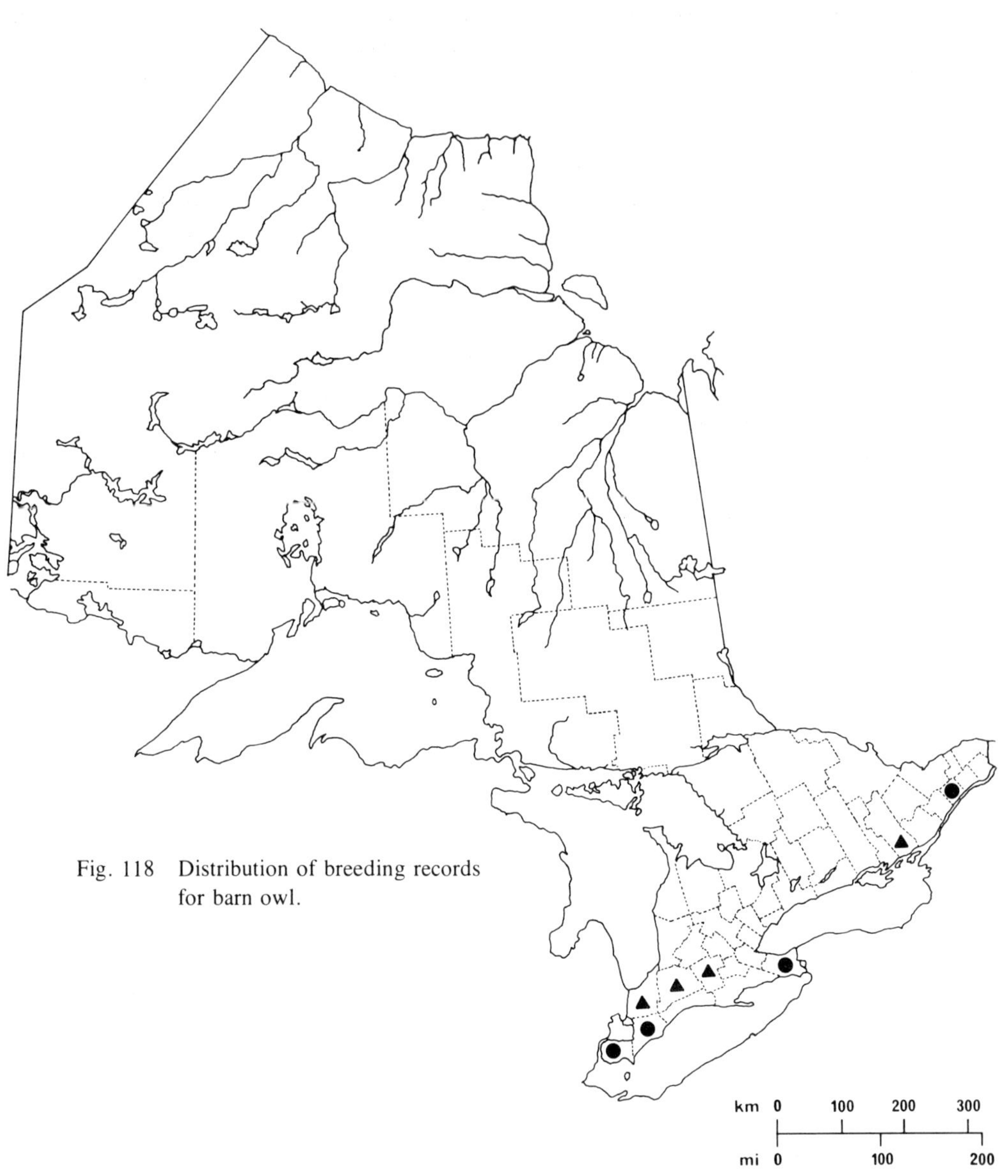

Fig. 118 Distribution of breeding records for barn owl.

Barn Owl, *Tyto alba* (Scopoli)

Nidiology

RECORDS 29 nests representing 7 provincial regions.
This rare breeding resident of southern Ontario usually nests in manmade structures. Most nest sites recorded were farmyard buildings (17 records), but other sites included water towers (2 records), an old flour mill (1 record), an airport hangar (1 record), a store (1 record), a canal lift bridge (1 record), and a woodland (1 record).

Nests were variously situated as follows: in barn lofts, where they were positioned on hay or straw, on boards and beams, in nest boxes placed inside the lofts, and in recesses between hay bales; and in silos, where they were positioned on boards of the attic, on top of silage, and in burrows in the silage up to 1 m (3 ft) in length. Single nests were positioned on the bottom of an empty water tower, on floorboards of the fourth floor of an old flour mill, in an unused chimney of a store, on a metal corner span of a lift bridge, and in a cavity 1.1 m (3.5 ft) deep in an old hackberry tree. Heights of 18 nests ranged from 3 to 33.5 m (10 to 110 ft), with 9 averaging 7.5 to 15 m (25 to 50 ft). The height of the lift-bridge nest was 7.5 m (25 ft) when the bridge was down, and 42.5 m (140 ft) when the bridge was up.

Nests were depressions in material such as hay, or the eggs were laid upon existing material. Nests were usually unlined. Pellets and a few feathers were often found near the eggs.

EGGS 27 nests with 1 to 10 eggs; 1E (2N), 2E (5N), **3E** (6N), **4E** (4N), **5E** (5N), **6E** (4N), 10E (1N).
Average clutch range 3 to 5 eggs (15 nests).

INCUBATION PERIOD No information.

EGG DATES 10 nests, 3 March to 15 September (12 dates); 5 nests, 18 May to 5 June. One record noted young ca 4 to 5 weeks of age in a nest on 17 November, which indicates a probable October egg date.

Breeding Distribution

In Ontario the barn owl is near the northern extremity of its range. Although captive-bred birds have been released in southern Ontario, this species is known to nest naturally in the province only in the counties of the extreme south of the Deciduous Forest region and, occasionally, in the Kingston area (Quilliam, 1973; Weir, 1976) and Dundas County. It is possible that a few birds occur naturally in the western Rainy River District.

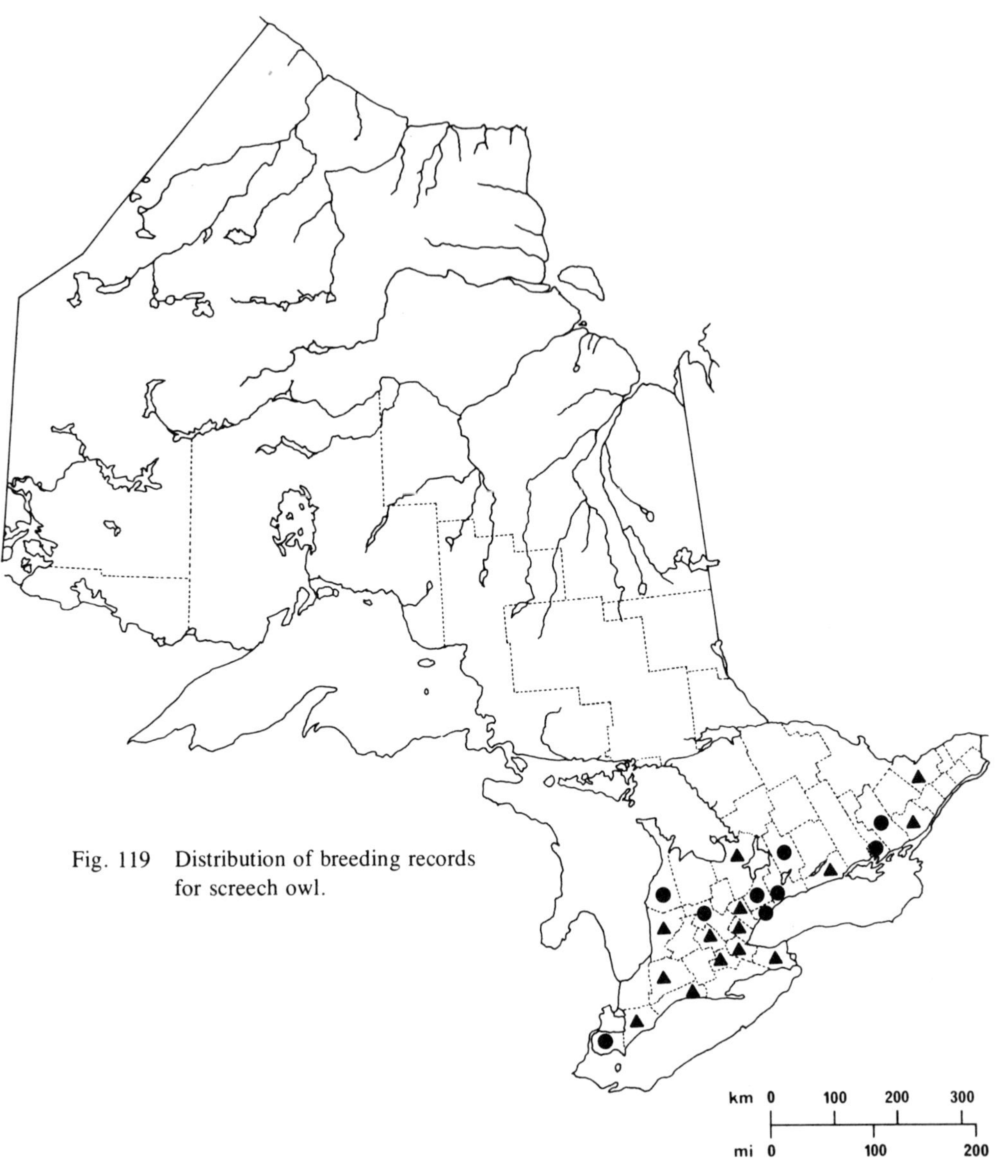

Fig. 119 Distribution of breeding records for screech owl.

Screech Owl, *Otus asio* (Linnaeus)

Nidiology

RECORDS 70 nests representing 19 provincial regions.
Breeds in mixed, deciduous, and coniferous woods and swamps; near ponds, marshes, and rivers (at least partly because of the availability of duck nest boxes in those areas); in gardens and parks in residential areas; and in orchards and barn buildings in rural areas.

Nests were situated in natural cavities (26 nests) and woodpecker cavities (5 nests) in dead and living trees, and in duck and other nest boxes (31 nests). One nest was on a hay-fork carriage in the peak of a barn roof. Deciduous trees (8 spp., 29 nests) were preferred to coniferous trees (1 sp., 4 nests). One nest was in the same stub as an active nest of starling, and another was in a colony of black-crowned night heron. Heights of 27 nests in natural tree cavities ranged from 1 to 13.5 m (3 to 45 ft), with 14 averaging 3.2 to 6 m (10.5 to 20 ft).

Depths of 8 tree cavities ranged from 15 to 152.5 cm (6 to 60 inches); diameters of 4 entrance holes were 6.5, 7.5, 7.5 × 18, and 15 cm (2.5, 3, 3 × 7, and 6 inches); diameter of 1 cavity at egg level was 30.5 cm (12 inches). Cavities contained pre-existing material such as leaves and wood chips, but the following additional materials were recorded as contributions of this species: fur, feathers, remains of prey, and sprigs of fresh cedar.

EGGS 58 nests, 1 to 6 eggs; 1E (3N), 2E (5N), **3E** (13N), **4E** (19N), **5E** (14N), **6E** (4N).
Average clutch range 3 to 5 eggs (46 nests).

INCUBATION PERIOD 1 nest, at least 24 days.

EGG DATES 44 nests, 20 March to 7 June (57 dates); 22 nests, 16 April to 3 May.

Breeding Distribution

The screech owl apparently breeds only in the wooded regions of southern Ontario (Fig. 153). Its numbers probably became much reduced as land was cleared, but it apparently still occurs on Manitoulin Island (Nicholson, 1972), and at least as far north as Ottawa in the east.

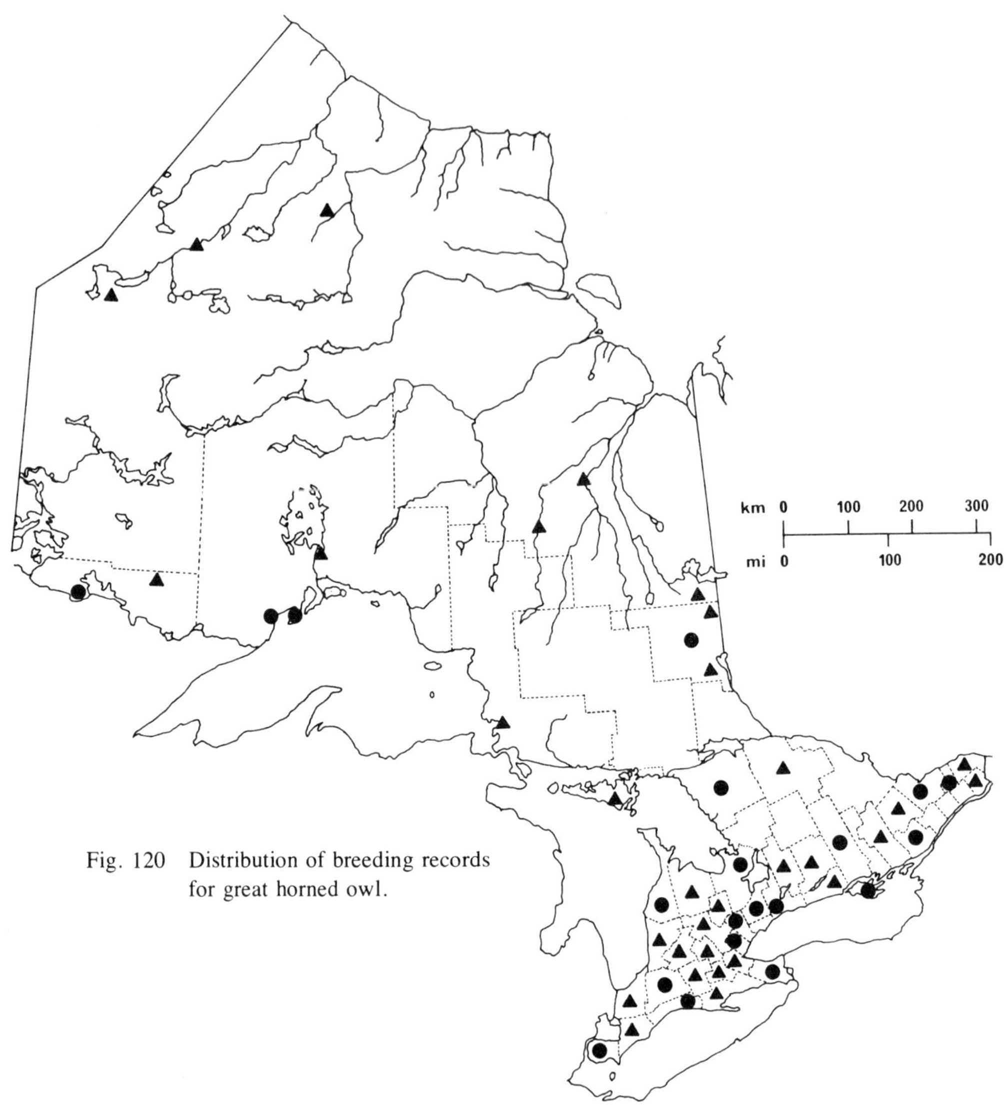

Fig. 120 Distribution of breeding records for great horned owl.

Great Horned Owl, *Bubo virginianus* (Gmelin)

Nidiology

RECORDS 448 nests representing 42 provincial regions.
Breeds most often in wooded areas of all kinds (dense or open, dry or swampy) and less often in agricultural and urban areas (in fields and pastures, in fence rows, in farm buildings, in gravel pits, on golf courses, and in cemeteries). Deciduous woods (108 nests) were

preferred to mixed (37 nests) and coniferous (9 nests) stands.

In considering 436 nest records, 416 nests (95.4%) were open tree nests; 13 nests (3%) were in tree cavities or broken-stub depressions; 3 nests (0.7%) were on hydro pylons; 2 nests (0.5%) were in barn and mine buildings; 1 nest (0.2%) was in a cliff crevice; and 1 nest (0.2%) was on a manmade wooden platform in a tree. Deciduous trees (17 spp., 239 nests) were preferred to coniferous trees (7 spp., 95 nests) as nest sites, and the most favoured trees were beech (79 nests), pine spp. (65 nests), elm spp. (42 nests), maple spp. (40 nests), oak spp. (25 nests), and hemlock (21 nests). Open-tree nests were probably always old nests of other species or squirrel dreys. Of 184 open-tree nests, most were built by other raptors: red-tailed hawk (80 nests), red-shouldered hawk (16 nests), unspecified hawk (13 nests), bald eagle (11 nests), and goshawk (3 nests); some were built by common crow (28 nests) and great blue heron (12 nests); and some were old squirrel dreys (21 nests). Nests were sometimes reused for consecutive seasons by this species but more often were reused on an alternating or intermittent basis with other raptors. Heights of 379 tree nests ranged from 2 to 27.5 m (7 to 90 ft), with 190 averaging 10.5 to 17 m (35 to 55 ft).

Open-tree nests were in main forks, in large crotches, or on horizontal limbs, situated at various levels, and either against the trunk or at various distances from it. These nests were of various sizes, structures, and materials, depending on the identity of their original builders. Basically they ranged from flat to well-hollowed platforms made of sticks and often containing leaves. Nests were variously lined with such pre-existing materials as bark chips, small twigs, leaves, grasses, conifer sprigs and needles, green plastic, binder twine, and animal fur. They were also often lined with down and feathers from the owl itself. When cavities or stub hollows were used, there was usually no nest material present. Outside diameters of 25 nests ranged from 25.5 to 152.5 cm (10 to 60 inches), with 12 averaging 53.5 to 76 cm (21 to 30 inches); outside depths of 9 nests ranged from 18 to 91.5 cm (7 to 36 inches); inside diameters of 8 nests ranged from 20.5 to 51 cm (8 to 20 inches); inside depths of 6 nests ranged from 4 to 12.5 cm (1.5 to 5 inches).

EGGS 226 nests with 1 to 3 eggs; **1E** (22N)—9.7%, **2E** (175N)—77.4%, **3E** (29N)—12.8%. *Average clutch range* 2 eggs (175 nests).

INCUBATION PERIOD 8 nests; 4 of ca 30 days, 1 of ca 31 days, 1 of 30 to 31 days, 1 of more than 33 days, 1 of ca 35 days. Incubation commenced with the first egg.

EGG DATES 136 nests, 31 January to 21 April (172 dates); 68 nests, 1 March to 18 March.

Breeding Distribution

The great horned owl apparently breeds throughout the forested portions of Ontario. Records are few in the north, where population levels of the species are probably lower than in the south.

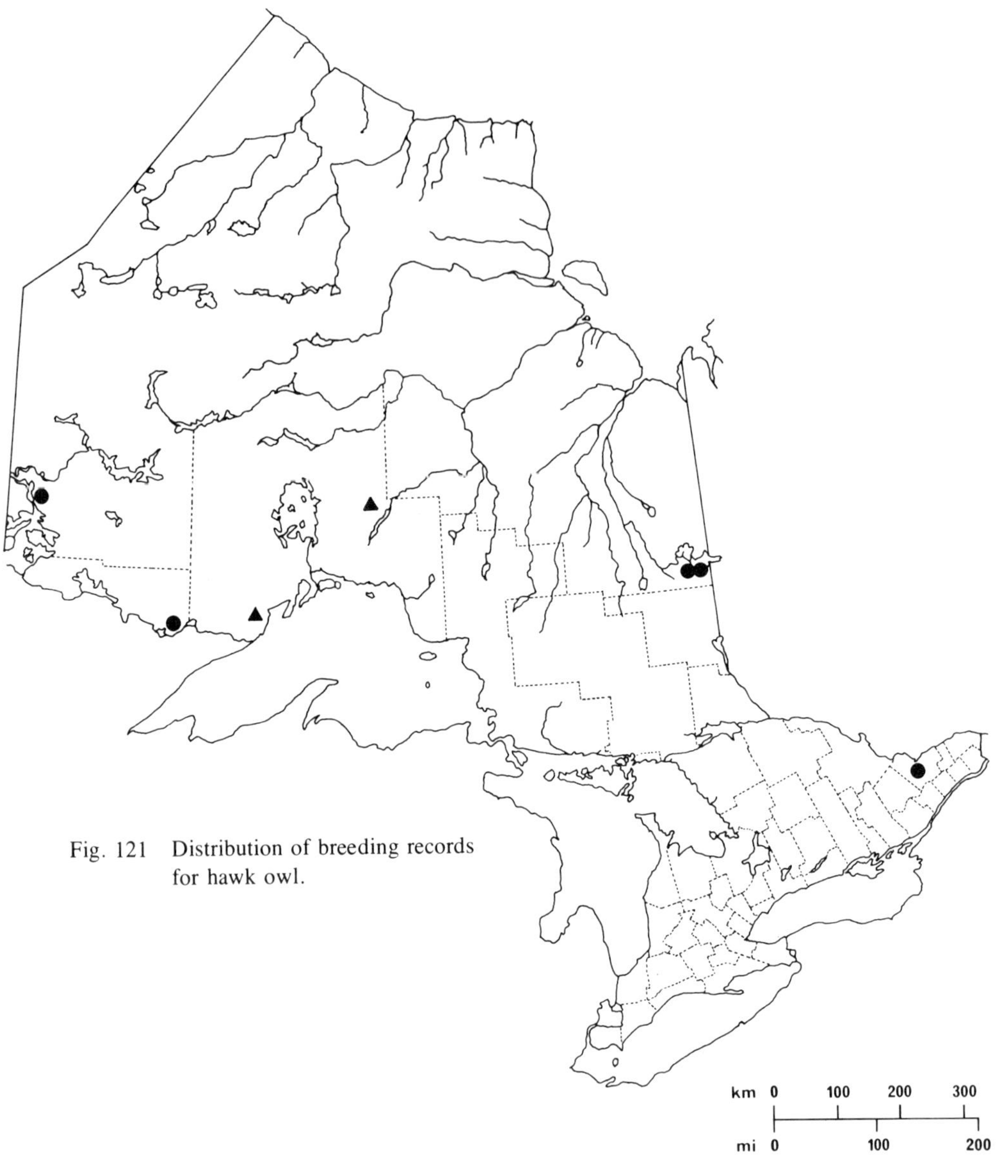

Fig. 121 Distribution of breeding records for hawk owl.

Hawk Owl, *Surnia ulula* (Linnaeus)

Nidiology

RECORDS 2 nests representing 2 provincial regions.
This species breeds in the boreal forest of Canada and Eurasia. It was first discovered nesting in Ontario in 1963, near Ottawa, south of its normal breeding range and in a somewhat typical but isolated habitat locally known as the "Mer Bleue" bog. A second nest was found in 1972 in a geographically more typical area, in the boreal forest in Thunder Bay District.

The first nesting habitat was a small, deciduous wood lot in a large peat bog, and the nest was situated in a tree 13.5 m (45 ft) inside the edge of the wood lot; the second nest was in a large 4-year-old burn with many standing dead trees.

The first nest was in a cavity in the broken top of a red maple, 7.5 m (24 ft) above ground. Internal diameter of the maple stub was 18 cm (7 inches); its cavity depth ranged from a few centimetres to 45.5 cm (18 inches) because of the diagonal break in the stub. The cavity floor was of rotten wood and was covered with damp fur and feathers 5 cm (2 inches) deep. The second nest was in the hollow top of a fire-killed birch at a height of 10.5 m (35 ft).

The first nest was found on 25 May and contained 4 downy young of various sizes and 1 egg. On 29 May the infertile egg was collected (NMC 2313), and the nest and 4 young were photographed (Smith, 1970). The second nest was discovered on 21 May with adults in attendance; its contents were not ascertained, but the adult owls were present throughout May and June.

Breeding Distribution

Although records are very few, the hawk owl probably breeds throughout the forested portions of northern Ontario (Fig. 161) as far south as Lake Superior and Cochrane. The record from near Ottawa indicates that it may breed farther south in suitable habitat. Despite the widespread occurrence of this species, its habit of nesting in cavities and the inaccessibility of much of its breeding range explain the fact that only two nests have been reported so far.

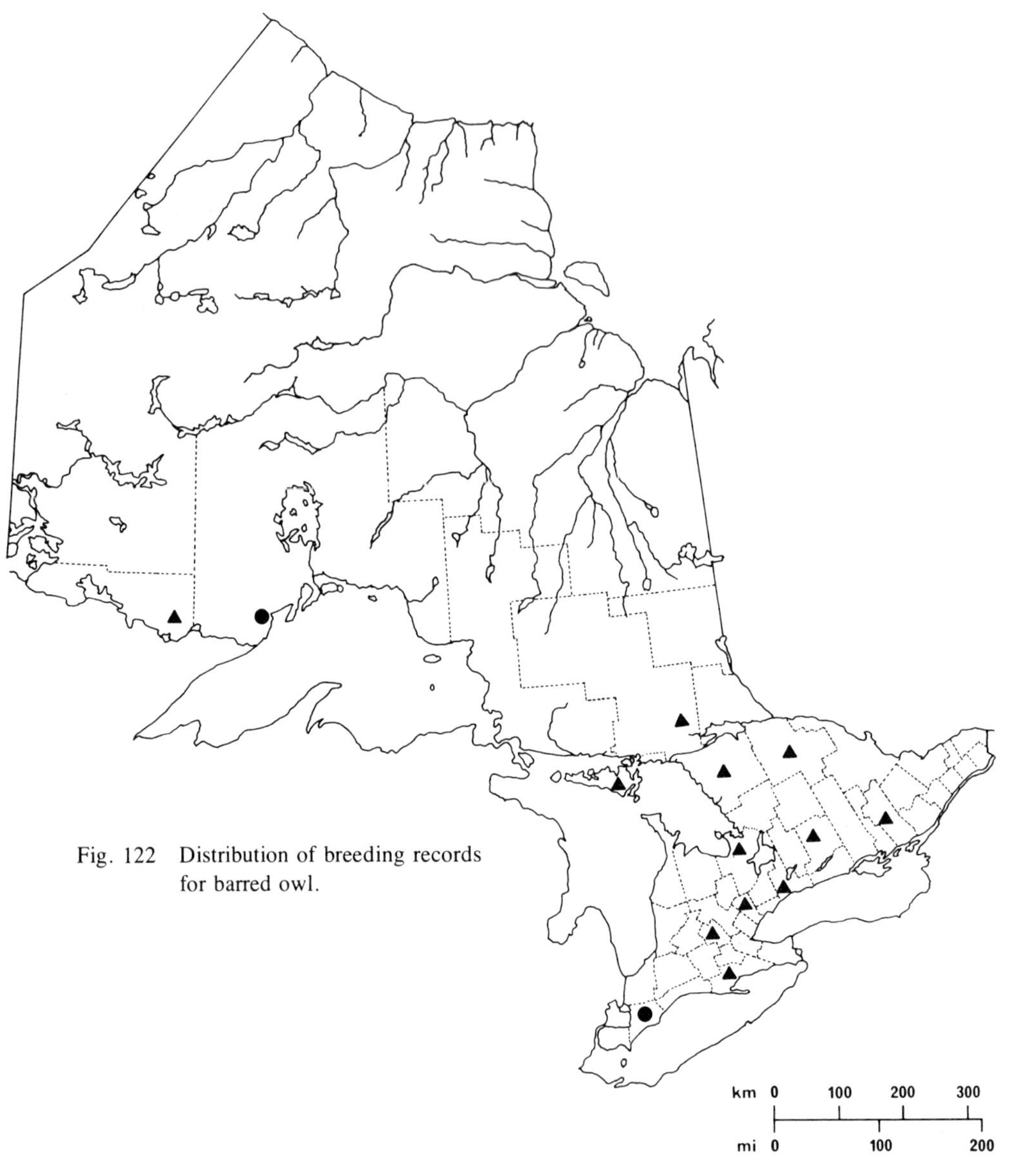

Fig. 122 Distribution of breeding records for barred owl.

Barred Owl, *Strix varia* Barton

Nidiology

RECORDS 10 nests representing 5 provincial regions.
The barred owl is a rare, permanent resident of wooded regions in Ontario. Few nests have been discovered so far. The collection of a set of eggs in Kent County in 1897 (Macoun, 1903) was not documented. The first documented record was on 3 May 1940, when a nest from which a young bird was collected (ROM 10960) was found in Thunder Bay District (Dear, 1940).

The species breeds in large mixed and deciduous woodlands. One record indicated an open type of deciduous woodland with little undergrowth near a large body of water.

Nests were positioned in living and dead trees: in natural cavities (8 nests), in a squirrel drey in a crotch (1 nest), and in a stick nest on a broken tree-top (1 nest). Nest trees were deciduous: balsam poplar (5 nests), beech (2 nests), and white birch (2 nests); 1 tree species was unspecified. Heights of 10 nests ranged from 4.5 to 10.5 m (15 to 35 ft), with 5 averaging 7.5 to 10.5 cm (25 to 35 ft); tree diameters of 5 cavity nests were 61 and 76 cm (24 and 30 inches).

An abandoned squirrel drey used by this species had some small pine boughs added to it; an adult was observed carrying nest material to a stick nest on a broken tree-top. One cavity nest contained a handful of feathers and the cavity depth was 1.5 m (5 ft). Nest sites were sometimes reused in successive years, and 1 site was 90 m (300 ft) from a site used 2 years previously. In 1 cavity, a second clutch was laid in 2 separate and successive years after the first clutch was collected.

EGGS 8 nests with 2 to 3 eggs; 2E (5N), 3E (3N).

INCUBATION PERIOD No information.

EGG DATES 6 nests, 4 April to 18 May (8 dates); 3 nests, 17 April to 20 April.

Breeding Distribution

Most breeding records of the barred owl in Ontario are based on reports of young birds, since its cavity nests are difficult to locate. It breeds very sparingly in the agricultural south but increases in numbers throughout the forested Canadian Shield. Its northern limits probably exceed considerably the records plotted on the map, but are poorly known. It may range as far north as the Albany River and the southern tip of James Bay.

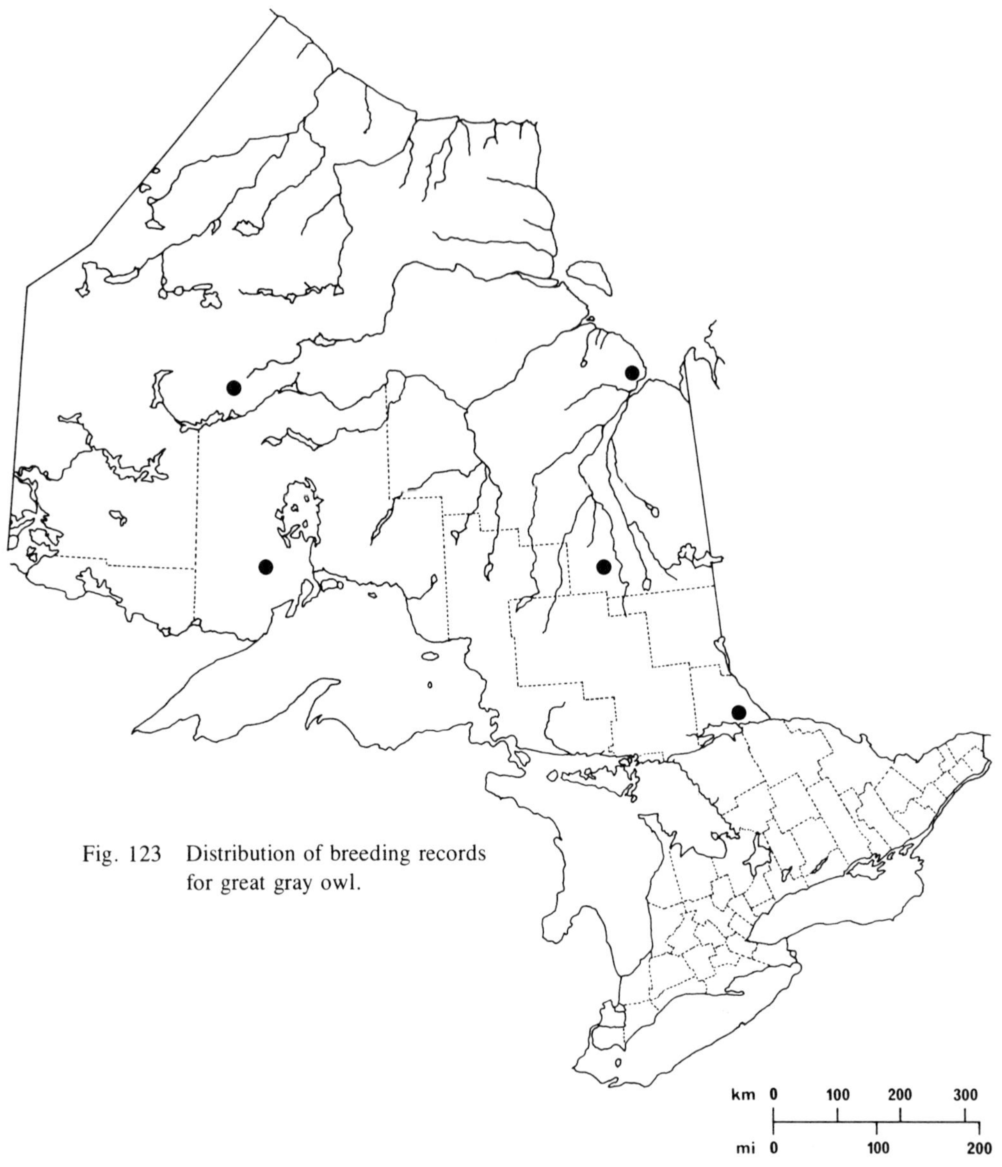

Fig. 123 Distribution of breeding records for great gray owl.

Great Gray Owl, *Strix nebulosa* Forster

Nidiology

RECORDS 4 nests representing 3 provincial regions.
This relatively rare owl of Canada's and Eurasia's boreal forest has been recorded as nesting in Ontario only 4 times. In 1977 a nest was discovered and documented near Central Patricia, Kenora District. Later a report was submitted for a documented nest that was actually found in 1976 northwest of Thunder Bay, Thunder Bay District. A third nest was discovered in 1980, just north of Moosonee, Cochrane District. The fourth nest, first reported in 1980, was actually found in August 1964 ca 50 km (30 mi) south of Smooth Rock Falls, Cochrane District.

The Kenora nest was in an aspen stand at the edge of a beaver meadow (Fig. 161). The Thunder Bay nest was in an area cut over in 1971–72, which had irregular clear cuts alternating with patches of poplar, birch, and lowland black spruce, and uncut forest and black spruce swamps nearby. The Moosonee nest was in a fen that was treed with stunted tamaracks and the occasional black spruce. This nest was 150 m (490 ft) from a small creek that was lined with black spruce. The Smooth Rock Falls nest was in a mature aspen grove.

The Kenora nest was in the main crotch of an aspen near the top-central portion of the tree, at a height of 20 m (65 ft); the Thunder Bay nest was in a crotch of branches in a poplar, at a height of ca 7.5 m (25 ft); the Moosonee nest was on a manmade nest platform in a tamarack, at a height of 5 m (16.5 ft) (the top of the tamarack had been removed to support the platform).

The 2 natural nests were composed of sticks, and each was photographed when it contained only 1 young owl (1977 nest, ROM PR 810–824; 1976 nest, ROM PR 1075–1076). In the Kenora nest, the young owl was seen and photographed on 6 June; in the Thunder Bay nest, in late June. The Kenora young owl was less than a week old when photographed. The Cochrane nest contained 3 eggs when observed on 29 April, 3 May, and 5 June; on the last 2 occasions an adult was sitting; on 18 July, the nest was empty and no owls were seen. The Smooth Rock Falls nest contained 2 young when found, and the first young left the nest only on 30 August when it was photographed (ROM PR 1168).

Breeding Distribution

An indication of breeding by the great gray owl (Fig. 162A) in Ontario was first secured in 1911, when young already able to fly were collected in Nipissing District (Baillie and Harrington, 1936). Nothing further was recorded until 1977, when the first nest was reported (James, 1977). This and the three other above-mentioned nests are the only breeding records available for the province, but the species is believed to be thinly scattered over the forested portions of northern Ontario, extending at least as far south as northern Lake Superior.

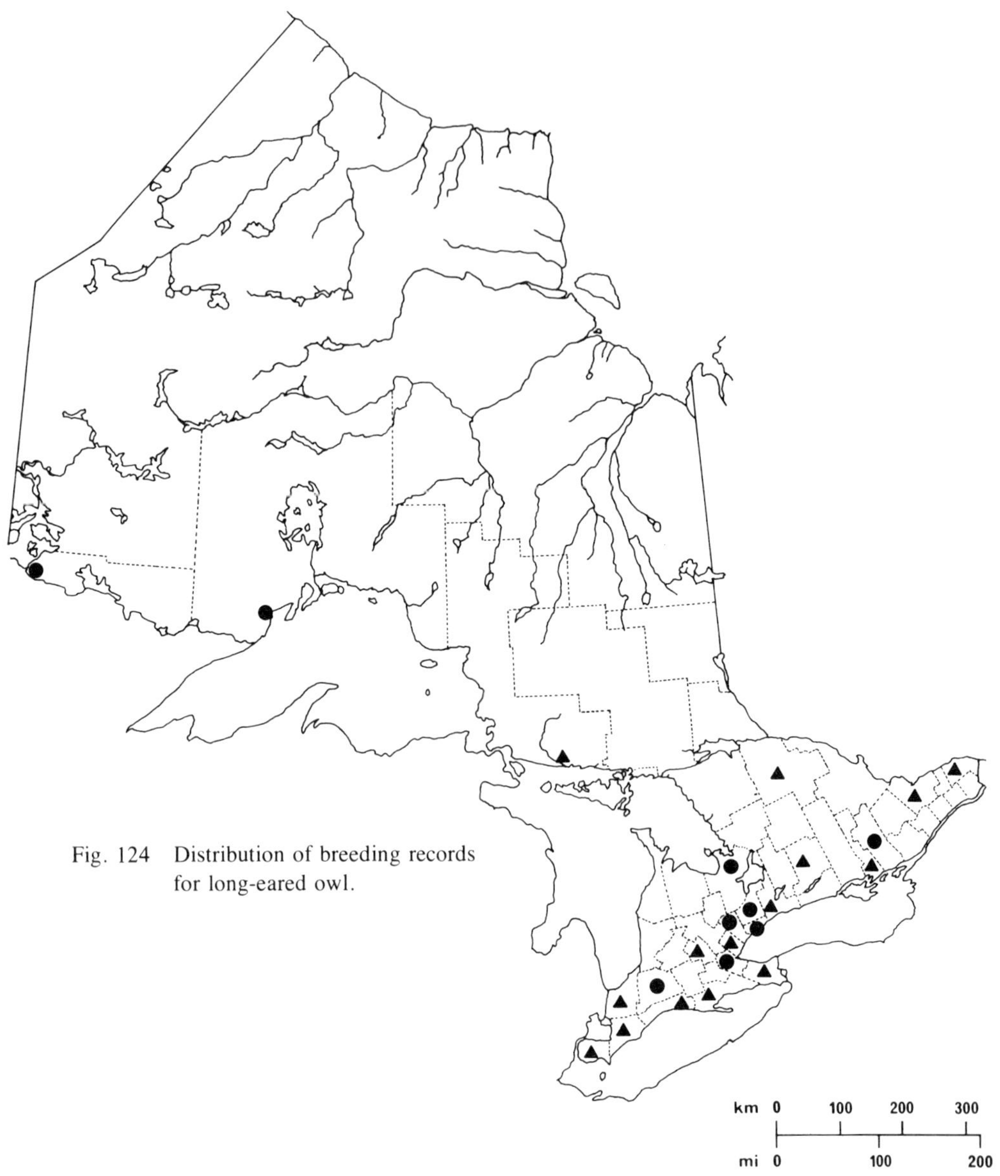

Fig. 124 Distribution of breeding records for long-eared owl.

Long-eared Owl, *Asio otus* (Linnaeus)

Nidiology

RECORDS 73 nests representing 22 provincial regions.
Breeds most often in dense coniferous woods and reforestation groves, which are more often wet than dry; breeds less often in mixed woods and occasionally in deciduous stands. A few nests were recorded in conifer rows in open wood lots, in farm fields, and in hawthorn pastures.

Nests were usually in coniferous trees (6 spp., 40 nests), which were nearly always living (1 coniferous nest tree was dead), and were sometimes in deciduous trees (4 spp., 8 nests). Pine spp. (18 nests), cedar and juniper spp. (11 nests), tamarack (4 nests), hawthorn spp. (3 nests), hemlock (3 nests), and spruce spp. (3 nests) were preferred.

Nests were usually positioned in main crotches near the trunk; occasionally they were on horizontal branches at a distance from the trunk. Heights of 57 nests ranged from 2.5 to 18.5 m (7.5 to 60 ft), with 29 averaging 5.5 to 9 m (18 to 30 ft).

This owl used old nests of other species such as crows (26 records) and also used squirrel dreys (4 records). Typically, occupied nests were twig and stick platforms, sometimes with leaves in the outer part of the nest. In order of frequency, nests were lined with feathers and down, leaves, bark, mosses, cedar sprigs, binder twine, grasses, pine needles, and rootlets. The above materials were usually gathered by the previous occupant. Outside diameters of 3 nests ranged from 25.5 to 45.5 (10 to 18 inches); outside depths of 2 nests were both 25.5 cm (10 inches); inside diameter of 1 nest was 15 cm (6 inches) and its inside depth was 4 cm (1.5 inches).

EGGS 61 nests with 2 to 6 eggs; 2E (6N), **3E** (12N), **4E** (16N), **5E** (23N), **6E** (4N).
Average clutch range 3 to 5 eggs (51 nests).

INCUBATION PERIOD 2 nests, 27 days.

EGG DATES 43 nests, 19 March to 24 May (49 dates); 22 nests, 15 April to 5 May.

Breeding Distribution

The northern extent of the breeding range of the long-eared owl in Ontario is uncertain. Various reports suggest that the species breeds as far north as Hudson Bay, but actual breeding records are lacking for almost the whole northern part of the province. It does, however, breed throughout the south.

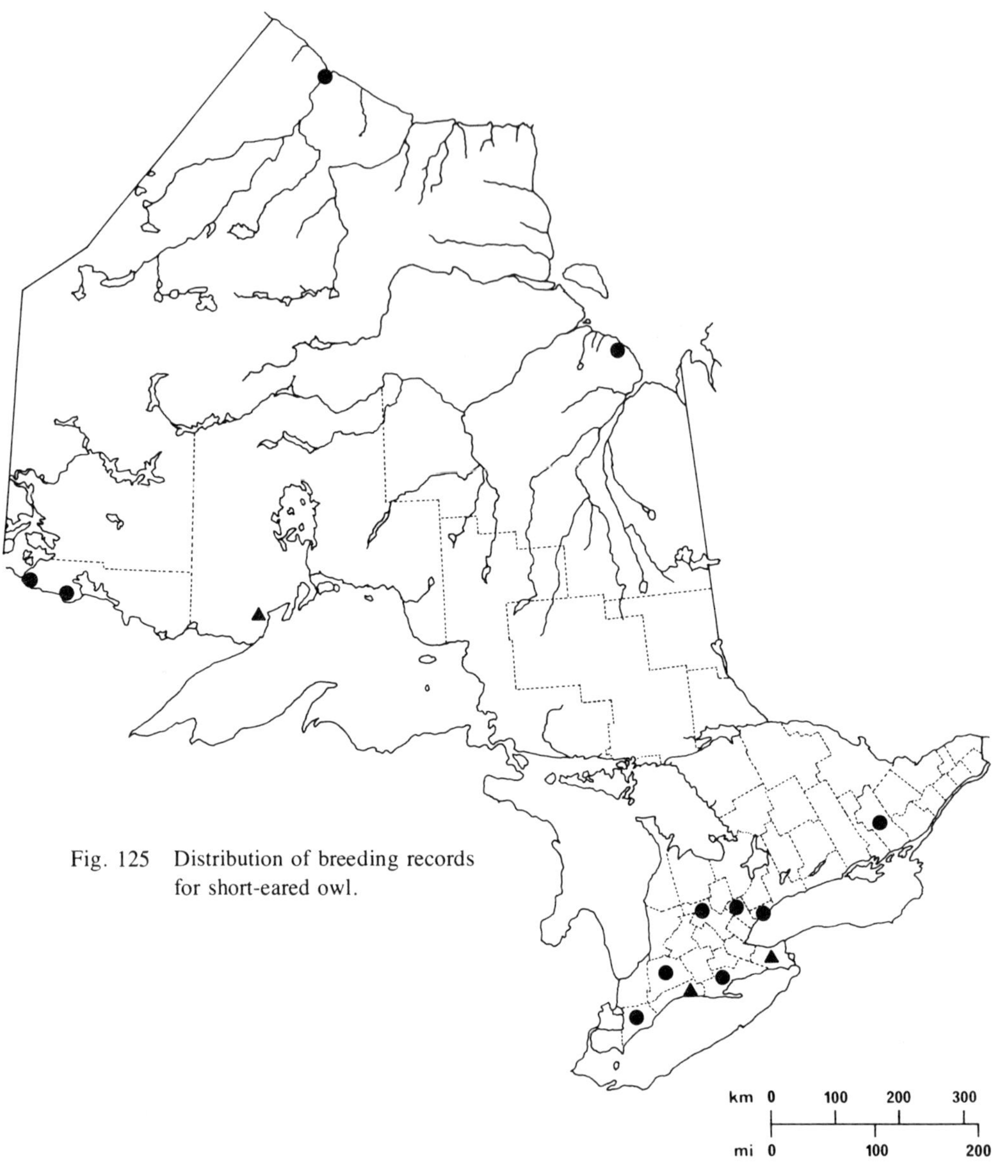

Fig. 125 Distribution of breeding records for short-eared owl.

Short-eared Owl, *Asio flammeus* (Pontoppidan)

Nidiology

RECORDS 14 nests representing 8 provincial regions.
This owl is a rare, local breeding species in Ontario and probably nests more frequently in the narrow tundra zone of the north than it does in grassland areas farther south.

Breeding sites recorded were in abandoned grassy fields in agricultural areas (5 nests), in heath bogs (3 nests), on airport fields of short grass (2 nests), in cattail marshes (2 nests), and on tundra (1 nest).

Nests were on the ground, usually in grass and often under shrubs such as cranberry, Labrador tea, and stunted willows. In wet areas, some nests were on hummocks and small knolls. One nest was within 128 m (420 ft) of an occupied nest of marsh hawk.

Some nests were slight depressions in the ground, and some were merely cups of dried weeds and/or flattened grasses. At least 1 nest was in a tuft of tall grasses, which formed a canopy over it. Linings sometimes included a few feathers.

EGGS 14 nests with 1 to 9 eggs; 1E (2N), 2E (1N), **4E** (3N), **5E** (1N), **6E** (2N), **7E** (4N), 9E (1N).
Average clutch range 4 to 7 eggs (10 nests).

INCUBATION PERIOD No information.

EGG DATES 10 nests, 14 April to 1 August (14 dates); 5 nests, 6 May to 19 May.

Breeding Distribution

The short-eared owl is believed to nest in all parts of Ontario. It breeds throughout the south and along the coasts of Hudson and James bays. The species is probably thinly scattered elsewhere in suitable habitats but may be virtually absent from the extensive Boreal Forest region.

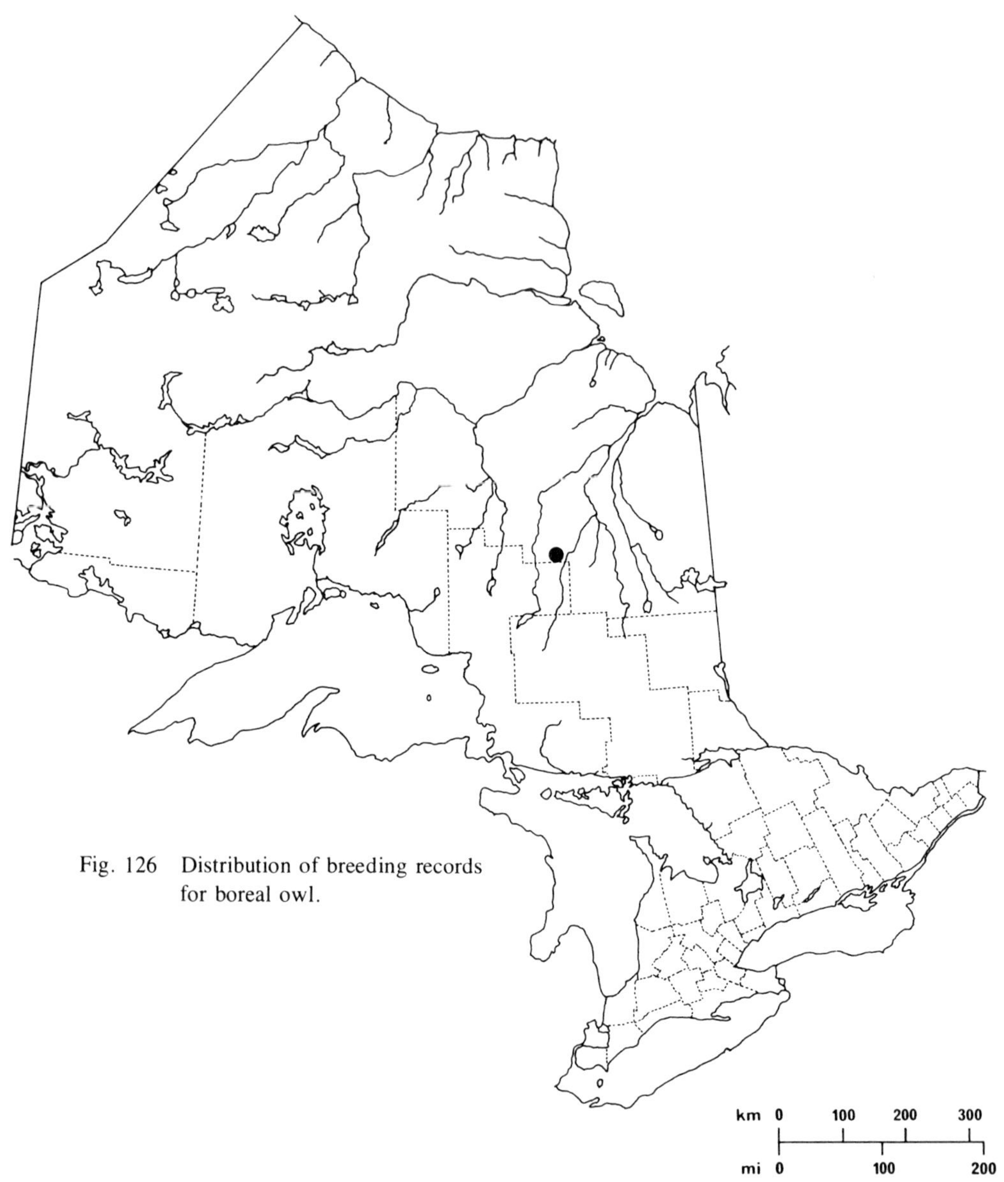

Fig. 126 Distribution of breeding records for boreal owl.

Boreal Owl, *Aegolius funereus* (Linnaeus)

Nidiology

RECORDS 1 nest representing 1 provincial region.
In the breeding season, this seldom seen, nocturnal owl is an inhabitant of the province's vast Boreal Forest region, and it undoubtedly nests more commonly than the single nest thus far discovered would indicate.

The nest was located on 2 May 1975 in a mixed stand of black and white spruce, balsam fir, trembling aspen, and white birch. The nest cavity was in a trembling aspen at a height of 17 m (55 ft). The cavity was in the main trunk in the centre of a branch scar. The hole measured 6 by 8.5 cm (2.4 by 3.3 inches) and the cavity, 20 by 20 by 25 cm (7.9 by 7.9 by 9.8 inches).

The cavity contained 3 eggs when first examined on 14 May, and 3 young hatched between 22 and 27 May (ROM PR 326–328).

Breeding Distribution

The boreal owl breeds across Canada, primarily in the Boreal Forest region (Godfrey, 1966). However, there were only a few summer records in Ontario and no reports of its nesting in the province until 1975, when the above nest was located in southern Cochrane District (Bondrup-Nielsen, 1976).

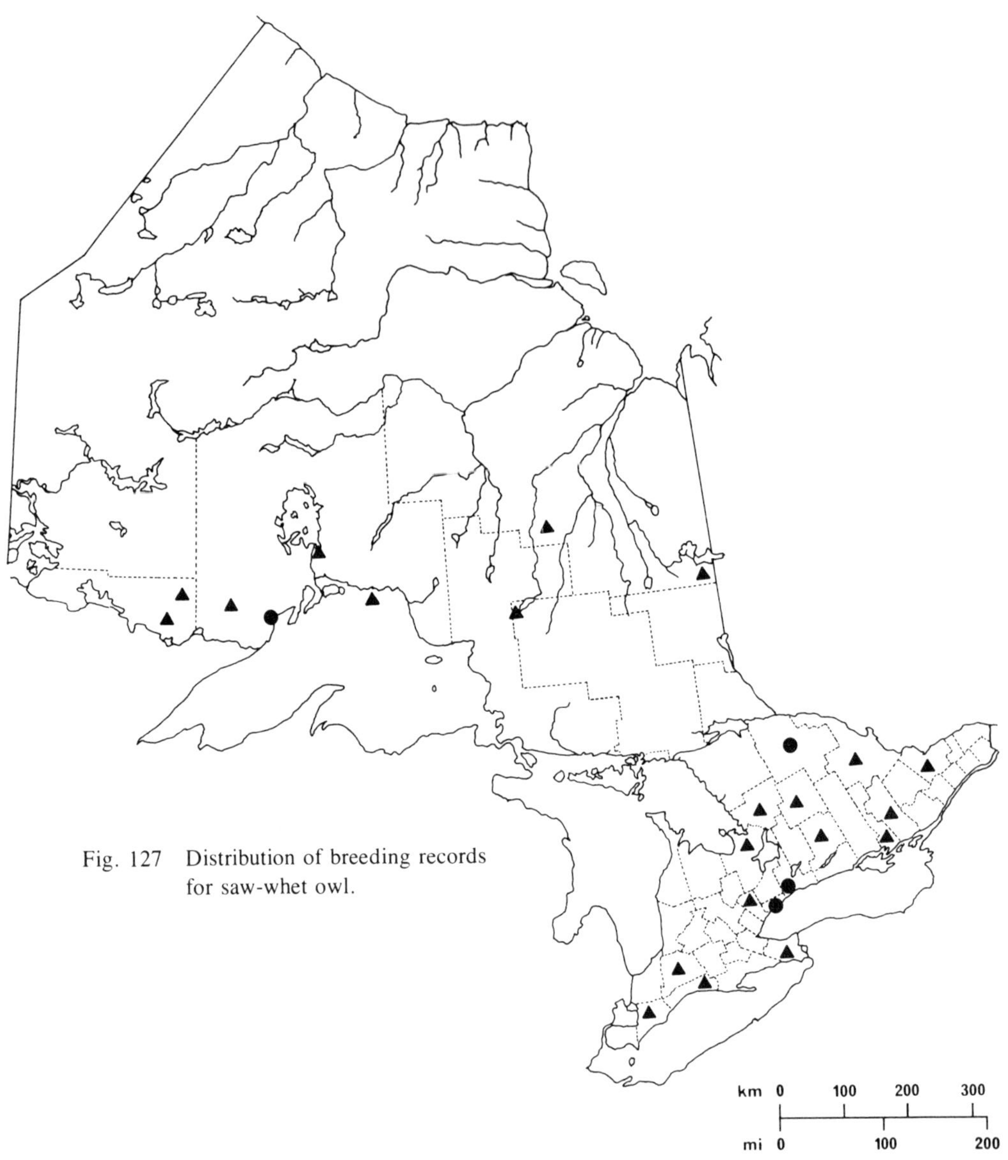

Fig. 127 Distribution of breeding records for saw-whet owl.

Saw-whet Owl, *Aegolius acadicus* (Gmelin)

Nidiology

RECORDS 18 nests representing 12 provincial regions.
Breeds in both mature and second-growth and both deciduous and mixed woodlands that are located either on the mainland or on islands and in either wet or dry situations.

Most nests were in tree cavities—old woodpecker cavities (9 nests) and unspecified cavities (5 nests)—but 1 nest was in a wood duck nest box. Nest trees were usually dead (11 nests), but 1 was living. Deciduous trees (4 spp., 10 nests) were more commonly selected than coniferous trees (1 sp., 2 nests). Heights of 13 tree nests ranged from 2.5 to 13.5 m (8 to 43.5 ft), with 7 averaging 3.5 to 6 m (12 to 20 ft).

Diameter of 1 nest hole was 9 cm (3.5 inches); inside diameters of 2 nest cavities were 7.5 and 9 cm (3 and 3.5 inches); depths of 4 nest cavities ranged from 22 to 45.5 cm (9 to 18 inches). Nest cavities were usually unlined but some contained grasses, twigs, bark, feathers, and the hair and bones of rodents. Some of this material may have been placed in the cavities by previous occupants.

EGGS 14 nests with 1 to 6 eggs; 1E (1N), 2E (1N), **4E** (2N), **5E** (3N), **6E** (7N).
Average clutch range 4 to 6 eggs (12 nests).

INCUBATION PERIOD 1 nest, at least 28 days. Incubation commenced before the clutch was complete.

EGG DATES 13 nests, 1 April to 27 July (15 dates); 7 nests, 10 April to 17 May.

Breeding Distribution

The saw-whet owl breeds throughout southern Ontario, but because it is seldom seen during the breeding season, we have no definite idea so far of the limits of its breeding range in northern Ontario. The species has been found nesting at widely scattered localities, as far north as Lake Nipigon and southern Cochrane District, but it may range as far north as the latitude of southern James Bay.

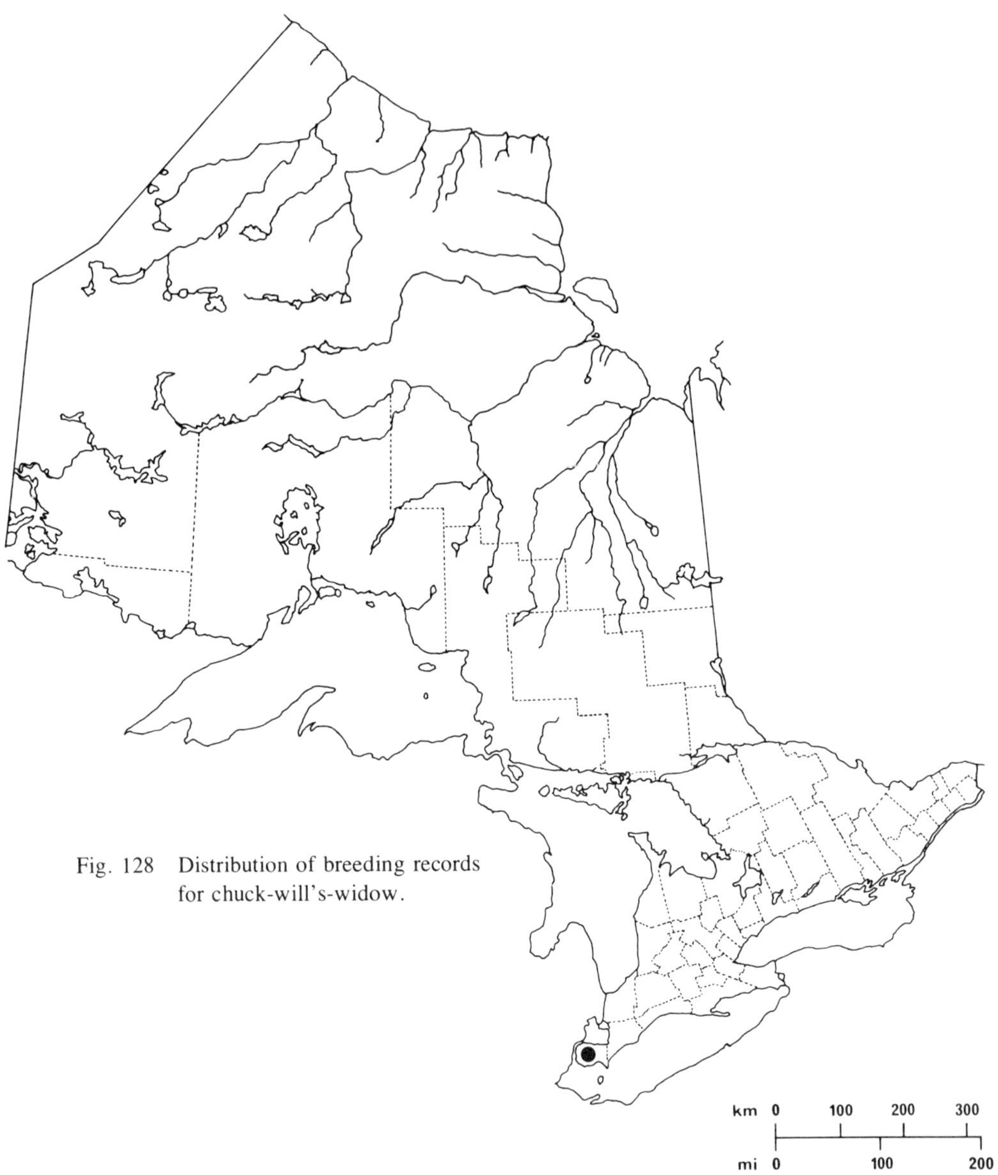

Fig. 128 Distribution of breeding records for chuck-will's-widow.

Chuck-will's-widow, *Caprimulgus carolinensis* Gmelin

Nidiology

RECORDS 1 nest representing 1 provincial region.
This southern caprimulgid was recorded in summer on several occasions recently at Point Pelee and Pelee Island, Essex County.

To date, however, only 1 nest has been discovered. It contained 2 eggs and was found on 5 June 1977, in a wood lot in Point Pelee National Park. This nest was in a clearing with low, dense understory and was photographed (ROM PR 826–828) on 6 June, by A. Wormington.

The eggs were deposited on a bed of oak leaves, pine needles, and grasses.

Breeding Distribution

In 1906 a chuck-will's-widow was collected at Point Pelee, Essex County, but this was considered only an extralimital occurrence. The next record of this species in Ontario was a sighting in 1964. Since 1975 a few individual birds have been seen and/or heard each year at Point Pelee and at least once on Pelee Island. In 1977, following reports of more than one chuck-will's-widow remaining in the area, the above nest was found at Point Pelee (Fig. 146A). Although birds continue to occur each spring in the extreme southwest of Ontario. no additional nests have been found.

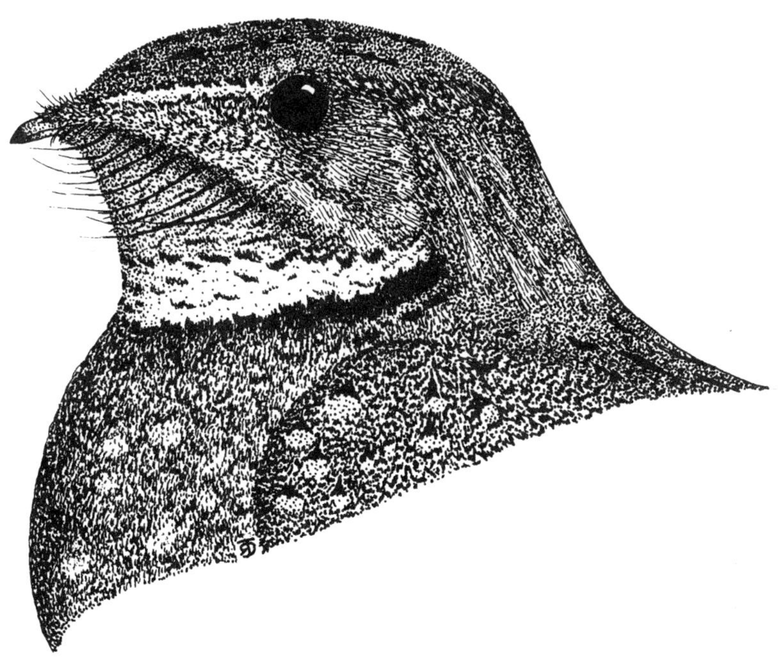

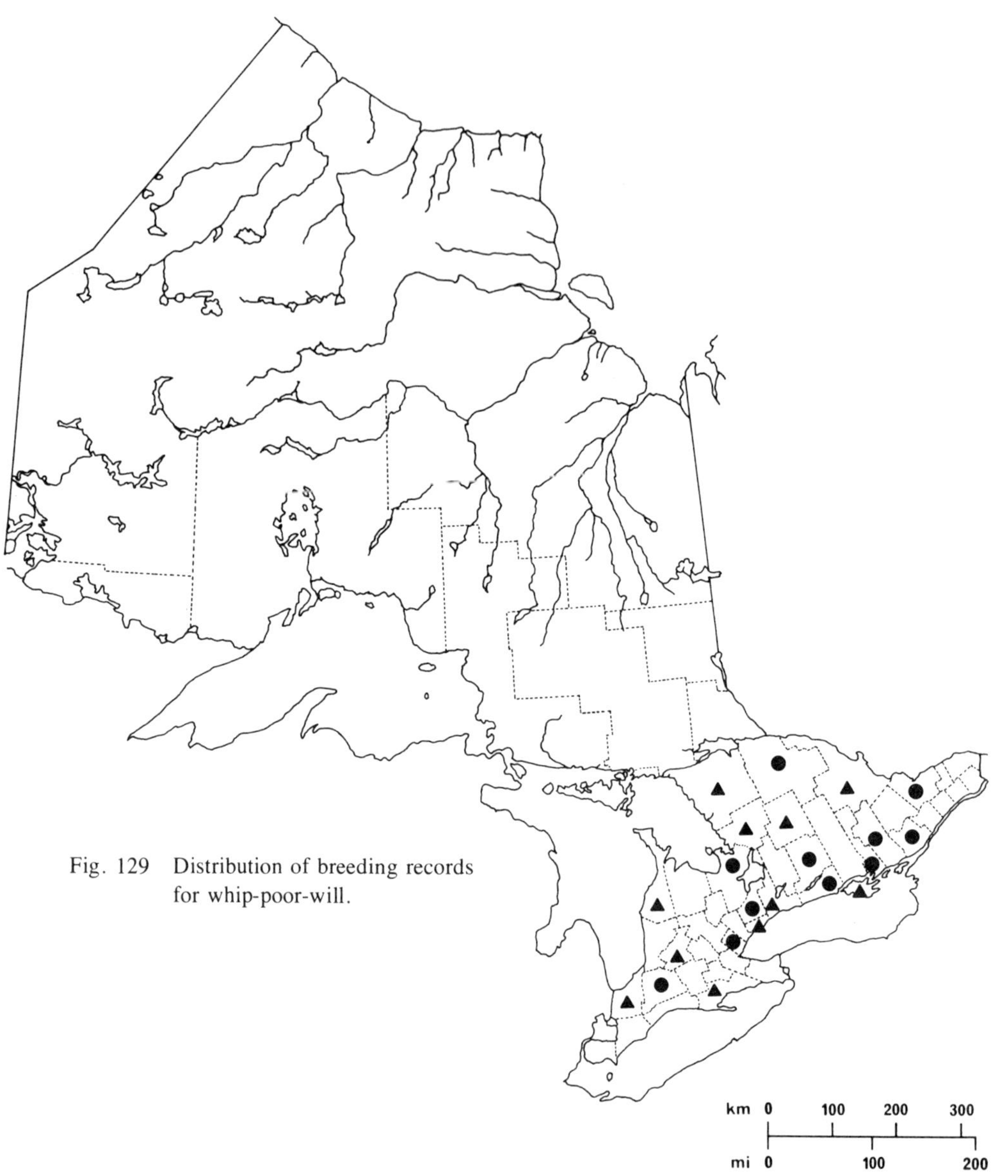

Fig. 129 Distribution of breeding records for whip-poor-will.

Whip-poor-will, *Caprimulgus vociferus* Wilson

Nidiology

RECORDS 33 nests representing 15 provincial regions.
Breeds in both dense and open areas, in deciduous, mixed, or coniferous woods. Nesting habitats were large forests, small wood lots in agricultural areas, pine plantations, and tree-covered sand dunes. Some nests were on hillsides and hilltops.

Nests were invariably on the ground, and were situated beneath trees, beside fallen trees or branches, under or near bushes, or in more open areas without ground cover.

Most nests were depressions. Eggs were often laid on dead leaves or pine needles, sometimes on bare ground or sand, and once on decayed wood.

EGGS 32 nests with 1 to 2 eggs; 1E (1N), **2E** (31N).
Average clutch range 2 eggs (31 nests).

INCUBATION PERIOD 2 nests, at least 16 days.

EGG DATES 26 nests, 21 May to 8 July (34 dates); 13 nests, 9 June to 30 June.

Breeding Distribution

At one time the whip-poor-will nested commonly throughout southern Ontario, but it has become quite scarce in agricultural areas, where forests have largely been removed. No records of this species' nesting in northern Ontario are available, but summer records indicate that it probably breeds as far north as Kenora and Kapuskasing on occasion.

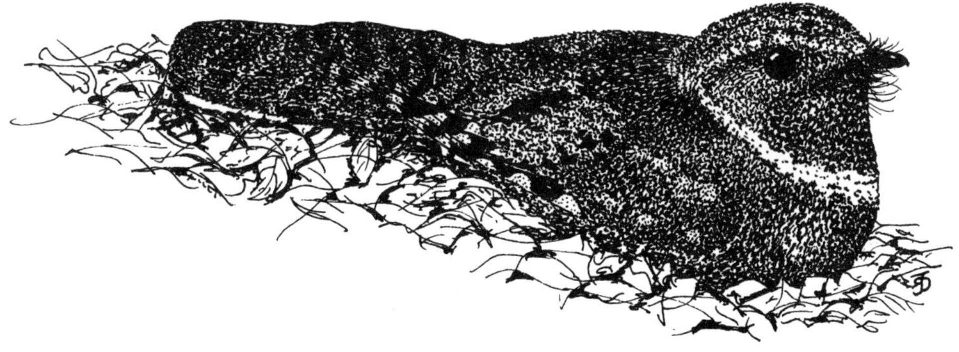

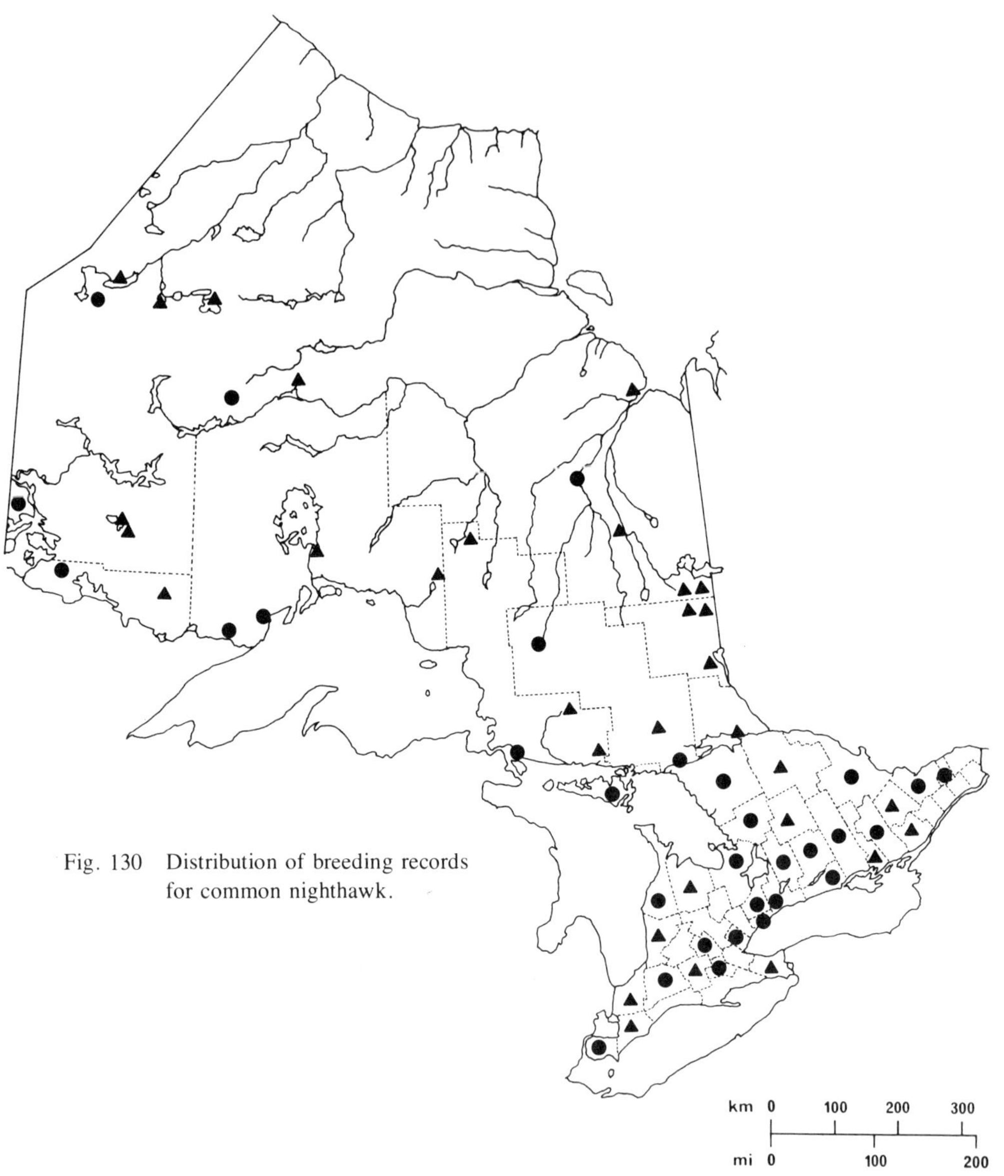

Fig. 130 Distribution of breeding records for common nighthawk.

Common Nighthawk, *Chordeiles minor* (Forster)

Nidiology

RECORDS 215 nests representing 36 provincial regions.
Breeds in open and dense woods, burns, bogs, woodland clearings, and rocky outcroppings, both bare and vegetated, as well as in urban and in agricultural areas. In urban areas, nest sites were commonly on flat, gravel roofs of houses, schools, factories, and other buildings; and occasionally on railroad beds, footpaths, and coal piles. In rural areas, nests were in grass and crop fields and pastures; on limestone prairies; on compounds and yards of parks, lumber mills, and military bases; in gravel pits; on rock piles; and on airfields.

Nests were positioned on the ground (147 nests) or on gravel roofs (45 nests). Ground nests were variously situated on bare rock, on gravel, or on rocky surfaces among lichen, mosses, leaves, and conifer needles; on bare ground, or on ground among leaves, rotted wood, bark chips, mosses, conifer needles, lichen, wood ash, grasses, and twigs; in sand; in sawdust and wood chips; and on coal pieces. Most ground nests were in exposed situations, but a few (9 nests) were under bushes and trees.

Almost all nests were merely scrapes in existing surface material, or eggs were laid on bare rock or gravel. However, 2 records reported the use of compacted organic material (conifer needles, dead twigs, and bark chips) apparently formed into a rudimentary structure. Some nest sites were reused in successive seasons for a number of years.

EGGS 206 nests with 1 to 4 eggs; 1E (32N), **2E** (172N), 4E (2N).
Average clutch range 2 eggs (172 nests).
The 4-egg clutches were probably the product of 2 females.

INCUBATION PERIOD 2 nests: 1 of 19 days, 1 of 20 days. Incubation commenced with the first egg.

EGG DATES 182 nests, 26 May to 13 August (207 dates); 91 nests, 11 June to 29 June.

Breeding Distribution

The common nighthawk probably breeds throughout Ontario. Summer records extend as far north as the Hudson Bay coast, but nest reports are lacking in that region at the present time.

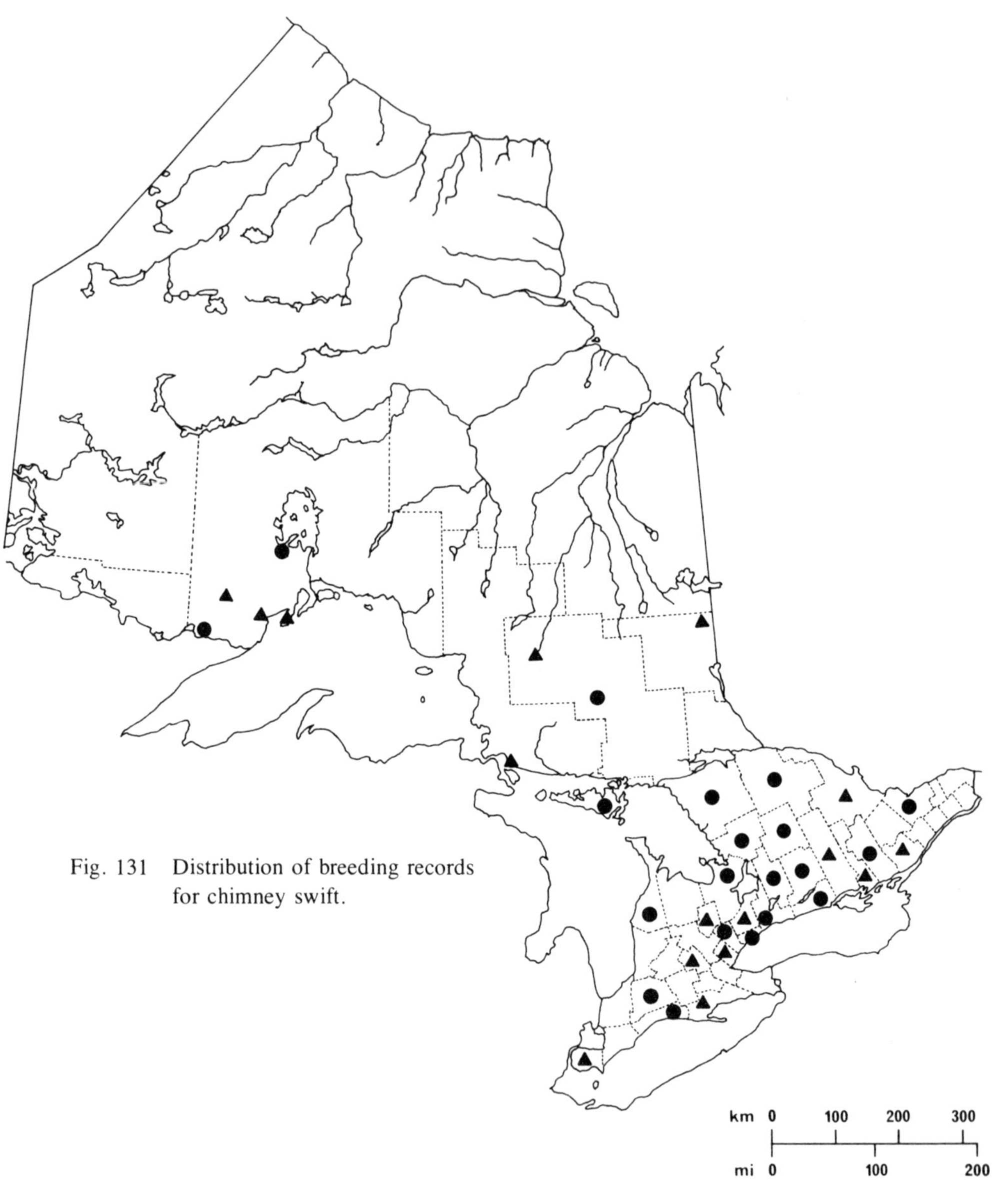

Fig. 131 Distribution of breeding records for chimney swift.

Chimney Swift, *Chaetura pelagica* (Linnaeus)

Nidiology

RECORDS 92 (95 nests) representing 28 provincial regions.
Breeds wherever suitable nests sites are available. Nest sites recorded were in rural (42 nests), urban (22 nests), and park or cottage (19 nests) areas.

Almost invariably nests were located in houses, barns and silos, sheds and other unspecified buildings, and lighthouses. They were placed on walls, rafters, and gables of buildings (41 nests); in chimneys (40 nests); and in air-vent shafts (2 nests). One nest was attached to a fireplace fireguard at the bottom of a chimney; another was in a hollow elm stub. Nests in chimneys were situated from 1.2 to 3.5 m (4 to 12 ft) below chimney tops (11 nests); nests in rooms were situated from 1 to 1.5 m (3 to 5 ft) below ceilings (10 nests). Heights of 29 nests ranged from 2 to 15 m (6 to 50 ft), with 15 averaging 4.5 to 10.5 m (15 to 35 ft).

Nests were frail, semicircular, basket-shaped structures. Measurements of 4 nests ranged as follows: lateral inside diameter, 5.5 to 7 cm (2.25 to 2.75 inches); front-to-back inside diameter, 5 to 6.5 cm (2 to 2.5 inches); outside depth, 4 to 4.5 cm (1.5 to 1.75 inches). Nests were made of small dead twigs that were of uniform thickness and 2.5 to 4 cm (1 to 1.5 inches) in length. These twigs were cemented together and to the side of the supporting site with a glue-like saliva. Nests were almost flat on top and were unlined.

EGGS 67 nests with 2 to 5 eggs; 2E (5N), **3E** (10N), **4E** (29N), **5E** (23N).
Average clutch range 4 to 5 eggs (52 nests).

INCUBATION PERIOD 1 nest, 16 days.

EGG DATES 61 nests, 24 May to 4 August (66 dates); 31 nests 23 June to 7 July.

Breeding Distribution

The chimney swift breeds throughout Ontario ranging about as far north as Lake Nipigon and Lake Abitibi.

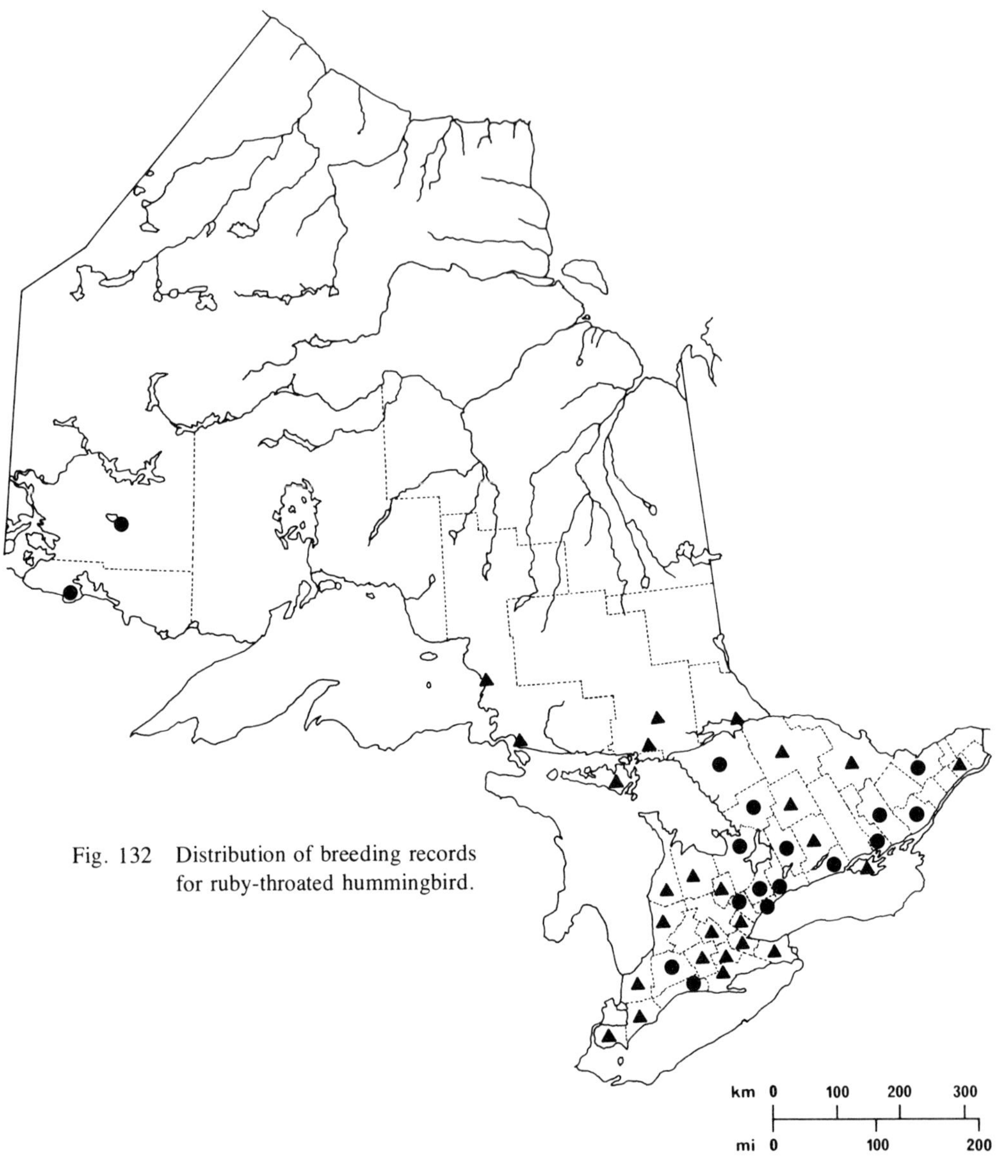

Fig. 132 Distribution of breeding records for ruby-throated hummingbird.

Ruby-throated Hummingbird, *Archilochus colubris* (Linnaeus)

Nidiology

RECORDS 109 nests representing 38 provincial regions.
Breeds in a variety of habitats including mixed and deciduous woods and coniferous groves, cottage areas on the shores or on islands in lakes and rivers, swamps and beaver ponds, farmland and orchards, and golf courses and residential gardens. Nests were sometimes in dense woods but more often in open areas (in gardens, over roads and paths, over or near water, and near dwellings). The preponderance of open sites seems to indicate a preference for clearings near the nest tree.

Almost all nests were in trees (1 nest was in a vine on the wall of a house); deciduous trees (12 spp., 60 nests) were preferred to conifers (4 spp., 18 nests). Nests were often saddled on bare, horizontal branches or in forks, and less often on downward-sloping branches, which had diameters of 1 and 6.5 m (0.4 to 2.5 inches). Nest branches were usually living and sometimes dead; occasionally nests were canopied by overhanging leaves. Nests were almost invariably from 1 to 3.5 m (3 to 12 ft) away from the tree trunk, often near the end of the branch. Heights of 85 nests ranged from 1.2 to 15 m (4 to 50 ft), with 43 averaging 3 to 6 m (10 to 20 ft).

Nests were tiny, well-constructed cups that resembled knots. They were characteristically made of lichens, bud scales, bark fragments and fibres, and fine grasses, all held in place by spiders' webs on the outside. Nests were lined with plant down and fibre. Other materials occasionally used were mosses, hair, weed seeds, small leaves, and feathers. Outside nest diameters ranged from 3.8 to 5 cm (1.5 to 2 inches); inside diameter of 1 nest was 2 cm (0.75 inch). The cups of 2 nests were quite shallow and the eggs were resting on the bare limbs.

EGGS 77 nests with 1 to 3 eggs; 1E (10N), **2E** (66N), 3E (1N).
Average clutch range 2 eggs (66 nests).

INCUBATION PERIOD 2 nests, at least 13 to 15 days.

EGG DATES 61 nests, 25 May to 2 September (68 dates); 31 nests, 16 June to 12 July.

Breeding Distribution

The ruby-throated hummingbird breeds throughout southern Ontario, and its range extends into the north about as far as Lac Seul and Cochrane.

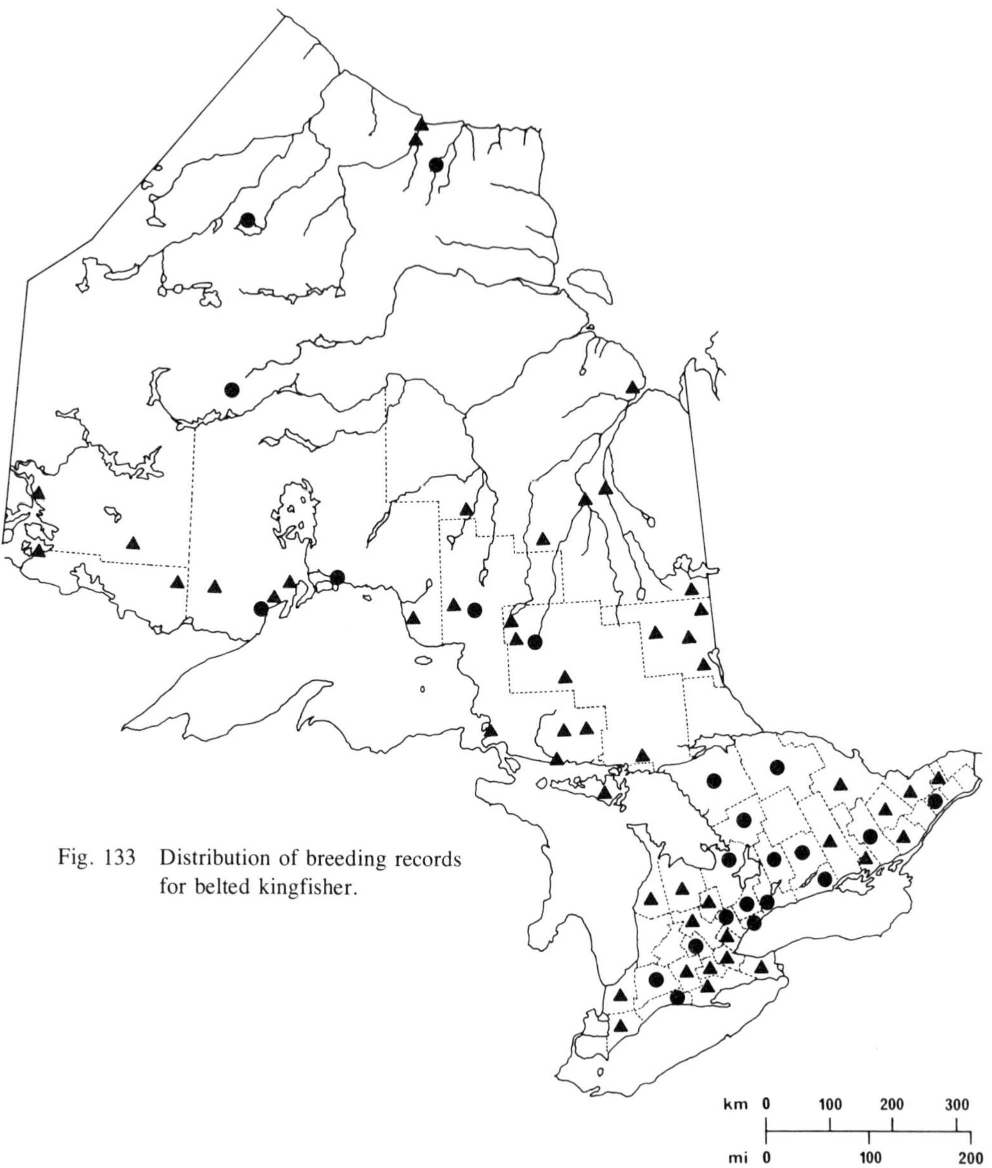

Fig. 133 Distribution of breeding records for belted kingfisher.

Belted Kingfisher, *Megaceryle alcyon* (Linnaeus)

Nidiology

RECORDS 271 nests representing 40 provincial regions.
Breeds most often in woodland and farmland areas, at lake and river shores, in roadside and railroad cuts, and in gravel and sand pits; breeds less commonly in marsh and sand-dune areas, at pit mines, at sawmills, and in city parks.

Nesting was often solitary, but was sometimes in or near bank swallow and rough-winged swallow colonies. Nests were almost invariably burrows in banks of sand, gravel, earth, and clay, although 1 nest was in a cavity in an aspen tree at 10.5 m (35 ft). Nests were also recorded in sawdust piles, in mine tailings, and once in a drainpipe. Nests were often over or near water but sometimes up to 0.8 km (0.5 mi) distant from it. Heights of 163 nests ranged from ground level to 215 m (700 ft) in cliffs, with 82 averaging 1.5 to 6 m (5 to 20 ft). Situations of 63 nests ranged from 0.3 to 3 m (1 to 10 ft) below the top of the bank or other supporting site, with 31 averaging 0.3 to 0.6 m (1 to 2 ft) below the top. Occasionally burrow entrances were obscured by overhanging roots.

Diameters of openings to nest tunnels ranged from 7.5 to 30.5 cm (3 to 12 inches) but averaged ca 10 cm (4 inches). Tunnels ran horizontally or slightly upward, and many tunnels curved, sometimes to the left or more often to the right. Lengths of 71 nest tunnels ranged from 0.3 to 2.5 m (1 to 8 ft), with 35 averaging 1.1 to 1.8 m (3.5 to 6 ft). Tunnels expanded into nest chambers, which ranged from 20.5 to 23 cm (8 to 9 inches) in diameter.

Nest chambers were sometimes unlined but frequently contained fish bones and scales, crayfish and shell fragments, feathers, grasses, straw, twigs, and mosses.

EGGS 98 nests with 2 to 8 eggs; 2E (2N), 3E (11N), 4E (7N), **5E** (13N), **6E** (29N), **7E** (34N), 8E (2N).
Average clutch range 5 to 7 eggs (76 nests).

INCUBATION PERIOD 1 nest, 22 days.

EGG DATES 71 nests, 4 May to 2 July (76 dates); 35 nests, 23 May to 9 June.

Breeding Distribution

The belted kingfisher breeds throughout Ontario.

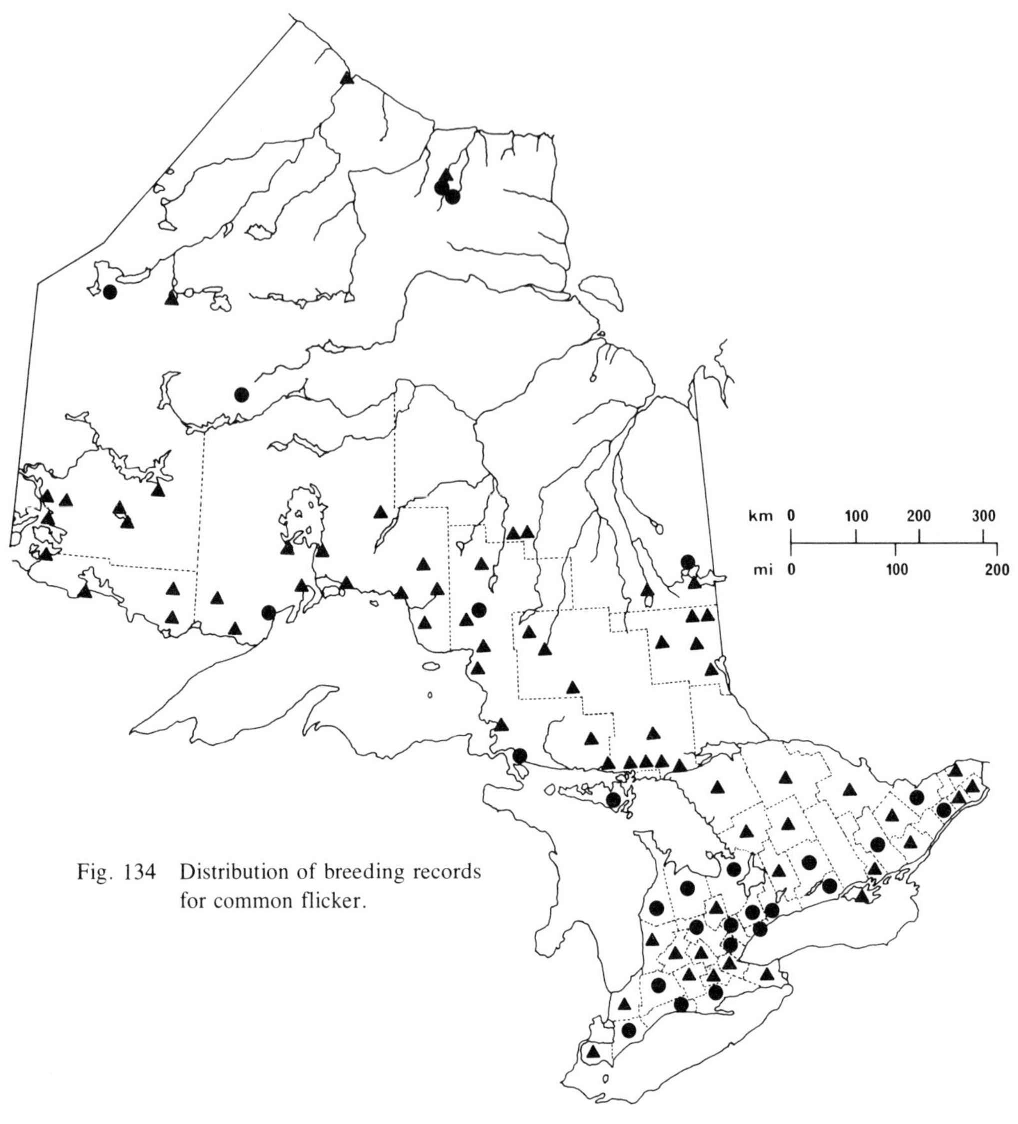

Fig. 134 Distribution of breeding records for common flicker.

Common Flicker, *Colaptes auratus* (Linnaeus)

Nidiology

RECORDS 757 nests representing 49 provincial regions.
Breeds in a variety of habitats. Those recorded included mixed (135 nests), deciduous (122 nests), and coniferous (44 nests) woodlands; shorelines and forest edges (88 nests); open agricultural land (64 nests); residential areas (42 nests); beaver ponds and marsh areas

(25 nests); burns (14 nests); and campgrounds (6 nests). Dry woodland areas were preferred to wet.

Most nests were in cavities, and were excavated usually in trees (584 nests)—most often in trunks and occasionally in limbs—sometimes in hydro poles (24 nests) and fence posts (14 nests), and once in a stump of driftwood. Nest boxes (21 nests) were sometimes used. In those records where nest trees were described, dead trees (302 nests) were preferred to living trees (20 nests). Nests were more often in deciduous trees (24 spp., 288 nests) than in coniferous trees (8 spp., 19 nests), and nest trees most commonly selected were elm spp. (102 nests), poplar spp. (95 nests), apple spp. (17 nests), maple spp. (16 nests), and pine spp. (9 nests). This species usually excavated new nest cavities, but occasionally it reused old excavations of the same or of other species, and more rarely it used natural cavities. Cavities faced in all directions, but southerly exposures (28 of 57 records—49%) were preferred. Some nest trees were previously used by yellow-bellied sapsucker, hairy woodpecker, eastern bluebird, and starling; some nest trees were simultaneously shared with wood duck, American kestrel, pileated woodpecker, yellow-bellied sapsucker, hairy woodpecker, starling, and house sparrow. One nest cavity was reused for 8 seasons. Diameters (DBH) of 10 nest trees ranged from 19 to 46 cm (7.5 to 18 inches), with 5 averaging 23 to 31 cm (9 to 12 inches); bole diameters at 17 nest cavities ranged from 15 to 31 cm (6 to 12 inches), with 9 averaging 19 to 31 cm (7.5 to 12 inches); heights of 616 nests ranged from 0.3 to 24.5 m (1 to 80 ft), with 308 averaging 2.5 to 7.5 m (8 to 25 ft). In 1 nest the eggs were just above ground level in a cavity of a 0.6 m (2 ft) stump.

Most nest cavities were unlined, but 1 adult was observed carrying grass into a cavity. Measurements of 30 cavity entrances ranged from 5 to 18 cm (2 to 7 inches), with 15 averaging 7.5 cm (3 inches); depths of 41 cavities ranged from 20 to 91.5 cm (8 to 36 inches), with 21 averaging 25.5 to 40.5 cm (10 to 16 inches). One nest cavity had 2 entrances.

EGGS 245 nests with 1 to 13 eggs; 1E (7N), 2E (9N), 3E (13N), **4E** (35N), **5E** (39N), **6E** (51N), **7E** (54N), **8E** (14N), **9E** (8N), 10E (3N), 11E (1N), 13E (1N).
Average clutch range 4 to 7 eggs (189 nests).
In 1 nest, 15 eggs were collected at intervals, then a further 5 eggs were incubated and young hatched. One nest contained 5 eggs of common flicker and 1 egg of hairy woodpecker; an adult hairy woodpecker was flushed from this nest once.

INCUBATION PERIOD 7 nests: 1 of ca 12 days, 1 of less than 13 days, 4 of ca 13 days, 1 of less than 15 days. Incubation commenced in 1 nest before the clutch was complete.

EGG DATES 218 nests, 5 May to 20 July (245 dates); 109 nests, 27 May to 9 June.

Breeding Distribution

The common flicker breeds throughout Ontario.

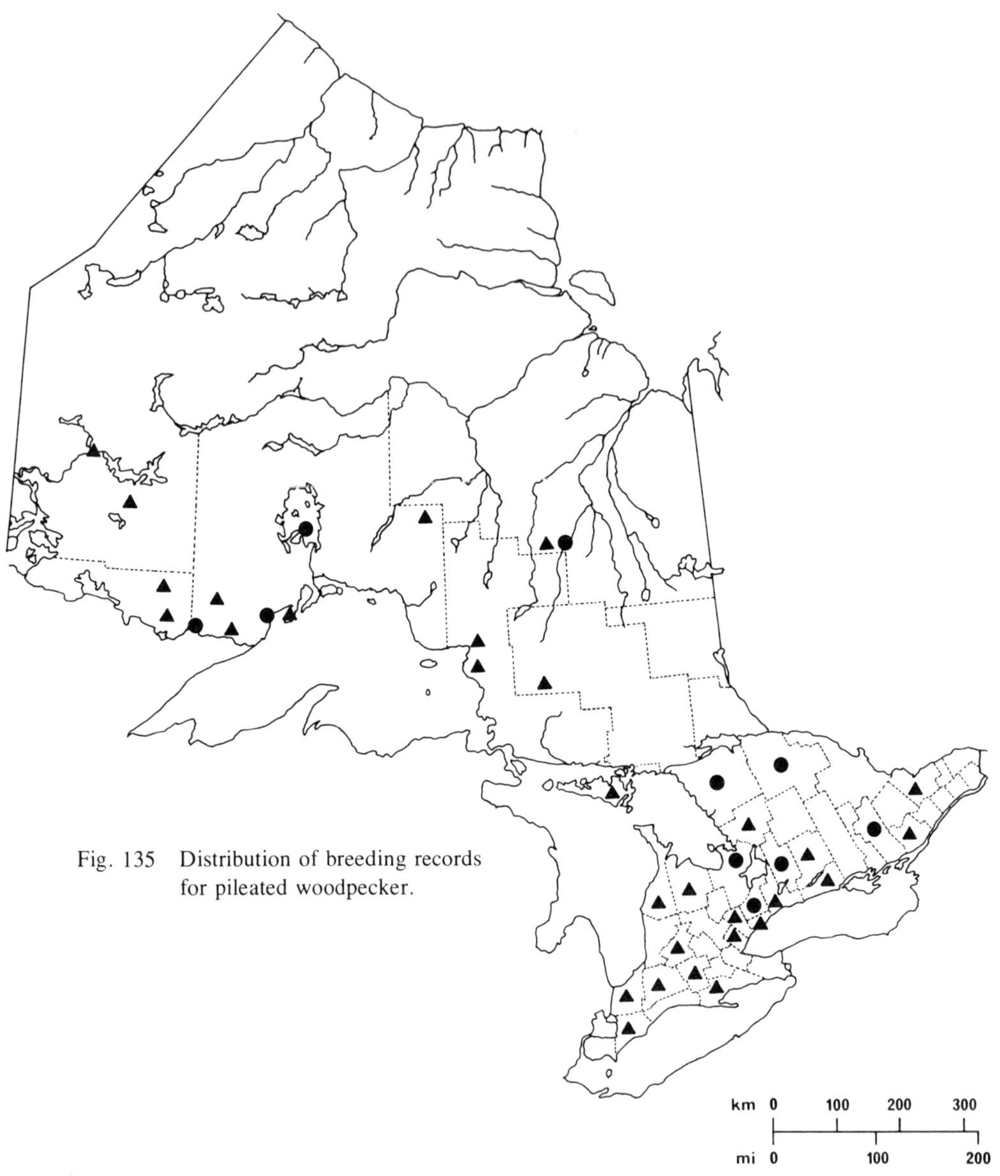

Fig. 135 Distribution of breeding records for pileated woodpecker.

Pileated Woodpecker, *Dryocopus pileatus* (Linnaeus)

Nidiology

RECORDS 89 nests representing 28 provincial regions.
Breeding habitats recorded were usually woodlands—mixed (25 nests), deciduous (18 nests), coniferous (3 nests)—and occasionally more open agricultural and park areas. As well, some nests were in wooded swamps, beaver ponds, and flood plains.

All nests were in cavities excavated in trees that were usually dead or less often living and were most often deciduous (7 spp., 59 nests) and sometimes coniferous (2 spp., 8 nests); occasionally old excavations were used. The most commonly selected nest trees were poplar spp. (16 nests), elm spp. (15 nests), birch spp. (10 nests), beech (7 nests), and maple spp. (7 nests). Nest cavities were in the trunk or less often in large limbs. Diameters (DBH) of 3 nest trees ranged from 35.5 to 51 cm (14 to 20 inches); bole diameter at 1 nest cavity was 23 cm (9 inches). Nest holes faced in all directions, although southerly exposures (8 nests) seemed to be slightly preferred to all other directions (5 nests). Heights of 69 nests ranged from 3.5 to 27.5 m (12 to 90 ft), with 35 averaging 7.5 to 12 m (24 to 40 ft). Two cavities were reused in 2 successive years, and 1 nest tree was simultaneously used by common flicker.

Nest holes were upright and oval in shape. Horizontal diameters of 7 nest holes ranged from 10 to 12.5 cm (4 to 5 inches); their vertical diameters ranged from 12.5 to 23 cm (5 to 9 inches); cavity depths of 4 nests ranged from 35.5 to 76 cm (14 to 30 inches).

EGGS 27 nests with 1 to 6 eggs; 1E (1N), 2E (6N), **3E** (11N), **4E** (7N), **5E** (1N), 6E (1N).
Average clutch range 3 to 4 eggs (18 nests).

INCUBATION PERIOD No information.

EGG DATES 16 nests, 4 May to 28 June (18 dates); 8 nests, 9 May to 30 May.

Breeding Distribution

The pileated woodpecker breeds throughout the forested parts of Ontario, about as far north as southern James Bay and the Albany River in the east, and possibly somewhat farther in the west.

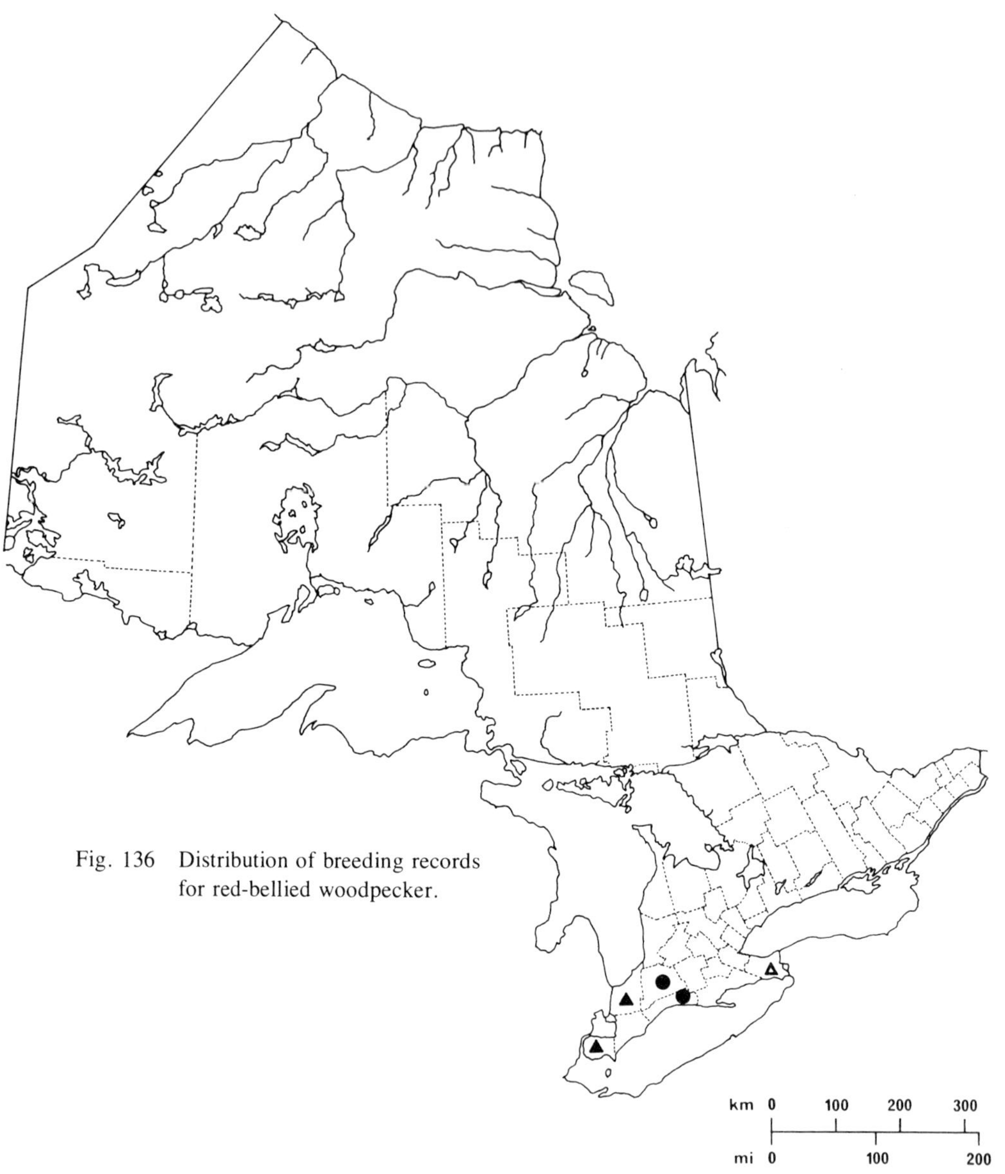

Fig. 136 Distribution of breeding records for red-bellied woodpecker.

Red-bellied Woodpecker, *Melanerpes carolinus* (Linnaeus)

Nidiology

RECORDS 9 nests representing 5 provincial regions.
A rare, local, permanent resident of the extreme southwest of Ontario, this woodpecker breeds in deciduous forests (Fig. 145) and wooded residential areas.

Nest cavities were in dead or partially dead trees such as beech (1 record), elm sp. (1 record), and hickory sp. (1 record). Heights of 6 nests ranged from 10 to 20 m (35 to 65 ft), with 3 averaging 12 m (40 ft).

EGGS 4 nests with 2 to 7 eggs; 2E (2N), 5E (1N), 7E (1N).

INCUBATION PERIOD No information.

EGG DATES 4 nests, 1 May to 27 May.

Breeding Distribution

In southern Ontario, the red-bellied woodpecker (Fig. 146B) is at the northern limit of its breeding range. It has been reported as nesting in only four counties in the extreme southwest of the province. At one time its range may have been larger, and possibly it nested as far north as Wellington and Halton counties (Baillie and Harrington, 1937), but the only specific record of more widespread nesting is in Niagara RM (Beardslee and Mitchell, 1965).

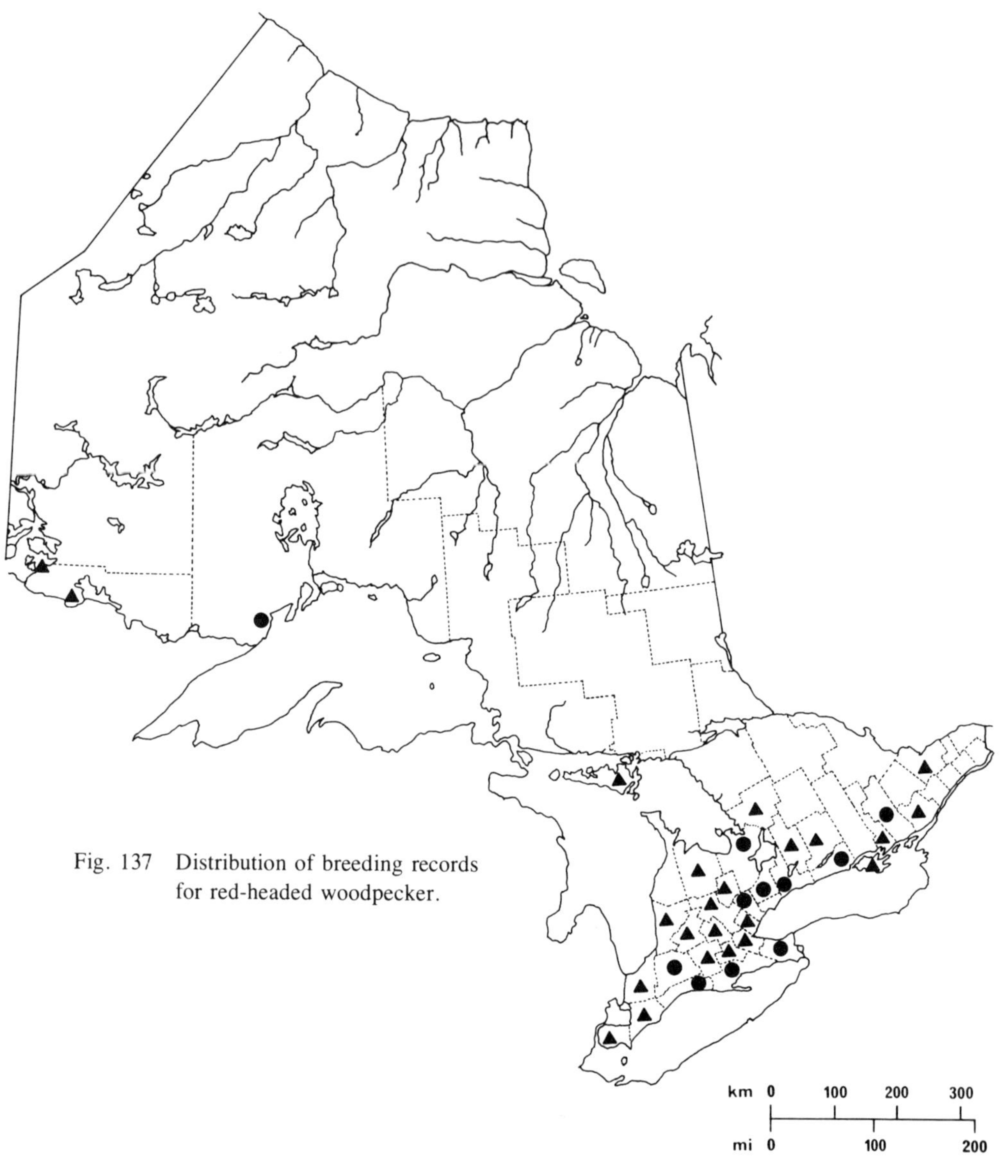

Fig. 137 Distribution of breeding records for red-headed woodpecker.

Red-headed Woodpecker, *Melanerpes erythrocephalus* (Linnaeus)

Nidiology

RECORDS 161 nests representing 32 provincial regions.
Breeds in the following habitats: both in and at the edge of woodlands that are predominantly open and most often deciduous but occasionally mixed; in areas flooded by beavers; in fields and pastures and in fence rows and roadsides on agricultural land; in city parks, ravines, golf courses, and residential gardens; and at the edges of ponds, rivers, and river flood plains.

Nest cavities were usually in trees (145 nests), but 1 cavity was in a hydro pole and another in a hollow post. Nest trees selected were usually dead or partially dead (110 nests); 3 nest trees were living. Nest trees were invariably deciduous (11 spp.); those most often used were elm spp. (63 nests), maple spp. (13 nests), oak spp. (11 nests), and birch spp. (7 nests). Nest cavities were in the trunk or in main limbs, and ranged in position from low heights to near the top of the tree. The holes faced in all directions, although southerly exposures (7 nests) were apparently preferred to other directions (4 nests). Heights of 120 nests ranged from 2 to 24.5 m (7 to 80 ft), with 60 averaging 7.5 to 12 m (24 to 40 ft).

The species usually excavated its own nest, although 1 nest excavated by common flicker was used. Hole diameters of 4 nests ranged from 5 to 7.5 cm (2 to 3 inches); cavity depths of another 4 nests were all 25.5 cm (10 inches).

Two broods were raised in 1 nest in the same season; 2 nests were reused the following year; 1 nest was reused for a number of successive years.

EGGS 31 nests with 1 to 7 eggs; 1E (2N), 2E (2N), 3E (3N), **4E** (12N), **5E** (8N), **6E** (1N), 7E (3N).
Average clutch range 4 to 5 eggs (20 nests).

INCUBATION PERIOD No information.

EGG DATES 27 nests, 14 May to 21 July (28 dates); 14 nests, 31 May to 17 June.

Breeding Distribution

The red-headed woodpecker nests throughout the south but less commonly in the Canadian Shield. Although its breeding range is mainly outside the areas of coniferous forest in Ontario, the species nests as far north as Manitoulin Island, the north shore of Lake Huron, and the area between Thunder Bay and western Rainy River District. Although its numbers have decreased markedly in the past 50 years, the species still occupies much of its original range.

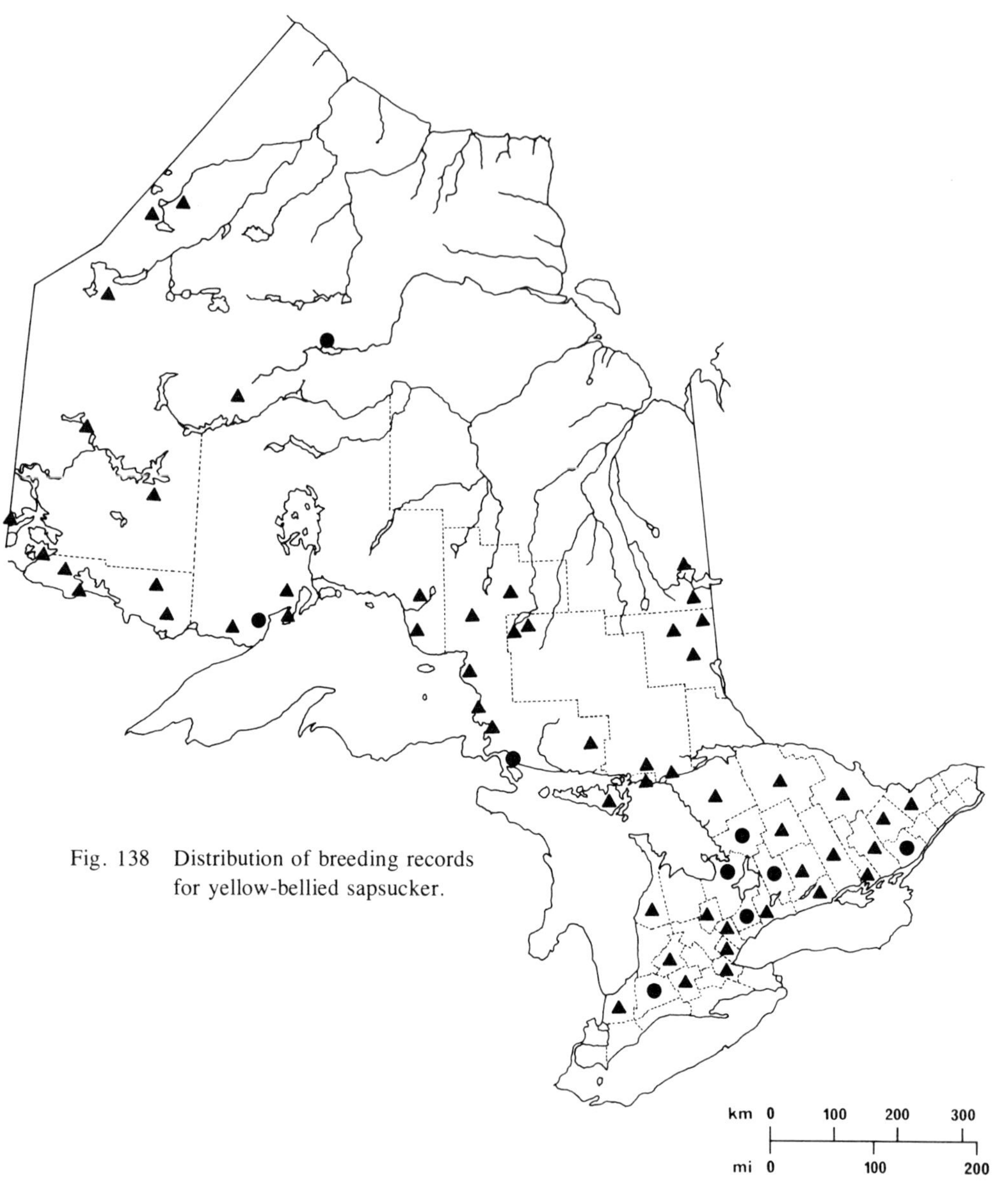

Fig. 138 Distribution of breeding records for yellow-bellied sapsucker.

Yellow-bellied Sapsucker, *Sphyrapicus varius* (Linnaeus)

Nidiology

RECORDS 342 nests representing 35 provincial regions.
Breeds usually in, or at the edge of, open and closed woodlands and prefers dry terrain to wet. Woodland habitats recorded were mixed tree stands (142 records) and deciduous stands (67 records); no coniferous stands were recorded. Nest sites also included campgrounds (7 nests), roadsides (3 nests), pastures (2 nests), residential areas (2 nests), and sand-dune areas (1 nest).

All nests were in excavated tree cavities. Deciduous trees (11 spp., 293 nests) were greatly preferred to coniferous trees (4 spp., 6 nests), and the trees most commonly used were poplar spp. (178 nests), birch spp. (39 nests), maple spp. (38 nests), beech (12 nests), and elm spp. (11 nests). Of the most frequently selected nest trees (poplar spp.), living trees were preferred, but dead trees were often recorded as well. Most nest cavities were newly excavated, but this species often reused old cavities and sometimes used the excavations of common flicker (2 records), hairy woodpecker (2 records), or downy woodpecker (1 record). One cavity was previously used by starling. Other species nesting in the same tree simultaneously were pileated and downy woodpeckers and starling. Nests were recorded close to active nests of goshawk and hairy woodpecker. Nest cavities faced in all directions, but southerly (26 nests) and easterly (25 nests) exposures were slightly more common than westerly (19 nests) and northerly (17 nests) directions. Diameters (DBH) of 31 nest trees ranged from 21.5 to 51 cm (8.5 to 20 inches), with 15 averaging 25.5 to 33 cm (10 to 13 inches). Heights of 307 nests ranged from 1.7 to 23 m (5.5 to 75 ft), with 154 averaging 5.5 to 10.5 m (18 to 35 ft).

Nest cavities were usually in the tree trunk but occasionally in large upright branches. Bole diameters at 15 nest cavities ranged from 12.5 to 30.5 cm (5 to 12 inches), with 8 averaging 19 to 25.5 cm (7.5 to 10 inches). Cavities were unlined and entrances were usually circular. Diameters of 21 holes ranged from 2.5 to 7.5 cm (1 to 3 inches), with 11 averaging 3 to 4 cm (1.25 to 1.6 inches); depths of 10 cavities ranged from 20.5 to 38 cm (8 to 15 inches), with 5 averaging 23 to 30.5 cm (9 to 12 inches); diameters of 3 cavities ranged from 6.5 to 15 cm (2.5 to 6 inches).

EGGS 43 nests with 2 to 7 eggs; 2E (5N), 3E (6N), **4E** (14N), **5E** (12N), **6E** (4N), 7E (2N).
Average clutch range 4 to 5 eggs (26 nests).

INCUBATION PERIOD 4 nests, ca 12 to 13 days.

EGG DATES 42 nests, 15 May to 12 July (47 dates); 21 nests, 29 May to 7 June.

Breeding Distribution

The yellow-bellied sapsucker breeds throughout most of the forested portions of Ontario as far north as Big Trout Lake and Moosonee. In the agricultural south, it is found only where suitable habitat exists.

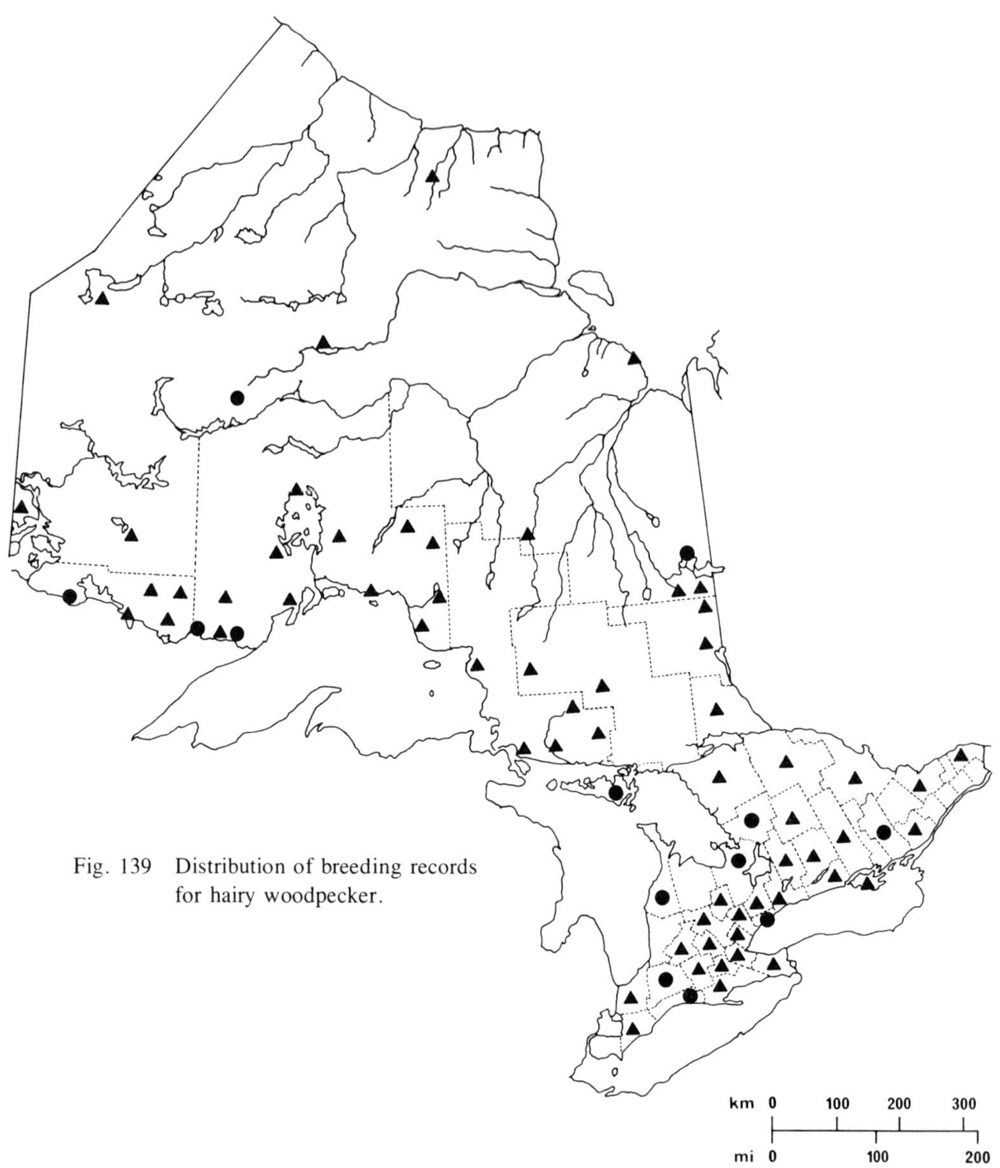

Fig. 139 Distribution of breeding records for hairy woodpecker.

Hairy Woodpecker, *Picoides villosus* (Linnaeus)

Nidiology

RECORDS 186 nests representing 40 provincial regions.
Breeds usually in woodlands and slightly more often in dry habitats than in wet habitats (swamps and beaver ponds). Woodlands were usually mixed (59 nests), and deciduous (28 nests), and sometimes coniferous (4 nests). Other habitats recorded included rural roadsides (6 nests), campgrounds (3 nests), and orchards (1 nest).

Almost all nests were in excavated tree cavities, but 1 nest was in a wood duck nest box and another in a cavity in a telephone pole. Deciduous trees (10 spp., 152 nests) were much preferred to coniferous trees (4 spp., 6 nests), and nest trees most often selected were poplar spp. (68 nests), maple spp. (29 nests), birch spp. (21 nests), ash spp. (12 nests), elm spp. (10 nests), and beech (8 nests). Whereas the downy woodpecker preferred dead trees, this species, when specified in records, selected living trees as often as dead trees; 66 records were unspecified. However, since the majority of the unspecified records were assumed to be of living trees, this species probably preferred living trees. Almost all cavities were newly excavated. One tree was used for 3 successive years. One record described nesting in a tree used by yellow-bellied sapsucker during the previous season, but in a different cavity. Nests were recorded near active nests of common flicker and yellow-bellied sapsucker. Nest holes faced in all directions but southerly exposures (12 of 33 records—36.4%) were slightly preferred to other directions. Heights of 163 nests ranged from 1.2 to 24.5 m (4 to 80 ft), with 82 averaging 4.5 to 10.5 m (14 to 35 ft).

Nest cavities were much more often in the main trunk (56 records) than in limbs (6 records). Three cavities were noted, respectively, directly beneath a bracket fungus, a knot, and a branch. Two cavities slanted upwards at first. Diameters (DBH) of 3 nest trees ranged from 18 to 31 cm (7 to 12.2 inches); bole diameters at 7 nest cavities ranged from 12.5 to 25.5 cm (5 to 10 inches). Nest cavities were unlined and usually entrances were circular. Diameters of 22 holes ranged from 2.5 to 6.5 cm (1 to 2.5 inches), with 11 averaging 3.8 to 5 cm (1.5 to 2 inches); depths of 11 cavities ranged from 25.5 to 45.7 cm (10 to 18 inches), with 6 averaging 25.5 to 35.5 cm (10 to 14 inches); diameters of 2 nest cavities were 10 and 12.5 cm (4 and 5 inches).

EGGS 49 nests with 1 to 5 eggs; 1E (2N), 2E (3N), **3E** (13N), **4E** (23N), **5E** (8N).
Average clutch range 3 to 4 eggs (36 nests).

INCUBATION PERIOD 1 nest, at least 11 days.

EGG DATES 39 nests, 16 April to 8 June (45 dates); 20 nests, 5 May to 17 May. This species nested noticeably earlier than both yellow-bellied sapsucker and downy woodpecker.

Breeding Distribution

The hairy woodpecker breeds throughout the forested portions of Ontario, but becomes very rare in the far north of the province.

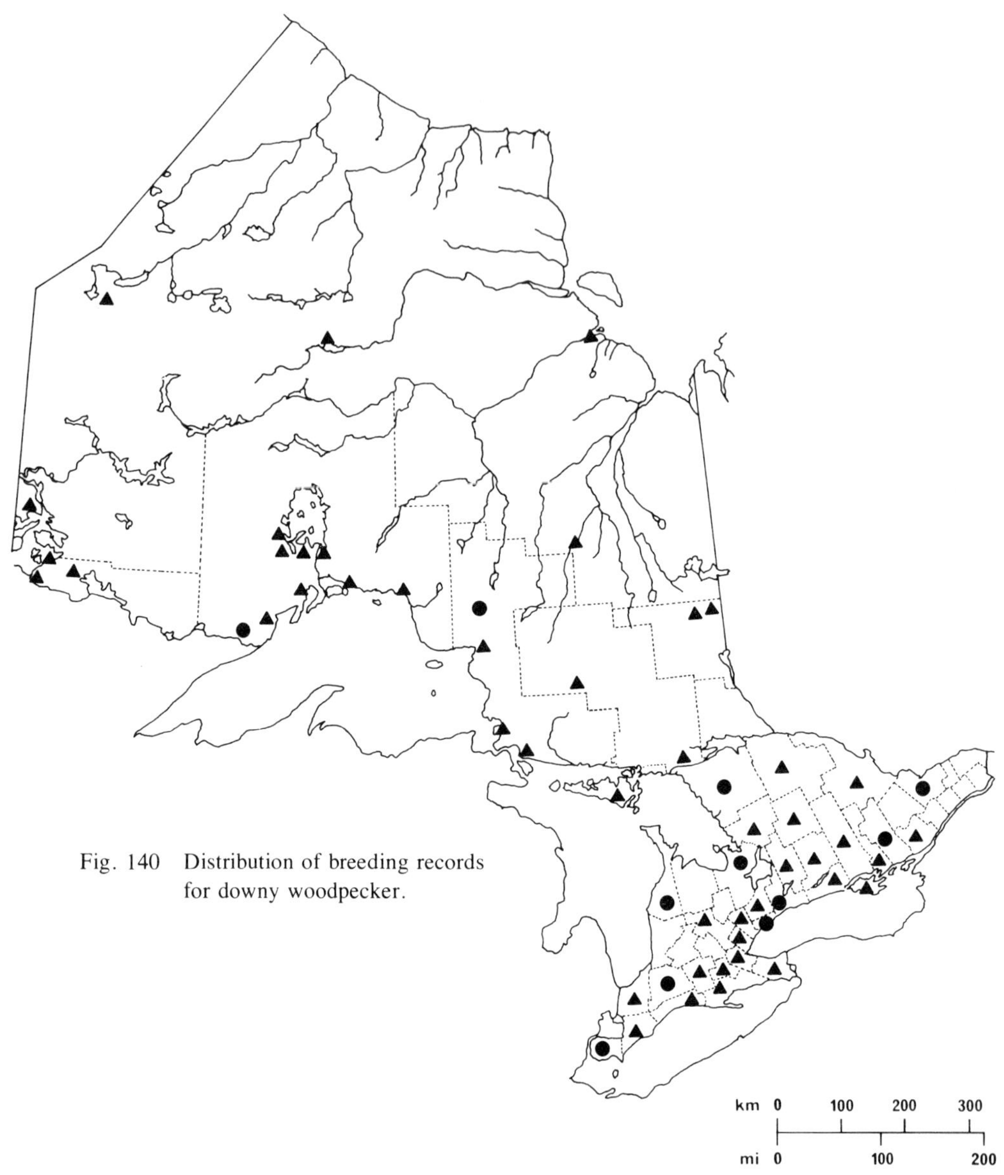

Fig. 140 Distribution of breeding records for downy woodpecker.

Downy Woodpecker, *Picoides pubescens* (Linnaeus)

Nidiology

RECORDS 201 nests representing 36 provincial regions.
Breeding habitats recorded were usually woodlands—deciduous (53 nests), mixed (25 nests), pure coniferous stands (3 nests). Other habitats recorded included pastures (5 nests), residential gardens (4 nests), an old burn (1 nest), a cemetery (1 nest), and a golf course (1 nest). Like the hairy woodpecker, this species bred slightly more often in dry habitats than in wet ones (beaver ponds with standing dead trees).

All nests were in excavated tree cavities; deciduous trees (13 spp., 181 nests) were greatly preferred to coniferous trees (2 spp., 4 nests), and nest trees most often selected were poplar spp. (46 nests), maple spp. (44 nests), elm spp. (26 nests), and birch spp. (25 nests). Dead trees or dead portions of living trees (132 nests) were much preferred to living trees (12 nests). Most nest cavities were newly excavated, but 1 nest was reused over a 4-year period. One nest was recorded in the same tree as a nest of yellow-bellied sapsucker; some nests were located near active nests of common flicker, house wren, and starling. Nest holes faced in all directions, but southerly exposures (9 of 21 records—42.9%) were slightly preferred to other directions. Heights of 148 nests ranged from 0.6 to 18.5 m (2 to 60 ft), with 74 averaging 3.5 to 9 m (12 to 30 ft).

Nest cavities were more often in the main tree trunk (82 nests) than in limbs (20 nests). Bole diameters at 7 nest cavities ranged from 12.5 to 45.5 cm (5 to 18 inches). Cavities were unlined and entrances were circular. Diameters of 17 holes ranged from 2.5 to 7.5 cm (1 to 3 inches), with 9 averaging 3 to 3.8 cm (1.2 to 1.5 inches); depths of 11 cavities ranged from 15 to 25.5 cm (6 to 10 inches), with 6 averaging 17 to 23 cm (6.6 to 9 inches); diameters of 5 nest cavities ranged from 7.5 to 10 cm (3 to 4 inches).

EGGS 64 nests with 1 to 7 eggs; 1E (1N), 2E (4N), **3E** (8N), **4E** (14N), **5E** (22N), **6E** (13N), 7E (2N).
Average clutch range 4 to 5 eggs (36 nests).

INCUBATION PERIOD 3 nests, ca 12 days.

EGG DATES 69 nests, 2 May to 1 July (74 dates); 35 nests, 23 May to 4 June.

Breeding Distribution

The downy woodpecker breeds throughout the forested portions of Ontario as far north as Fort Albany and Favourable Lake, and perhaps farther.

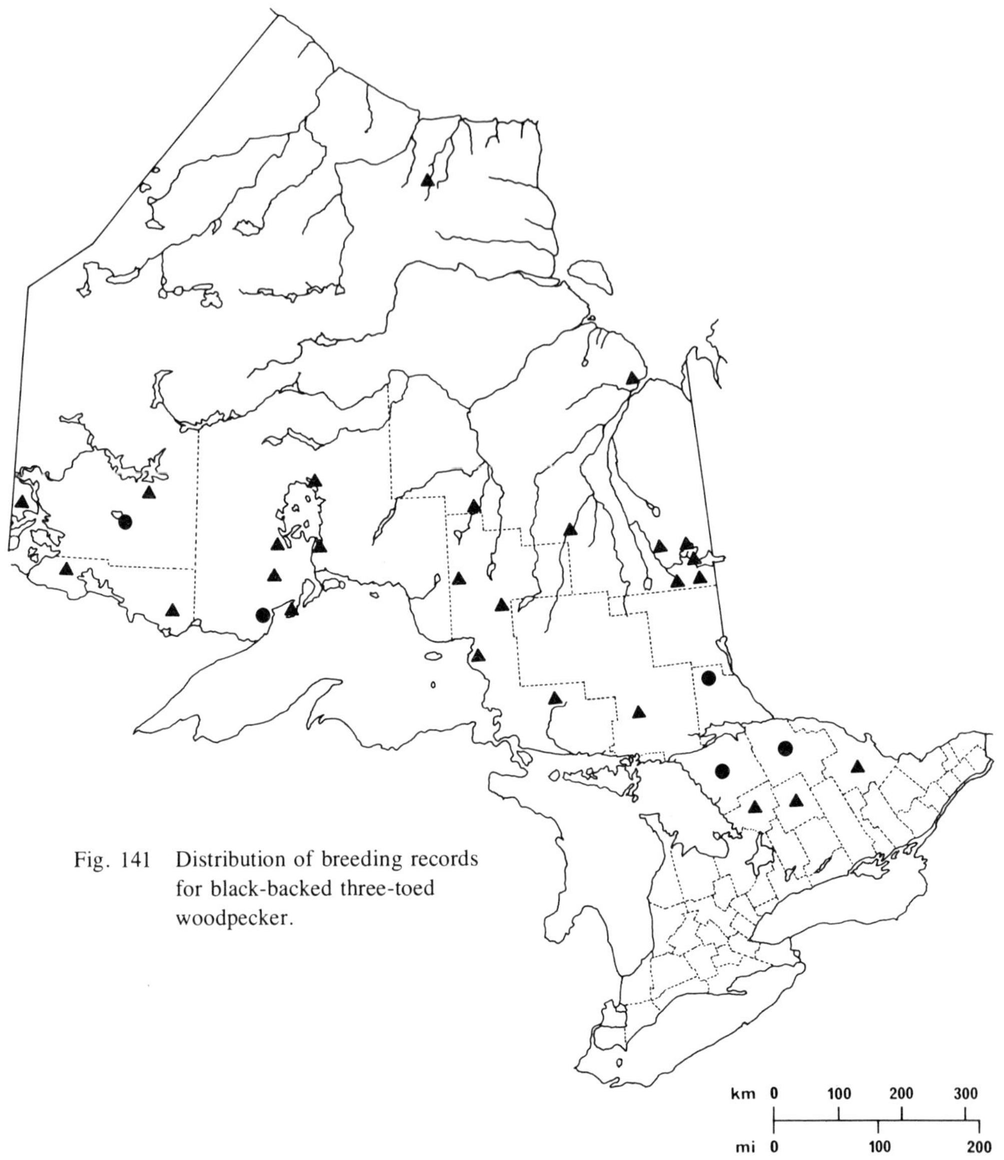

Fig. 141 Distribution of breeding records for black-backed three-toed woodpecker.

Black-backed Three-toed Woodpecker, *Picoides arcticus* (Swainson)

Nidiology

RECORDS 58 nests representing 11 provincial regions.
Breeding habitats recorded were usually tree stands—mixed (29 nests), coniferous (16 nests), and deciduous (1 nest). Nest sites were near or over water (on shores of lakes and rivers and at beaver ponds and bogs) (27 nests), beside roads (10 nests), and in cut-over and burnt-over areas (6 nests). The location of these sites seems to indicate a preference of the species for open areas near the nest.

Dead trees (27 nests) and hydro poles (8 nests) were preferred to living trees (12 nests), and coniferous trees (4 spp., 32 nests) were preferred to deciduous trees (3 spp., 12 nests). Nest trees most commonly selected were spruce spp. (13 nests), pine spp. (12 nests), poplar spp. (8 nests), and cedar (6 nests). Heights of 50 nests ranged from 0.5 to 12 m (1.5 to 40 ft), with 25 averaging 2.5 to 4.5 m (8 to 15 ft).

Nest cavities were excavated by the species and were in the main trunk of the tree. Cavities faced in all directions, but southerly exposures (5 nests) were apparently slightly preferred to all other directions (4 nests). Diameters (DBH) of 4 nest trees ranged from 19.3 to 30.5 cm (7.6 to 12 inches); hole diameters of 3 nests ranged from 4 to 9 cm (1.6 to 3.5 inches); cavity depths of 2 nests were both 25.5 cm (10 inches). The entrance hole of 1 nest was 4 cm (1.6 inches) wide by 5.8 cm (2.3 inches) high; the lip or sill of the hole measured 11.5 cm (4.5 inches); its cavity depth was 25.5 cm (10 inches); and its diameter at the level of the cavity floor was 9 cm (3.5 inches). Bark was removed all around the cavity entrance of 2 nests.

EGGS 19 nests with 2 to 6 eggs; 2E (2N), **3E** (9N), **4E** (6N), **5E** (1N), 6E (1N).
Average clutch range 3 to 4 eggs (15 nests).

INCUBATION PERIOD No information.

EGG DATES 9 nests, 18 May to 18 June; 5 nests, 18 May to 30 May.

Breeding Distribution

The black-backed three-toed woodpecker (Fig. 160B) is found throughout the forested regions of northern Ontario, frequently in burned areas. It extends into southern Ontario at least as far south as Muskoka and Haliburton districts.

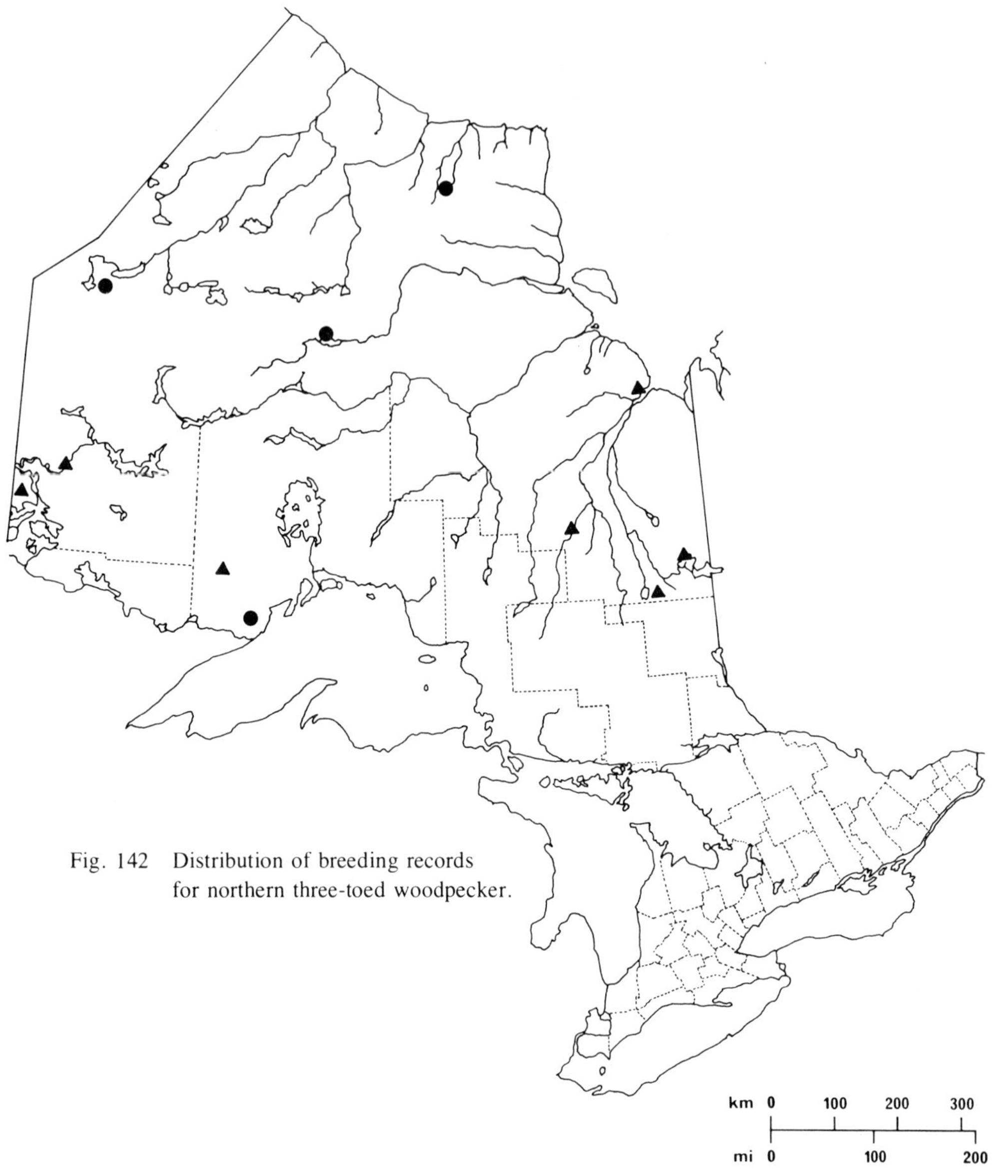

Fig. 142 Distribution of breeding records for northern three-toed woodpecker.

Northern Three-toed Woodpecker, *Picoides tridactylus* (Linnaeus)

Nidiology

RECORDS 10 nests representing 3 provincial regions.
An uncommon, permanent resident of the Boreal Forest region, this woodpecker breeds in open areas of black spruce, which are at times associated with deciduous trees as birch spp., willow spp., and alder. Of the habitats recorded, wet, lowland bog areas of sphagnum moss were often selected and some cedar was among the tree species present. One nest site near a lakeshore was in a mixed wood of large trees.

Nests were in cavities in trees, which were more often dead (6 nests) but occasionally living (3 nests). Spruce spp. (6 nests) were preferred to other tree species such as jack pine (2 nests) and birch sp. (1 nest). Heights of 10 nests ranged from 1.2 to 6 m (4 to 20 ft), with 5 averaging 1.8 to 4.6 m (6 to 15 ft).

Nest cavities were in the main trunk of the nest tree. Trunk diameter of each of 2 nest trees was 18 cm (7 inches); DBH of another tree was 25 cm (10 inches). One nest hole was beneath a fungus growth. Entrance-hole diameter of 1 nest was 6.5 cm (2.5 inches); entrance-hole diameter of a second nest was 4 by 4.5 cm (1.6 by 1.8 inches), and its sill width was 4 cm (1.5 inches); cavity depth of a third nest was 25.5 cm (10 inches); hole diameter of a fourth nest, which faced north, was 7.5 cm (3 inches), and its cavity depth was 15 cm (6 inches).

EGGS 4 nests with 3 to 5 eggs; 3E (2N), 4E (1N), 5E (1N).

INCUBATION PERIOD No information.

EGG DATES 2 nests, 27 May to 6 June.

Breeding Distribution

Because there are only a few nesting records of the northern three-toed woodpecker, knowledge of the extent of its breeding range is very incomplete. However, this seldom-seen species seems to nest in northern Ontario from Thunder Bay and Wawa as far north as the Hudson Bay coast. Baillie and Harrington (1937) were apparently unaware that the first provincial breeding record had been established in 1904, when eggs were collected at Thunder Bay (Dear, 1940).

Hypothetical Breeders

The following four species have been reported as breeding in Ontario, but there is no material evidence to support these reports.

Yellow-crowned Night Heron, *Nyctanassa violacea* (Linnaeus)

A sighting of a single bird on 23 May 1954 near a nest in a colony of great blue and black-crowned night herons on East Sister Island in Lake Erie (Baillie, 1955b) led to the erroneous inclusion of this species on the Ontario breeding-bird list. Nothing was ever seen to indicate that the bird sighted was actually breeding. The species ordinarily breeds in the southern United States. However, it has been known to nest as far north as Michigan and should be watched for in Ontario in future.

Brant, *Branta bernicla* (Linnaeus)

Brant ordinarily breed in the high arctic, but during the summer of 1953 a large number apparently remained on Lake Abitibi. Again in 1954 a number stayed in Ontario at several locations. On Kelly Lake, at Copper Cliff (near Sudbury), 1 pair was reported to have raised a brood of 5 young (Baillie, 1955a, 1963). Two other unsubstantiated breeding records of this species from Prescott and Stormont counties are on file at the ROM, but it is difficult to be sure that these were not escaped captive birds.

Canvasback, *Aythya valisineria* (Wilson)

This duck breeds widely in western Canada. It was reported as breeding in Ontario as long ago as 1897 (Baillie, 1962), but in this century there are no reports until 1948. There are sight records from Lambton County (1948 and 1952), Kent County (1953 and 1954) (Baillie, 1962), and Luther Marsh (Goodwin, 1965). Most of these observations were of females with ducklings.

Short-billed Dowitcher, *Limnodromus griseus* (Gmelin)

A single downy chick believed to be of this species was captured and released near Winisk in 1963 (Tuck, 1968). No other reports of nesting or breeding have subsequently been made, although this species breeds in similar habitat in Manitoba.

Acknowledgements

No one is more aware than the authors that this book could not have been written without the cooperation and work of many amateur and professional naturalists and ornithologists. We express our sincere appreciation to the hundreds of contributors who turned in nest cards to the ONRS or made available their field notes and journals, all of which represent thousands of man-hours spent in the field.

A number of volunteers gave generously of their time to transpose onto cards the records of the ROM nest and egg collection and many other available nests records both published and unpublished; others made field trips to areas where more coverage was necessary, in order to gather more nest records; still others contributed artwork and assisted in various ways in the publication of annual ONRS reports. Our particular thanks are extended to the following for such invaluable assistance: J. Armstrong, D. Baldwin, J. Baillie, G. Bennett, M. Blackshaw, C. Blomme, D. Chamberlain, R. Cubitt, E. Czerwinski, C. Dobson, A. Farraway, P. Geraghty, R. Grant, T. Gretsky, C. Hanrahan, C. Harris, S. Hibbard, H. James D. Kozlovic, S. Lane, L. Lawrence, P. Livingston, G. Logan, D. Loncke, C. Long, H. Lumsden, R. MacMunn, W. Mansell, A. McColl, P. McKane, P. McLaren, P. Milton, L. Montgomerie, R. Montgomerie, B. Mordy, P. Nasmith, G. O'Reilly, B. Parker, C. Peck, M. Peck, S. Peck, S. Postupalsky, B. Ranford, J. Richards, L. Rudkin, J. Satterly, P. Singleton, L. Skene, F. Starr, D. Sutherland, D. Thompson, R. Ussher, S. Vice, P. Waggett, J. Wallace, S. White, B. Wilkins, and C. Young.

We are grateful for the field assistance rendered during ONRS and ROM trips by staff of the Canadian Wildlife Service, the Ontario Ministry of Natural Resources, the Ontario Parks Service, various Conservation Authorities, the Department of Ornithology at the ROM, and the following individuals: C. Campbell, V. Crich, V. Dunston, P. Geraghty, C. Harris, J. Harris, D. Howes-Jones, D. Hussell, C. Long, M. McLaren, P. McLaren, A. Mills, R. Montgomerie, B. Mordy, J. Mordy, P. Mordy, C. Peck, M. Peck, S. Peck, W. Rapley, J. Richards, M. Stasuilis, P. Stepney, R. Tasker, P. Waggett, and U. Watermann.

R. C. Long, research associate in the ROM Department of Ornithology, helped greatly in the summarization of nest card information for a number of species in the present volume.

We appreciate the labours of Mrs. P. Urquhart, secretary in the Department of Ornithology at the ROM, who typed the manuscript and assisted in numerous tasks involving the ONRS. We are grateful for the assistance of J. C. Barlow and A. J. Baker, curators in the Department of Ornithology. W. Robertson and the staff of the Photography Department of the ROM kindly assisted in the preparation of the illustrations for this volume.

We wish to thank C. A. Campbell, W. E. Godfrey, R. C. Long, H. G. Lumsden, and J. M. Richards for reading the manuscript and for their constructive criticisms and suggestions.

The financial assistance we received from the Canadian National Sportsmen's Fund and the Canadian Wildlife Service to further this project has been of great assistance.

Plant Species Mentioned in the Text

Alder *Alnus*
  speckled *A. rugosa*
Algae ALGAE
Apple *Malus*
Arrowhead *Sagittaria*
Ash *Fraxinus*
Aspen *Populus*
  trembling *P. tremuloides*
Avens
  mountain or arctic *Dryas integrifolia*
Beech *Fagus grandifolia*
Birch *Betula*
  bog, swamp, or dwarf *B. pumila*
  white *B. papyrifera*
Blueberry *Vaccinium angustifolium*
Bracken see Fern
Buckbean *Menyanthes trifoliata*
Bulrush *Scirpus*
Bunchberry *Cornus canadensis*
Bur-reed *Sparganium*
Buttonbush *Cephalanthus occidentalis*
Cattail *Typha*
  narrow-leaved *T. angustifolia*
Cedar
  white *Thuja occidentalis*
  red *Juniperus virginiana*
Cherry *Prunus*
  chokecherry *P. virginiana*
Cinquefoil *Potentilla*
Common reed grass see Grass, common reed
Corn *Zea mays*
Cottonwood *Populus deltoides*
Cranberry *Vaccinium*
Dogwood *Cornus*
Elder *Sambucus*
Elm *Ulmus*
  white *U. americana*
Fern FILICALES
  bracken *Pteridium aquilinum*
Fir *Abies*
  balsam *A. balsamea*
Gale
  sweet see Sweet gale
Goldenrod *Solidago*
Grape *Vitis*
Grass GRAMINEAE
  common reed *Phragmites communis*
Hackberry *Celtis occidentalis*
Hawkweed *Hieracium*
Hawthorn *Crataegus*
Hazel *Corylus*
Heath ERICACEAE
Hemlock *Tsuga canadensis*
Hickory *Carya*
Honeysuckle
  bush *Diervilla lonicera*
Horsetail *Equisetum*
Ivy
  poison *Rhus radicans*
Juniper *Juniperus*
Labrador Tea *Ledum groenlandicum*
Laurel
  bog or swamp *Kalmia polifolia*
Leatherleaf *Chamaedaphne calyculata*
Lichen LICHENES
  reindeer *Cladonia rangiferina*
Lilac *Syringa vulgaris*
Lily
  water *Nymphaea*
Locust *Robinia*
Loosestrife *Lythrum*
Maple *Acer*
Milfoil
  water *Myriophyllum*
Moss MUSCI
  sphagnum *Sphagnum*
Mustard CRUCIFERAE
Nettles *Urtica*
Nightshade *Solanum*
Oak *Quercus*
  bur *Q. macrocarpa*
Pine *Pinus*
  jack *P. banksiana*
Pondweed *Potamogeton*
Poplar *Populus*
  balsam *P. balsamifera*
Raspberry *Rubus*
Rhododendron *Rhododendron*
Rice
  wild *Zizania aquatica*

Rush *Juncus*
Sedge CYPERACEAE
Carex
Serviceberry *Amelanchier*
Smartweed *Polygonum*
Spiraea *Spiraea*
Spruce *Picea*
black *P. mariana*
white *P. glauca*

Stonecrop *Sedum*
Sumac *Rhus typhina*
Sweet gale *Myrica gale*
Tamarack *Larix laricina*
Water Lily see Lily, water
Water Milfoil see Milfoil, water
Willow *Salix*
arctic *S. arctica*
dwarf *S. reticulata*

Literature Cited

ALISON, R.M.

1975 The Ring-necked Duck—a new species in Ontario? Ontario Fish and Wildlife Review 14:3–5.

1976 The history of the Wild Turkey in Ontario. Canadian Field-Naturalist 90:481–485.

ALLIN, A.E.

1940 The vertebrate fauna of Darlington Township, Durham County, Ontario. Transactions of the Royal Canadian Institute, 23:83-118.

AMERICAN ORNITHOLOGISTS' UNION

1957 Check-list of North American birds. 5th ed. Baltimore, American Ornithologists' Union. 691 pp.

1973 Thirty-second supplement to the American Ornithologists' Union check-list of North American birds. Auk 90:411–419.

1976 Thirty-third supplement to the American Ornithologists' Union check-list of North American birds. Auk 93:875–879.

ANGEHRN, P.A.M., H. BLOKPOEL, and P. COURTNEY

1979 A review of the status of the Great Black-backed Gull in the Great Lakes area. Ontario Field Biologist 33(2):27–33.

BAILLIE, J.L.

1939 Four additional breeding birds of Ontario. Canadian Field-Naturalist 53:130–131.

1947 The summer birds of Sudbury District, Ontario. Contributions of the Royal Ontario Museum of Zoology 28:1–32.

1955a Fall migration. Audubon Field Notes 9:23.

1955b The winter season. Audubon Field Notes 9:254.

1958 Six old yet new Ontario breeding birds. Ontario Field Biologist 12:1–7.

1960 New Ontario breeding birds. Ontario Field Biologist 14:14–23.

1961 More new Ontario breeding birds. Ontario Field Biologist 15:1–9.

1962 Fourteen additional Ontario breeding birds. Ontario Field Biologist 16:1–15.

1963 The 13 most recent Ontario nesting birds. Ontario Field Biologist 17:15–26.

BAILLIE, J.L. and P. HARRINGTON

1936 The distribution of breeding birds in Ontario. Part I. Transactions of the Royal Canadian Institute 21:1–50.

1937 The distribution of breeding birds in Ontario. Part II. Transactions of the Royal Canadian Institute 21:199–283.

BAILLIE, J.L. and C.E. HOPE

1943 The summer birds of the northeast shore of Lake Superior, Ontario. Contributions of the Royal Ontario Museum of Zoology 23:1–27.

1947 The summer birds of Sudbury District, Ontario. Contributions of the Royal Ontario Museum of Zoology 28:1–32.

BEARDSLEE, C.S. and H.D. MITCHELL

1965 Birds of the Niagara Frontier region. Bulletin of the Buffalo Society of Natural Sciences 22:1–478.

BENT, A.C.

1919–68 Life histories of North American birds. Bulletins of the U.S. National Museum. [Reprinted: New York, Dover Publications]

BONDRUP-NIELSEN, S.

1976 First Boreal Owl nest for Ontario, with notes on development of the young. Canadian Field-Naturalist 90:477–479.

BOYER, G.F. and O.E. DEVITT

1961 A significant increase in the birds of Luther Marsh, Ontario, following fresh-water impoundment. Canadian Field-Naturalist 75:225–237.

BUERKLE, U. and W.D. MANSELL

1963 First nesting record of the Cattle Egret (*Bubulcus ibis*) in Canada. Auk 80:378–379.

CLARKE, C.H.D.

1954 The Bob-white in Ontario. Ontario Dept. of Lands and Forests, Technical Bulletin, Fish and Wildlife Series 2:1–10.

COOCH, F.G.

1963 Recent changes in the distribution of color phases of *Chen c. caerulescens*. Proceedings of the International Ornithological Congress 13:1182–1194.

COOKE, W.W.

1906 Distribution and migration of North American ducks, geese and swans. U.S. Dept. of Agriculture, Biological Survey Bulletin 26:19–80.

DEAR, L.S.

1940 Breeding birds of the region of Thunder Bay, Lake Superior, Ontario. Transactions of the Royal Canadian Institute 23:119–143.

DENIS, K.

1961 Birds of the Canadian Lakehead area. Thunder Bay Field Naturalists Club, Supplement 2. 8 pp. Mimeographed.

DEVITT, O.E.

1939 The Yellow Rail breeding in Ontario. Auk 56:238–243.

1967 The birds of Simcoe County, Ontario. Barrie, Brereton Field Naturalists Club. 192 pp.

DEVITT, O.E. and D.S. MILLER

1954 The Prairie Chicken nesting at Gore Bay, Manitoulin Island. Canadian Field-Naturalist 68:93.

DZUBIN, A.H., H. BOYD, and W.J.D. STEPHEN

1975 Blue and Snow Goose distribution in the Mississippi and central flyways, 1951–1971: a preliminary report. Canadian Wildlife Service, Progress Notes 54:1–34.

ELLIOT, R.D. and R.I.G. MORRISON

1979 The incubation period of the Yellow Rail. Auk 96:422–423.

ERSKINE, A.J.

1971 Buffleheads. Canadian Wildlife Service, Monograph Series 4. 242 pp.

GLEASON, H.A. and A. CRONQUIST

1963 Manual of vascular plants of Northeastern United States and adjacent Canada. Princeton, Van Nostrand. 810 pp.

GODFREY, W.E.

1966 The birds of Canada. National Museum of Canada, Bulletin 203:1–428.

GOODWIN, C.E.

1965 The nesting season. Audubon Field Notes 19:537–540.

HANSON, H.C., H.G. LUMSDEN, J.J. LYNCH, and H.W. NORTON

1972 Population characteristics of three mainland colonies of blue and lesser snow geese nesting in the southern Hudson Bay region. Ontario Ministry of Natural Resources Research Report 92:1–38.

HUSTICH, I.

1957 On the phytogeography of the subarctic Hudson Bay Lowland. Acta Geographica 16:1–48.

JAMES, E.

1956 A narrative of the captivity and adventures of John Tanner during thirty years residence among the Indians in the interior of North America. [Reprint of the 1830 ed.] Minneapolis, Ross and Haines. 428 pp.

JAMES, R.D.

1977 First nesting of the Great Gray Owl in Ontario. Ontario Field Biologist 31:55.

JAMES, R.D., P.L. McLAREN, and J.C. BARLOW

1976 Annotated checklist of the birds of Ontario. Life Sciences Miscellaneous Publications, Toronto, Royal Ontario Museum. 75 pp.

KELLEY, A.H., D.S. MIDDLETON, and W.P. NICKELL

1963 Birds of the Detroit-Windsor area. Bloomfield Hills, Cranbrook Institute of Science. 119 pp.

KRUG, H.H.

1956 The Great Black-backed Gull nesting on Little Haystack Island, Lake Huron. Auk 73:559.

LUMSDEN, H.G.

1951 Breeding diving ducks on Lake St. Clair, Ontario. Canadian Field-Naturalist 65:31–32.

1957 Notes on the wildlife from the Cape Henrietta Maria region of Ontario. Unpublished report of the Ontario Ministry of Natural Resources. In the Ministry of Natural Resources Research Library, Maple, Ontario. 20 pp.

1959a The status of waterfowl in the Cape Henrietta Maria region of Ontario. *In* Transactions of the Northeast Wildlife Conference, Tenth Annual Meeting. Montreal, Office of Biology of the Dept. of Games and Fisheries, University of Montreal, pp. 156–164.

1959b Mandt's Black Guillemot breeding on the Hudson Bay coast of Ontario. Canadian Field-Naturalist 73:54–55.

1964 Golden Eagle nesting in Ontario. Auk 81:91.

1966 The Prairie Chicken in southwestern Ontario. Canadian Field-Naturalist 80:33–45.

1971 The status of the Sandhill Crane in northern Ontario. Canadian Field-Naturalist 85:285–293.

1975 The Whistling Swan in James Bay and the southern region of Hudson Bay. Arctic 28:194–200.

MacLULICH, D.A.

1938 Birds of Algonquin Provincial Park, Ontario. Contributions of the Royal Ontario Museum of Zoology 13:1–47.

MACOUN, J.

1903 Catalogue of Canadian birds. Part II. Ottawa, Geological Survey of Canada, pp. 219–413.

MACOUN, J. and J.M. MACOUN

1909 Catalogue of Canadian birds. Ottawa, Canada Dept. of Mines, Geological Survey Branch. 761 pp.

MANNING, T.H.

1952 Birds of the west James Bay and southern Hudson Bay coasts. National Museum of Canada, Bulletin 125:1–114.

McILWRAITH, T.

1886 The birds of Ontario. Hamilton, The Hamilton Association. 303 pp.

McKEATING, G.B.

1978 Return of the Peregrine Falcon. Ontario Naturalist 17:4–11.

MITCHELL, M.H.

1935 The Passenger Pigeon in Ontario. Contributions of the Royal Ontario Museum of Zoology 7:1–181.

MORRISON, R.I.G. and T.H. MANNING
1976 First breeding records of Wilson's Phalarope for James Bay, Ontario. Auk 93:656–657.

MORRISON, R.I.G., T.H. MANNING, and J.A. HAGAR
1976 Breeding of the Marbled Godwit, *Limosa fedoa*, in James Bay. Canadian Field-Naturalist 90:487–490.

NICHOLSON, J.C.
1972 The birds of Manitoulin Island. Sudbury, Published by the author. 46 pp.
1974 Birds of the Sudbury District. Sudbury, Published by the author. 57 pp.

PALMER, R.S.
1962 Handbook of North American birds. Vol. 1. New Haven, Yale University Press. 567 pp.
1976 Handbook of North American birds. Vols. 2 and 3. New Haven, Yale University Press. 521 and 560 pp.

PECK, G.K.
1966 First published breeding record of Mute Swan for Ontario. Ontario Field Biologist 20:43.
1970 First Ontario nest records of Arctic Loon (*Gavia arctica*) and Snow Goose (*Chen hyperborea*). Ontario Field Biologist 24:25–28.
1972 Birds of the Cape Henrietta Maria Region, Ontario. Canadian Field-Naturalist 86:333–348.
1976 Recent revisions to the list of Ontario's breeding birds. Ontario Field Biologist 30:9–16.

PREVETT, J.P. and F.C. JOHNSON
1977 Continued eastern expansion of breeding range of Ross' Goose. Condor 79:121–123.

QUILLIAM, H.R.
1973 History of the birds of Kingston, Ontario. Kingston, Kingston Field Naturalists. 209 pp.

RICHARDS, J.M.
1977 A summary of nesting records for Ruddy Duck, *Oxyura jamaicensis,* in Ontario, with particular reference to the Regional Municipality of Durham. Ontario Field Biologist 31:45–47.

ROWE, J.S.
1959 Forest regions of Canada. Canada, Dept. of Northern Affairs and Natural Resources, Forest Branch, Bulletin 123:1–71.

SHUELER, F.W., D.H. BALDWIN, and J.D. RISING
1974 The status of birds at selected sites in northern Ontario. Canadian Field-Naturalist 88:141–150.

SCOTT, G.A.
1963 First nesting of the Little Gull (*Larus minutus*) in Ontario and in the New World. Auk 80:548–549.

SIMKIN, D.W.
1963 A Surf Scoter nesting record for northwestern Ontario. Canadian Field-Naturalist 77:60.
1968 Red-throated Loon nesting in northern Ontario. Canadian Field-Naturalist 82:49.

SMITH, D.A.
1970 Observations on nesting Hawk Owls at the Mer Bleue, near Ottawa, Canada. Canadian Field-Naturalist 84:377–383.

SMITH, W.J.
1957 Birds of the Clay Belt of northern Ontario and Quebec. Canadian Field-Naturalist 71:163–181.

SNYDER, L.L.
1928 The summer birds of Lake Nipigon. Transactions of the Royal Canadian Institute 16:251–277.
1930 A faunal investigation of King Township, York County, Ontario. Transactions of the Royal Canadian Institute 17:183–202.
1931 A faunal investigation of Long Point and vicinity, Norfolk County, Ontario. III. The birds of Long Point and vicinity. Transactions of the Royal Canadian Institute 18:139–227.
1938 A faunal investigation of western Rainy River District, Ontario. Transactions of the Royal Canadian Institute 22:181–213.
1941 The birds of Prince Edward County, Ontario, University of Toronto Studies, Biological Series 48:25–92.
1942 Summer birds of the Sault Ste. Marie region, Ontario. Transactions of the Royal Canadian Institute 24:121–153.
1953 Summer birds of western Ontario. Transactions of the Royal Canadian Institute 30:47–95.
1957 Changes in the avifauna of Ontario. *In* Urquhart, F.A., ed., Changes in the fauna of Ontario. Toronto, University Press, pp. 26–42.

SNYDER, L.L. and E.M. WALKER
1928 A faunal investigation of the Lake Abitibi region, Ontario. University of Toronto Studies, Biological Series 32:1–46.

TODD, W.E.C.
1963 Birds of the Labrador Peninsula and adjacent areas. Toronto, University of Toronto Press. 819 pp.

TOZER, R.G. and J.M. RICHARDS
1974 Birds of the Oshawa-Lake Scugog region, Ontario. Oshawa, Published by the authors. 384 pp.

TUCK, L.M.
1968 Dowitcher breeding in Ontario. Ontario Field Biologist 21:124.

WEIR, R.D.
1976 Autumn season 1 Aug.–30 Nov. 1976. Blue Bill 23:55.

Selected Bibliography

Some additional papers consulted for information on nidiology and breeding distribution.

BLOKPOEL, H.
1977 Gulls and terns nesting in northern Lake Ontario and the upper St. Lawrence River. Canadian Wildlife Service, Progress Notes 75:1–12.
1978 Fish-eating birds nesting in Canadian Lake Erie and adjacent waters. Canadian Wildlife Service, Progress Notes 87:1–12.

BREMNER, R.M.
1949 Observations on the birds of the Casummit-Birch Lakes region of northwest Ontario. Canadian Field-Naturalist 63:161–165.

BROOMAN, R.C.
1954 The birds of Elgin County. St. Thomas, Gilbert Press. 41 pp.

GOODWILL, J.E.V.
1942 The summer birds of the Madsen Area, Patricia District, Ontario. Canadian Field-Naturalist 65:131–133.

KELLEY, A.H.
1978 Birds of southeastern Michigan and southwestern Ontario. Cranbrook Institute of Science, Bulletin 57:1–99.

KENDEIGH, S.C.
1947 Bird population studies in the coniferous forest biome during a spruce budworm outbreak. Ontario Dept. of Lands and Forests, Division of Research, Biological Bulletin 1:1–100.

LEE, D.
1978 An annotated list of the birds of the Big Trout Lake area, Kenora District. Ontario Field Biologist 32:17–36.

LLOYD, H.
1923–36 The birds of Ottawa. Canadian Field-Naturalist 37:101–105, 125–127, 151–156; 38:10–16; 39:16–17; 46:123–127, 162–166; 50:143–144.

LLOYD, H. and R.G. LANNING
1948 Observations on the birds of Renfrew County, Ontario. Canadian Field-Naturalist 62:47–65.

LONG, R.C.
1965–72 An annotated list of the birds of Pickering Beach. Ontario Field Biologist 19:26–35; 20:25–34; 21:14–29; 22:8–24; 23:14–23; 24:19, 21, 22; 26:38–45.

O'REILLY, G.
1969 Polar Bear Provincial Park investigational survey, 1969. Unpublished report of the Ontario Dept. of Lands and Forests. Copy in Royal Ontario Museum Dept. of Ornithology. 52 pp.

PERUNIAK, S.
1969/1971 The birds of the Atikokan area, Rainy River District, Ontario. Ontario Field Biologist 23(1969):35–38, 25(1971):15–33.

RICKER, W.E. and C.H.D. CLARKE
1939 Birds of the vicinity of Lake Nipissing, Ontario. Contributions of the Royal Ontario Museum of Zoology 16:1–25.

SAUNDERS, W.E. and E.M.S. DALE
1933 History and list of the birds of Middlesex County, Ontario. Transactions of the Royal Canadian Institute 19:161–248.

SHEPPARD, R.W.
1960 Bird life of Canada's Niagara frontier. Niagara Falls, Niagara Falls Nature Club. 50 pp. Mimeographed.

SPEIRS, D.H. and J.M. SPEIRS
1947 Birds of the vicinity of North Bay, Ontario. Canadian Field-Naturalist 61:23–38.

SPEIRS, J.M.
1973–79 Birds of Ontario County. Toronto, Federation of Ontario Naturalists. 6 vols.

SPRAGUE, T.
1969 Birds of Prince Edward County. Picton, Prince Edward Region Conservation Authority. 215 pp.

TONER, G.C., W.E. EDWARDS, and M.W. CURTIS
1942 Birds of Leeds County, Ontario. Canadian Field-Naturalist 56:8–12, 21–24, 34–44, 50–56.

Index to Common and Scientific Bird Names

Fig. 143

Fig. 143 Great Lakes island habitat. Shown here is the southeastern edge of East Sister Island in Lake Erie, one of several completely or partially vegetated islands in the Great Lakes and in other large lakes and rivers in the province that are important nesting sites for many species. Islands in lakes Erie and St. Clair have heronries of two species whose ranges scarcely exceed the limits of these islands. (Photo by G. K. Peck)

Fig. 144A Nesting cattle egrets with four young, photographed on Pelee Island in 11 June 1976. This species is a relatively recent addition to the breeding avifauna of Ontario (1962) and now nests exclusively on islands in the southern Great Lakes. (Photo by M. K. Peck)

Fig. 144B White pelican colony on one of the Three Sister Islands in Lake of the Woods, photographed on 6 July 1976. This is the only known nesting area for white pelican in Ontario. Similar low, rocky islands with scant vegetation are important breeding sites for colonies of double-crested cormorant and several species of gulls and terns. (Photo by G. K. Peck)

Fig. 144A

Fig. 144B

Fig. 145

Fig. 146A

Fig. 146B

Fig. 145 Deciduous Forest habitat (Carolinian Zone). A variety of trees reach the northern limits of their distribution in this region of broad-leaved trees in the extreme south of Ontario. Similarly, a number of bird species, including barn owl and red-bellied woodpecker, are found almost exclusively in this region. (Photo by G. K. Peck)

Fig. 146A Incubating chuck-will's-widow on a nest with two eggs, photographed in the Deciduous Forest region at Point Pelee on 5 June 1977. This constitutes the first and only nesting record of this species in Canada. (Photo by A. Wormington)

Fig. 146B Red-bellied woodpecker is seen over more extended area in winter, but the small resident breeding population is confined to the Deciduous Forest region. (Photo by G. K. Peck)

Fig. 147

Fig. 147 Southern Ontario cattail marsh. This marsh at Mitchell's Bay in Lake St. Clair, Kent County, is typical of many in the agricultural portion of southern Ontario. Such marshes are important nesting sites for grebes, ducks, gallinules, coots, and terns. These birds become much rarer in the Canadian Shield, where marshes are scarce. King rail and Forster's tern breed exclusively in the southern marshes. (Photo by G. K. Peck)

Fig. 148A Nesting Forster's tern photographed at Mitchell's Bay on 30 May 1977. This tern was confirmed as a breeding species in Ontario only in 1976. (Photo by G. K. Peck)

Fig. 148B Nest and eggs of little gull photographed at Eastbourne Marsh (also known as "Cranberry Marsh") on 13 July 1971. This was the third nesting site for little gull not only in Ontario but also in North America. (Photo by G. K. Peck)

Fig. 148A

Fig. 148B

Fig. 149

Fig. 149 Lakeshore beach habitat. Such areas are important for many species; shorebirds of all types visit these lakeshores during migration, and a few species nest regularly on or near the beaches. However, human activity has been harmful to bird populations in such areas. The bald eagle, for example, formerly nested and hunted in considerable numbers along the shores of the Great Lakes but now has almost disappeared from southern Ontario. (Photo by R. D. James)

Fig. 150A Incubating piping plover on a typical beach. This species requires extensive undisturbed sandy beaches for nest sites. It has become an endangered species in Ontario largely as a result of human disturbance. (Photo by G. K. Peck)

Fig. 150B Nesting Wilson's phalarope with four eggs, photographed at Luther Marsh on 18 May 1970. This species nests among sedges and grasses adjacent to water. It has been expanding its range slowly in Ontario. (Photo by G. K. Peck)

Fig. 150A

Fig. 150B

Fig. 151

Fig. 152A

Fig. 152B

Fig. 151 Southern Ontario agricultural habitat. This area has been largely cleared of forests that were the habitats of species now disappeared such as passenger pigeon and turkey. Resulting farmlands have been of benefit to a number of species such as bobwhite and ring-necked pheasant. The more recent practice of clean farming, however, has in turn been detrimental to these and many other breeding species. (Photo by G. K. Peck)

Fig. 152A Upland sandpiper is a species that prefers pastures and abandoned fields, and it has moved into southern Ontario as land has been cleared. (Photo by G. K. Peck)

Fig. 152B Red-tailed hawk is perhaps the only Ontario raptor whose numbers have increased with the clearing of land and presence of man. This hawk hunts small rodents in open country. (Photo by G. K. Peck)

Fig. 153

Fig. 153 Niagara Escarpment. The largest forest tracts in the agricultural portion of southern Ontario are found on the Escarpment and attract a variety of hawks, grouse, cuckoos, owls, and woodpeckers (as well as many passerine birds). In addition, the cliffs and caves on the Escarpment provide nest sites for a number of species. Unverified reports indicate that golden eagle, which nests on similar cliffs in northern Ontario, may once have nested on this Escarpment. (Photo by R. D. James)

Fig. 154A Turkey vulture nests frequently on the Niagara Escarpment but prefers cleared land for foraging. (Photo by G. K. Peck)

Fig. 154B Peregrine falcon is no longer an Ontario breeding bird, but it once nested on Escarpment cliffs. Reintroduction efforts may re-establish this magnificent species in the province. (Photo by G. K. Peck)

Fig. 154A

Fig. 154B

Fig. 155

Fig. 155 Great Lakes-St. Lawrence Forest habitat. The mixture of deciduous and coniferous trees in this region is attractive to some species of hawks, grouse, cuckoos, owls, and woodpeckers. Mature forests in the southerly deciduous portions of the region (particularly those south and east of the Canadian Shield) are an important habitat for species such as red-shouldered hawk and red-headed woodpecker. The coniferous forests farther north attract boreal species, such as spruce grouse and black-backed three-toed woodpecker, that extend their ranges south of the Boreal Forest region. (Photo by G. K. Peck)

Fig. 156A Nesting goshawk photographed in Halton RM. In Ontario, this species, like other accipiters, probably breeds mainly in the mixed woods of the Great Lakes-St. Lawrence Forest region. (Photo by G. K. Peck)

Fig. 156B Canadian Shield habitat. Lakes with rocky shores found throughout the Shield are generally low in nutrients and productivity. They attract only a limited number of breeding species including common loon and red-breasted merganser. Osprey and bald eagle inhabit the larger lakes. (Photo by G. K. Peck)

Fig. 156A

Fig. 156B

Fig. 157

Fig. 157 Poor farmland habitat. Rocky land on the Canadian Shield and the Niagara Escarpment that has been cleared and subsequently abandoned quickly becomes overgrown with hawthorns and other shrubs. The shrubby fields and associated overgrown orchards, fence rows, and wood lots provide breeding habitats for a number of species, including red-tailed hawk, mourning dove, and several species of woodpeckers. (Photo by R. D. James)

Fig. 158A Nesting black-billed cuckoo with one egg and two young, photographed in a hawthorn meadow near Toronto on 25 July 1965. This species breeds regularly in shrubby, overgrown land. (Photo by G. K. Peck)

Fig. 158B American kestrel often breeds in dead trees in abandoned fields where it hunts for insects and small rodents. (Photo by G. K. Peck)

Fig. 158A

Fig. 158B

Fig. 159

Fig. 159 Boreal Forest habitat. Although this region dominated by spruce trees extends across Canada and covers more than a third of Ontario, comparatively few bird species live in the unbroken stretches of forest. A larger variety of species are found in openings that allow the sunlight to penetrate and the vegetation to diversify, be they openings created by fires, logging, and blow downs, or clearings naturally occurring near wet areas such as bogs and rivers. (Photo by G. K. Peck)

Fig. 160A Spruce grouse, capable of living on a diet of conifer needles, is well suited to life in the Boreal Forest. (Photo by G. K. Peck)

Fig. 160B Black-backed three-toed woodpecker in a nest with four eggs, photographed in Algonquin Park on 30 May 1971. This species breeds in the Boreal Forest and prefers dead trees for its nesting cavities. (Photo by G. K. Peck)

Fig. 160A

Fig. 160B

Fig. 161

Fig. 161 Beaver-meadow habitat. This beaver meadow near Pickle Lake, Kenora District, illustrates the typical mixture of sedges, grasses, aspen trees, and shrubs of many kinds that grow around old ponds where beaver dams have broken up. These oases of deciduous growth in the Boreal Forest develop into important environments that attract several species of woodpeckers. In turn, the abandoned nest cavities of these woodpeckers, enlarged by weathering and rotting, provide nesting sites for other species, such as common goldeneye, American kestrel, and hawk owl. (Photo by R. D. James)

Fig. 162A Great gray owl is a seldom-seen species. It was found nesting in the stand of overmature aspens adjacent to the beaver meadow shown in Figure 161. (Photo by G. K. Peck)

Fig. 162B Bufflehead is found during the breeding season on northern beaver ponds and lakes where overmature trees provide suitable cavity nesting sites. (Photo by G. K. Peck)

Fig. 162A

Fig. 162B

Fig. 163

Fig. 163 Northern wetland habitat. The very conspicuous open fens and bogs with standing water, common throughout the Hudson Bay Lowland and scattered throughout the Boreal Forest, provide important habitats for Canada geese, several species of ducks, and shorebirds such as solitary sandpiper. Large graminoid fens attract sharp-tailed grouse, marsh hawk, and short-eared owl. (Photo by R. D. James)

Fig. 164A Greater yellowlegs is a breeding inhabitant of northern open fens. (Photo by G. K. Peck)

Fig. 164B Sandhill crane breeds in Ontario mainly in the fens of the Hudson Bay Lowland. (Photo by G. K. Peck)

Fig. 164A

Fig. 164B

Fig. 165

Fig. 165 Snow goose colony in northern coastal tundra. This marshy area near the Brant River is typical of those along the Hudson Bay and James Bay coasts and around many adjacent lakes and ponds that provide important nesting sites for several species. Nests in snow goose colonies along the coast are usually in wet areas so that geese can graze easily on sedges and grasses found there. Other species found in coastal marshy areas are the secretive yellow rail and northern phalarope. (Photo by G. K. Peck.

Fig. 166A Nesting whimbrel with three eggs, photographed on wet tundra near radar site 415, Cape Henrietta Maria, on 25 June 1970. (Photo by G.K. Peck)

Fig. 166B Incubating arctic loon on a nest with two eggs, photographed beside a small tundra lake near radar site 416, Cape Henrietta Maria, on 29 June 1970. (Photo by G. K. Peck)

Fig. 166A

Fig. 166B

Fig. 167

Fig. 168A

Fig. 168B

Fig. 167 Dry heath-lichen tundra habitat. The most southern extension of this type of vegetation on a mainland site anywhere in the world is in northern Ontario. Maintained by the cold expanse of water in Hudson and James bays, the tundra conditions attract several species that would otherwise be found only much farther north. (Photo by G. K. Peck)

Fig. 168A Nesting American golden plover with four eggs, photographed near radar site 415, Cape Henrietta Maria, on 23 June 1970. The two nests found so far in Ontario were both at this site. This species is a long-distance migrant that returns to tundra areas only during the brief summer. (Photo by G. K. Peck)

Fig. 168B Incubating willow ptarmigan with 12 eggs, photographed on dry tundra near radar site 416, Cape Henrietta Maria, on 27 June 1970. This species is a permanent resident of the tundra and is able to survive the harsh winters by burrowing into snowbanks for protection and by living on a diet consisting of small twigs and buds. (Photo by G. K. Peck)

Fig. 169

Fig. 169 Wet tussock-tundra habitat. The sunshine on longer summer days melts the surface layers of permafrost, and the low relief and poor drainage in the area result in numerous small ponds, pools, and temporarily flooded areas. This wet tundra is particularly attractive to waterfowl. (Photo by G. K. Peck)

Fig. 170A Nesting parasitic jaeger with two eggs, photographed near radar site 415, Cape Henrietta Maria, on 26 June 1970. This species is a predator and scavenger that regularly nests on tundra. (Photo by G. K. Peck)

Fig. 170B Oldsquaw is one of a number of ducks to be found nesting in Ontario's wet tundra areas. Its main breeding area is farther north. (Photo by G. K. Peck)

Fig. 170A

Fig. 170B